HUMAN-ANIMAL RELATIONS IN BRONZE AGE CRETE

Archaeologists have long admired the naturalistic animal art of Minoan Crete, often explaining it in terms of religion or a love of the natural world. In this book, Andrew Shapland provides a new way of understanding animal depictions from Bronze Age Crete as the outcome of human-animal relations. Drawing on approaches from anthropology and human-animal studies, he explores the stylistic development of animal depictions in different media, including frescoes, ceramics, stone vessels, seals and wall paintings, and explains them in terms of 'animal practices' such as bull-leaping, hunting, fishing and collecting. Integrating zooarchaeological finds, Shapland highlights the significance of objects and their associated human-animal relations in the history of the palaces, sanctuaries and tombs of Bronze Age Crete. His volume demonstrates how looking at animals opens up new perspectives on familiar sites such as Knossos and some of the most famous objects of this time and place.

Andrew Shapland is the Sir Arthur Evans Curator of Bronze Age and Classical Greece at the Ashmolean Museum, University of Oxford. He was previously Greek Bronze Age Curator at the British Museum, where he was co-curator of the exhibition *Troy: Myth and Reality*.

HUMAN-ANIMAL RELATIONS IN BRONZE AGE CRETE

A HISTORY THROUGH OBJECTS

ANDREW SHAPLAND
Ashmolean Museum, University of Oxford

CAMBRIDGE
UNIVERSITY PRESS

CAMBRIDGE
UNIVERSITY PRESS

University Printing House, Cambridge CB2 8BS, United Kingdom

One Liberty Plaza, 20th Floor, New York, NY 10006, USA

477 Williamstown Road, Port Melbourne, VIC 3207, Australia

314–321, 3rd Floor, Plot 3, Splendor Forum, Jasola District Centre, New Delhi – 110025, India

103 Penang Road, #05–06/07, Visioncrest Commercial, Singapore 238467

Cambridge University Press is part of the University of Cambridge.

It furthers the University's mission by disseminating knowledge in the pursuit of education, learning, and research at the highest international levels of excellence.

www.cambridge.org
Information on this title: www.cambridge.org/9781009151542
DOI: 10.1017/9781009151535

First published 2022

Printed in the United Kingdom by TJ Books Limited, Padstow Cornwall

A catalogue record for this publication is available from the British Library.

ISBN 978-1-009-15154-2 Hardback

CONTENTS

FIGURES

TABLE

PREFACE

This book has been a long time in the making; it originated in a 2004 Master's dissertation followed by a PhD, completed in 2009, at University College London (UCL) Institute of Archaeology, funded by the Arts and Humanities Research Board and Arts and Humanities Research Council respectively. I was fortunate to have been supervised in these by Cyprian Broodbank, Louise Martin, Jeremy Tanner and Todd Whitelaw, and am grateful for their advice and guidance during my time at UCL and beyond. Todd also provided me with an introduction to Knossos, through an invitation to take part in the Knossos Urban Landscape Project, a synergasia between the Heraklion Ephorate of Antiquities and the British School at Athens. This provided a wonderful hands-on introduction to the archaeology of Crete, and indeed some of its animals; one morning when I probably should have been looking at the ground for potsherds I noticed a distinctive bird sitting on a fence nearby, a hoopoe, instantly recognisable from the Caravanserai fresco now in Heraklion Museum. Yet it is this romantic moment when past and present seem to merge, as a bird seems to hop out of a Minoan fresco, that this book tries to resist by drawing attention to the gulf between present-day human-animal relations and those of the Bronze Age. The wild goats of Crete, agrimia, are easy to spot among souvenirs on the tourist stalls of Knossos but I have never seen one in the flesh, restricted as they are now to the mountains of West Crete and a few islands after almost being hunted to extinction in the twentieth century. Nor do I want to confuse my own, essentially touristic, experience of Crete with that of the present-day local inhabitants or those of the Bronze Age. Each will have encountered different animals in different ways. One of the reasons for highlighting the views of visitors to Crete in this book, particularly those of Sir Arthur Evans and his contemporaries, is to reflect on how outsiders have shaped Cretan archaeology with their various standpoints. My subsequent career path, as a curator at the British Museum and now the Ashmolean Museum, has resulted in an acute sense of the historical background of the discipline of Aegean Bronze Age archaeology. I have been lucky to work daily with objects removed from Crete over the course of more than a century which are both traces of Bronze Age lives and of colonial-era appropriation and exchange. At times, a long history of scholarship and interpretation helps to trace some of the Bronze Age associations of these objects, but at other times they have become caught up in modern-day concerns. Perhaps human-animal studies is also a modern-day concern, but it helps to highlight the more-than-human relationships of which these objects were once a part, besides their long and complicated route to the collections I work with.

Before turning my attention to animals and objects, I would like to thank the many human collaborators who have helped with this book. Beatrice Rehl of Cambridge University Press first suggested that I turn my doctorate into a book, and has shown great patience as it has gone through a number of rewrites. These, I should say, were the result of the judicious comments and criticisms of two sets of anonymous reviewers. I would like to thank them for helping me move away from a thesis and write a book instead. Ellen Adams, Laura Morley (Preston) and Nico Momigliano kindly read drafts of the book. Andrew Bevan and Carl Knappett examined my doctoral thesis, provided many useful comments and have continued to show an interest in its development. My colleagues at the British Museum and the Ashmolean Museum have also been supportive, but I owe particular thanks to Lesley Fitton, †Ian Jenkins and Paul Roberts, for whom this book has been a familiar item in review meetings.

This book has benefited directly or indirectly from discussions on Aegean Bronze Age archaeology, many of them at Knossos or at the London Mycenaean Seminar, with many people. Those not mentioned above include Lisa Bendall, John Bennet, Camilla Briault, Lesley Bushnell, Gerald Cadogan, Kostis Christakis, †Jo Cutler, Oliver Dickinson, Debi Harlan, Yannis Galanakis, Senta German, Angeliki Karagianni, Thomas Kiely, Antonis Kotsonas, Olga Krzyszkowska (who also generously shared her knowledge of seals with me), Matthew Haysom, Borja Legarra Herrero, Colin Macdonald, Erin McGowan, Margarita Nazou, Kostas Paschalidis, Conor Trainor and Peter Warren. In addition to the Mycenaean Seminar, more formal discussions of the ideas in this book have also taken place in seminars and conference sessions organised by the following: Kristin Armstrong Oma (University of Oslo Seminar on Animals), Jacqui Mulville (International Council for Archaeozoology International Conference), Marcus Brittain, Andy Needham, Nick Overton and Penny Spikins (Theoretical Archaeology Group conference), Chris Watts (Society for American Archaeology Annual Meeting), Dušan Borić and James Whitley (Cardiff University seminar), Branko van Oppen (Animals in Ancient Material Cultures Conference, Allard Pierson Museum), Tatiana Theodoropoulou, Jim Wright and Jenifer Neils (Harvesting the Sea conference, Wiener Laboratory of the American School of Classical Studies at Athens), Yannis Galanakis (University of Cambridge Classical Archaeology Seminar), Irene Lemos and Gian Piero Milani (University of Oxford Greek Archaeology Group and Prehistoric and

Early Greece Graduate Seminar), Robert Laffineur and Tom Palaima (ZOIA conference, originally to be held at the University of Texas at Austin) and Jan Driessen and Quentin Letesson (UCLouvain TALOS project webinar).

I am also grateful to a large number of people for sharing ideas, feedback or publications: Emily Anderson, Katerina Athanasaki, Emily Bonney, Ilaria Caloi, Georgia Flouda, Jack Fuller, Effie Gemi-Iordanou, Yannis Hamilakis, Kerry Harris, Eleni Hatzaki, Valasia Isaakidou, Robert Koehl, Dimitra Mylona, Jacke Phillips, Laerke Recht, David Reese, Alessandro Sanavia, Anna Simandiraki-Grimshaw, Katy Soar, Nancy Thomas, Judith Weingarten, David Wengrow and Marika Zeimbekis. My apologies to those I have neglected to mention. For help with images and other matters I would like to thank Calliopi Christophi (French School at Athens), Louise Dennis (Museum of Design in Plastics), Amalia Kakissis (British School at Athens), Stella Mandalaki (Heraklion Archaeological Museum), Diamantis Panagiotopoulos (CMS Archive), Angela Singer, Vassiliki Sithiakaki (Heraklion Ephorate), Amy Taylor (Ashmolean Museum Picture Library), Anastasia Tzigounaki (Rethymnon Ephorate) and British Museum Images.

The writing of this book would not have been possible without the forbearance, love and support of Susie, and the companionship of Penguin and Wanda. This book is dedicated to them.

ONE

INTRODUCTION

The meal furnished by the hospitality of my Samarióte guide consisted chiefly of the flesh of a wild-goat, killed by him on an expedition from which he had only just returned. I obtained from him three pairs of the animal's horns.

(Pashley 1837b: 271)

The 'Burgon ring' shows two agrimia, as the wild goats of Crete are also known, wide-mouthed and vigorously mating. The finely engraved detail on the circular bezel, only 1.6 cm in diameter, clearly delineates the male agrimi's bristly coat, beard and his long, curved, knobbly horns arching over his back (Figure 1.1). Beneath him, her head raised, stands a female goat, her ridged horns shorter. Their hooves merge into a rock formation below. This gold ring was part of a much larger collection accumulated by Thomas Burgon while he was a merchant in the Aegean from 1809 to 1814. He proved unable to make a living, perhaps because of his taste for antiquities, and joined the British Museum, to which he also sold his collection in 1842. The ring is described rather laconically in a subsequent British Museum catalogue as 'two wild goats; beneath them, rocks are indicated' and assigned to the Mycenaean period (Marshall 1907: 4). It was claimed for the Cretan Bronze Age by Sir Arthur Evans, who illustrated it in his seminal work *The Palace of Minos at Knossos*, where he described the scene as two agrimia 'in the act of coition' (Evans 1935: 510). The ring has been absorbed into the modern entity known as Minoan Crete, assembled from archaeological excavations and museum collections. This book aims to retrace the place of such animal depictions among the collective of animals, humans and things of which they were once part.

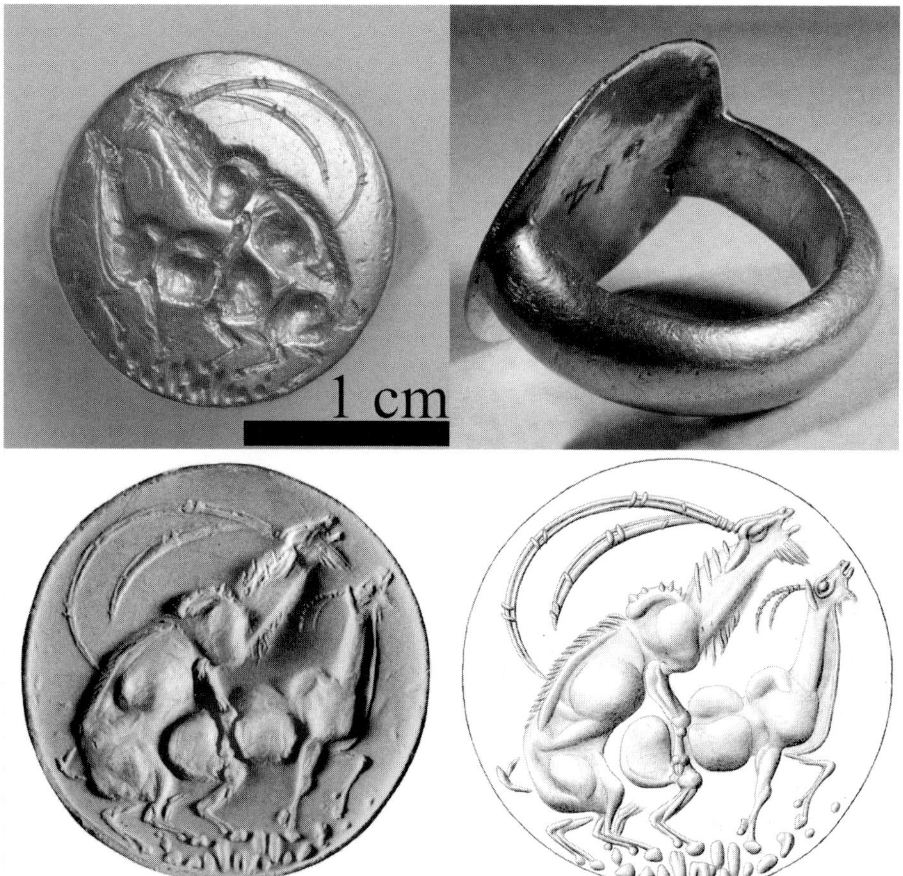

1.1. The Burgon ring: gold ring showing wild goats mating, MMIII–LMI. Clockwise from top left: photograph of bezel (GR 1842,0728.127). © The Trustees of the British Museum; photograph of hoop, drawing of impression, photograph of impression (CMS VII 68). © CMS Heidelberg

Three pairs of agrimi horns obtained on Crete in 1834 by Robert Pashley, a Cambridge academic, were similarly absorbed into a modern understanding of Cretan fauna which would have differed from that of the Samariote hunter he bought them from. On his return to Cambridge, Pashley's horns were transformed from hunting trophies to scientific specimens when his Cambridge colleague, Mr Rothman, identified them as belonging to the species *Capra aegagrus*. As Rothman noted:

> it is not the bouquetin, to which however it bears considerable resemblance, but the real wild-goat, the capra aegagrus. Pallas. the supposed origin of all our domestic varieties. The horns present an anterior trenchant edge, characteristic of this species. The discovery of the aegragus in

> Crete is perhaps a fact of some zoological interest, as it is the first well-authenticated European locality of this animal. (Pashley 1837b: 271)

The sixteenth-century French naturalist Pierre Belon had visited Crete, recorded his observations of the Cretan wild goats and also obtained horns (Belon 1555: 14). He called them boucestein, using the same name as the wild goat he would have known from the French Alps, now more commonly called the bouquetin or ibex (*Capra ibex*). It was this identification which Rothman was challenging, although the credit of naming the species of the Cretan wild goat went to another zoologist. Heinrich Schinz, relying largely on Belon's account, distinguished the Cretan wild goat from other worldwide species of wild goats, labelling it *Capra cretica* (Schinz 1838: 10).[1] Although his identification was accepted, until recently the Cretan wild goat was classified as *Capra aegagrus cretica* (Schinz 1838), a wild goat subspecies.

The classification of the Burgon ring also relied on the assembly of specimens against which it could be compared, which were assembled into a broader cultural group. It was only with the excavation of the Mycenae Shaft Graves on the Greek mainland in 1876 that rings of a similar type were found in an archaeological context, hinting that the Burgon ring might belong to the newly identified Bronze Age of Greece. By 1900 there were enough similar depictions on rings and sealstones for the German art historian Adolf Furtwängler to publish its impression alongside other examples of 'Mycenaean' pictures of animals. Here the 'truth to nature' of the goats' movements and expressions was praised (Furtwängler 1900: 14, pl. 3). In the same year, Arthur Evans began his excavations at Knossos in Crete, at first using the name 'Mycenaean' to describe his Bronze Age finds, but he increasingly adopted the term 'Minoan' to contrast them with the culture of the mainland. He published the Burgon ring alongside other animal depictions which he attributed to the 'finest' Middle Minoan III style (Evans 1935: 510). For Evans this period was distinguished by a 'naturalistic spirit' in art, exemplified by the Burgon ring, with its finely detailed depiction of animal bodies. This view still holds, and it has recently been described by one authority as 'an outstanding example of Minoan naturalism' (Krzyszkowska 2005: 128).

Among the finds from the first season at Knossos were fired-clay lumps with seal[2] impressions on them, many showing animals. Evans (1900: 69) identified these sealings as 'a class of object never before observed in any excavation of a Mycenaean site'. These were sometimes found alongside clay tablets inscribed with written records, some in a 'hieroglyphic' script and others in a 'linear' script (Evans 1900: 29, 56). This proved that objects like the Burgon ring were used for making sealings, used as a material trace of transactions, some of which were also recorded on clay tablets. One group of tablets inscribed with the script which Evans later called 'Linear B' was found in a building known as the

'Armoury' at Knossos, excavated in 1904. Evans was unable to decipher Linear B but could recognise a pictorial sign recording wild goats' 'long curving horns with the characteristic protuberances' (Evans 1935: 832) (Figure 1.2). He recalled the bow of Pandarus in the *Iliad*, made from the huge horns of a goat that the hero had himself hunted, and suggested that these horns had been brought by 'individuals who had succeeded in obtaining the horns by their own prowess' to be made into bows (Evans 1935: 835). With the decipherment of Linear B it is now apparent that the horns recorded on this tablet were collected as taxation from the area of Rethymnon, to the west of Knossos. This shows that the first records of agrimi horns were made not by zoologists but by Bronze Age administrators. Long before they were transformed into scientific objects, agrimi horns were being turned into weapons.

It was still possible to hunt agrimia on Crete when Evans was excavating at Knossos in the early twentieth century, although the introduction of modern weapons had started to affect their numbers: 'In Crete it is now mainly confined to the White Mountains, though it is still found occasionally on Ida and, more frequently, on the Lasithi ranges East of Knossos' (Evans 1935: 833). The food shortages resulting from subsequent conflicts, notably the Second World War, led to an increase in hunting and there were perhaps fewer than 100 agrimia remaining on the island following this period (Farmar 1952). Numbers have since recovered, but the island's wild goats are now only found in the White Mountains of western Crete. They are largely confined to the Samaria Gorge National Park, founded in 1962, now a popular tourist trail and UNESCO

1.2. Linear B tablet showing agrimi horns, LMIIIA. Length: 13.2 cm. Above: photograph (AN1910.217). © Ashmolean Museum, University of Oxford. Below: drawing (after Evans 1935: fig. 813d)

Biosphere Reserve. Hunting agrimia on the island is now illegal and as a result the practice, which has defined relations between humans and agrimia for millennia, has all but come to an end.

At the same time, debates over identification have continued as a result of recent morphological and genetic analysis showing that agrimia are likely to be descendants of domestic goats which became feral soon after they were brought to Crete by humans during the Neolithic (Shackleton 1997; Bar-Gal et al. 2002; Horwitz and Bar-Gal 2006). As a result they do not appear on the IUCN Red List of Threatened Species because they are classed there as domestic goats (*Capra hircus*) and not 'true' wild goats (*Capra aegagrus*) (Weinberg and Ambarli 2020). Nevertheless, attempts continue to save them not just from hunters but from contact with domestic goats because of fears about hybridisation (Spanos et al. 2008). Now often referred to as kri-kri, they have become a symbol of Crete, rarely seen in real life, but widely visible as tourist souvenirs.[3]

Pashley's agrimi horns and Burgon's gold ring have become distanced from the human-animal relations of which they were once a part. They instead became part of a modern way of understanding the world as they entered institutions such as the University of Cambridge or the British Museum and were published in accounts of travels and scholarly catalogues. The digitisation of these books, and of museum collections, has now made them instantly accessible from anywhere in the world. The horns, originally the product of a hunt, when brought to England became part of the description of the natural world of Crete, involving the collecting and listing of specimens. The gold ring has been recognised as a masterpiece of Minoan culture, expressing an appreciation of the natural world. Yet this natural world and the Minoan culture that appreciated it are products of a modern understanding of Crete. This book instead seeks to reintegrate agrimi horns and agrimi depictions into the relations between humans, animals and objects using the concepts of animal practices and animal things. Animal practices are context-specific relations between humans and animals, of which animal things are the material trace. Such animal things can endure long after the animal practice of which they were once part has disappeared and can be co-opted into new sets of relations. Now part of a display about Minoan Crete in the British Museum, the Burgon ring was once an animal thing which, at the moment it was used to make an impression in clay, created a set of material links between its user, the palace and the animals which inhabited the mountains beyond the settlements of Bronze Age Crete. This book will explore a variety of animal things from Bronze Age Crete and the animal practices that produced them. But first it is necessary to trace the development of the increasingly fragile concepts of Minoan culture and Cretan nature.

MINOAN CULTURE

Like other travellers before him, Robert Pashley paid a visit to the site of Knossos and observed 'the few shapeless heaps of masonry, which alone recal to the remembrance of the passing traveller its ancient and bygone splendour' (Pashley 1837a: 209). Finds there of Greek coins with the design of a labyrinth had shown that this was the location of the Greek and Roman city of Knossos and the mythical home of the Minotaur. Excavations began at the site in 1878, instigated by a local antiquarian named Minos Kalokairinos (Kopaka 1995, 2015). His discovery of a number of large storage jars in stone-built magazines attracted attention and he was soon asked to stop by his peers, who were concerned that finds from the site might be claimed by the Ottoman authorities. This was the decade in which Heinrich Schliemann had made spectacular finds at two other sites famous in Greek myth, Troy and Mycenae, and he made an unsuccessful bid to buy land at Knossos in order to dig there next (Schliemann 1878, 1880). Following his excavation at Mycenae, 'Mycenaean' was soon used as an adjective to describe a type of early pottery and associated finds from a number of sites across the Aegean, and rapidly became the name of a prehistoric mainland culture (Furtwängler and Loeschcke 1886). It was widely anticipated that Crete too would produce such material once the island gained its independence.

These nineteenth-century finds and publications helped structure what Arthur Evans found at Knossos and the way he interpreted his finds. Evans first came to Crete in 1894 in search of an early form of writing which he had recognised on sealstones said to come from the island (Evans 1894, 1897). He too visited Knossos, saw Kalokairinos's excavations and succeeded in buying a share of the land on which the site was located. He was particularly interested in the building Kalokairinos had discovered because some of its stone blocks were incised with what appeared to be written signs. The political situation precluded excavation but Evans was able to travel around Crete, observing ancient remains and purchasing sealstones with further evidence for the script he was looking for (Brown 2001). He rapidly published his discoveries in a paper on the 'Prae-Phoenician Script' of Crete, in which he used the term 'Minoan' Crete as a counterpart to Mycenaean Greece (Evans 1894: 367).[4] He regarded some of the seals as earlier than Schliemann's finds from Mycenae, noting that: 'We see before us the prototypes of more than one of the characteristic forms of Mycenaean times' (Evans 1894: 372). Even before Evans had excavated at Knossos, he was already trying to establish Minoan Crete as distinct from, and antecedent to, the Mycenaean culture of the mainland.

Arthur Evans began his excavations at Knossos in 1900, by which time Crete was effectively independent from the Ottoman Empire. The local authorities, particularly the Heraklion Syllogos led by Joseph Hazzidakis, were favourable to his purchase of the rest of the land around Knossos with a view to excavation (MacGillivray 2000b: 164–6; Brown 2001: xxiii). Almost immediately, Evans started to find the clay tablets which demonstrated the use of an early writing system at the site. He also rapidly came down on a room with a stone seat still *in situ* against walls painted with frescoes. In his report of the first year's excavations Evans named this the 'Throne-Room' and the large building he had uncovered was called a palace. Although he toyed with the idea of a matriarchy, he argued that a king sat on the throne (Evans 1900: 42). With the restoration of the relief fresco of a young man, apparently wearing a feathered crown, Evans had soon found an image of the ruler, whom he regarded as a 'priest-king' (Sherratt 2000). For Evans, the Palace of Minos, as he soon called the building, was both the residence of kings and queens and a religious centre.[5]

With the aid of his assistant, Duncan Mackenzie, Evans came to understand the complex stratigraphy of the Kephala Mound on which the palace was located. They realised that the palace was on top of a Neolithic tell, an artificial mound built up over thousands of years of occupation. The Bronze Age levels were distinguished on the basis of changes in architecture and pottery, and divided into a number of periods termed Early, Middle and Late Minoan. Their excavations suggested that in the Early Minoan period the buildings on the summit of the mound came to coalesce around a central courtyard until a palace was formed at the start of the Middle Minoan period. This building was severely damaged by an earthquake towards the end of this period, but was rapidly rebuilt and elaborated. It was finally destroyed by fire during the Late Minoan period, preserving the Linear B tablets that were found in large numbers across the site. Distinctions within periods were charted through changes in pottery styles; each period was divided into three phases and sometimes further subphases. These phases were associated with episodes of destruction and building, dated using the pottery scheme. This tripartite scheme was predicated on evolutionary ideas of birth, florescence and decay. This scheme, promoted by Evans (1906a), was soon adopted by archaeologists across the Aegean and is still in use today (Table 1.1). Although the Neolithic sequence at Knossos was divided in the same way by Evans and Mackenzie, this has been substantially revised as a result of subsequent excavation and study, and in order to integrate it with other Cretan and Aegean sequences (Tomkins 2007: 13–21).

The easing of political tensions on Crete at the very end of the nineteenth century opened the way for the excavation of sites across the island at the same time as Evans was working at Knossos (Figure 1.3). Italian excavations at Phaistos in southern Crete also began in 1900, where Luigi Pernier and Federico Halbherr found another 'palace' like that at Knossos, and another

TABLE 1.1 *Chronological table showing periods and approximate absolute dating*

Cultural Period	Palatial Period	Ceramic Period	Approximate Absolute Dates (BCE)
Neolithic		Initial Neolithic	7000–6500
		Early Neolithic	6500–5900
		Middle Neolithic	5900–5300
		Late Neolithic	5300–4500
		Final Neolithic	4500–3100
Early Minoan (EM)	Prepalatial	EMI	3100–2700
		EMII	2700–2200
		EMIII	2200–2000
Middle Minoan (MM)		MMIA	2000–1900
	Protopalatial	MMIB	1900–1800
		MMII	1800–1700
	Neopalatial	MMIII	1700–1600
		LMIA	1600–1500
		LMIB	1500–1450
Late Minoan (LM)	Final Palatial	LMII	1450–1400
		LMIIIA	1400–1350
	Postpalatial	LMIIIB	1350–1200
		LMIIIC	1200–1050

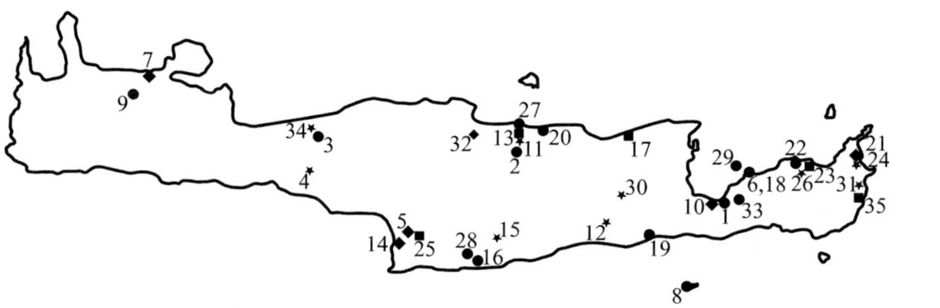

1.3. Map of Crete showing sites discussed in the text. Circle: settlement; square: palace site; diamond: other palatial centre; star: sanctuary site; (1) Alatzomouri-Pefka; (2) Archanes-Phourni; (3) Armenoi; (4) Atsipadhes; (5) Ayia Triada; (6) Chalinomouri; (7) Chania; (8) Chryssi; (9) Debla; (10) Gournia; (11) Juktas; (12) Kato Syme; (13) Knossos; (14) Kommos; (15) Kophinas; (16) Koumasa; (17) Mallia; (18) Mochlos; (19) Myrtos Fournou Kourifi; (20) Nirou Chani; (21) Palaikastro; (22) Papadiokampos; (23) Petras; (24) Petsophas; (25) Phaistos; (26) Piskokephalo; (27) Poros; (28) Porti; (29) Pseira; (30) Psychro; (31) Traostalos; (32) Tylissos; (33) Vronda/Kastro; (34) Vrysinas; (35) Zakros

substantial building at the nearby site of Ayia Triada two years later (Pernier and Banti 1935, 1951). In keeping with the desire to accommodate such buildings within a scheme of elite European architecture, this became known as a 'villa' (Halbherr 1903: 7). The Cretan Stephanos Xanthoudides worked in the same area, the southern Mesara plain, excavating a number of Early Bronze Age tombs (Xanthoudides 1924). American archaeologists worked in the east of the island; Harriet Boyd Hawes and her assistants, Richard Seager and Edith Hall, between them excavated a number of sites, including the settlement of Gournia, from 1901 (Boyd Hawes et al. 1908). At the same time British archaeologists were exploring further east. Neither Robert Carr Bosanquet at Petras nor David Hogarth at Zakros discovered palaces, although they did find Bronze Age material (Hogarth 1901; Bosanquet 1902b); palaces were subsequently discovered by Greek archaeologists at both sites in the later twentieth century (Platon 1971; Tsipopoulou 2012). Bosanquet (1902a) also excavated at Palaikastro, where a large settlement was uncovered, although, after more than a century, a palace has not yet been found. In 1915 another palace was found at Mallia, on the north coast of Crete to the east of Knossos, by Joseph Hazzidakis, and subsequently excavated by French archaeologists (Hazzidakis 1915; Chapouthier and Charbonneaux 1928).

Although Hazzidakis and Halbherr had been exploring sites on Crete during the period of Ottoman control, the burst of excavation at the start of the twentieth century meant that these archaeologists of various nationalities were simultaneously excavating Bronze Age sites. They informed one another of their discoveries and noted similarities in architecture and pottery as they were trying to understand their sites. What emerged was an idea that these sites were all essentially part of the same culture, and the term 'Minoan' came to be used to describe all of them. This was the period in which Minoan Crete came into being and, as another excavator remarked 100 years later, 'it is still the case that all modern research derives from, even if it is at times in antithesis to, the way the discoveries at the big sites were formulated in the first heady years of Cretan freedom in the early twentieth century' (Cadogan 2000: 17). It is no accident that the way in which Minoan Crete was established as a European civilisation, with affinities to prehistoric Greece, aligned closely with the cause of Cretan freedom, and its subsequent union with Greece in 1913. The modernist vision of the Cretan past resulted from a collaboration between Western European and American archaeologists and the local scholars, led by Hazzidakis and Xanthoudides, who encouraged them to excavate across the island (Carabott 2006; Varouchakis 2017). Since many of the still-dominant theories and interpretations of Minoan Crete are derived from this period, examining their origins provides a basis for evaluating them.

Whereas other archaeologists published site reports, Arthur Evans chose to follow his preliminary reports of his excavations at Knossos with an altogether different type of publication. *The Palace of Minos at Knossos* appeared in four volumes between 1921 and 1935 and set out Evans's vision of Minoan Crete intertwined with the results of his excavations. It opens with a claim that the Greek myth of Theseus and the Minotaur is unfair to 'this early civilisation of Crete'. Evans's excavations had shown that: 'The ogre's den turns out to be a peaceful abode of priest-kings, in some respects more modern in its equipments than anything produced by classical Greece' (Evans 1921b: 1). There has long been a feeling that the sense of modernity at Knossos owes much to Evans rebuilding the site with the help of reinforced concrete, one of the earliest uses of this material (Gere 2009). These 'reconstitutions', as Evans termed them, were often necessary to preserve the remains of the building, particularly on the east side where several storeys were found. And, like Evans's interpretations, these modern additions to the palace cannot simply be dismantled and discarded. More than any other site on Crete the sometimes fanciful concrete reconstructions open a dialogue between the visitor and the Bronze Age (Duke 2007). They are a metaphor for the Minoans, a twentieth-century vision of an ancient civilisation built on the material traces of the Bronze Age.

One of the most striking features of the modern-day palace at Knossos is the replica relief painting of the charging bull in front of an olive tree at the North Entrance (Figure 1.4). Based on fresco fragments recovered from the entrance passage below, this reconstruction was nevertheless an argument cast in concrete. The bull allowed Evans to connect the labyrinthine building he had excavated with the Minotaur of Greek myth by suggesting that this relief had inspired the tale: 'The monumental reliefs within its sea-gate – visible, it would appear, to a much later date – representing bull-catching scenes and, still more, the fresco panels with feats of the bull-ring in which girls as well as youths took part, go far to explain the myth' (Evans 1921b: 1–2). There is no evidence that

1.4. Relief fresco of a charging bull at the North Entrance of the palace at Knossos, MMIII–LMI. Left: photograph. © Hellenic Ministry of Culture and Sports (N. 3028/2002); right: reconstruction drawing based on photograph of surviving fresco fragment (Evans Fresco Drawing P/2 c). © Ashmolean Museum, University of Oxford

a picturesque ruin of this kind continued to stand at Knossos, although the remains of the palace did become a focus of later cult activity (Prent 2005: 514–18).[6] But even if the concrete reconstruction and direct associations with the Minotaur are stripped away, the stucco relief fragment of a bull's head found in the North Entrance passage is a remarkable part of a life-size depiction of a charging bull, his eye bulging, his tongue hanging out. It is clear why Evans regarded it as one of the highpoints of naturalistic art from Minoan Crete and made it a focal point of his restorations.

From the first excavations of Bronze Age Cretan sites, animal depictions were found in abundance in a variety of media including sealstones, stone vessels, clay and metal figurines and frescoes (Oulié 1926; Vanschoonwinkel 1996). These objects, and the often effusive descriptions of them in early publications, resulted in the modern interpretation of the Minoans as 'nature lovers' (Starr 1984), although Evans's view was more nuanced than this and connected with the evolution of Minoan art. Naturalism is a term used frequently in *The Palace of Minos at Knossos* to describe both the style and the prevalence of depictions of animals and plants. Evans placed the peak of naturalism at the end of the Middle Minoan period, with the charging bull a fine example: 'The brilliant naturalism of the grand Transitional Epoch that links the Middle with the Late Minoan Age reaches its acme in the high reliefs of painted stucco at Knossos, in the frescoes of Hagia Triada and such works as the "harvester vase"' (Evans 1921b: 28). This also illustrates how Evans drew in finds from other sites to act as comparanda for Knossos, so that *The Palace of Minos at Knossos* became a grand synthesis of Minoan civilisation. Given his position at the centre of Minoan archaeology for its first four decades, his insights are still important, but the thinking behind them needs to be examined critically.

Yannis Hamilakis (2002b: 4) has suggested that there are 'persistent and popular interpretative schemes' which archaeologists have inherited from Evans and continue to apply to the Bronze Age of Crete. The scheme which is most relevant here can be termed the 'nature-loving Minoans'. Over time in the popular – and scholarly – imagination, Minoan depictions of animals and plants came to be seen as joyfully expressing a love of the natural world and inextricably linked with religion (Forsdyke 1931: 29; Groenewegen-Frankfort 1951: 206–16). Those studying animal depictions in particular tend to regard 'nature' as a source of artistic inspiration.[7] Evans (1921b: 447) explained some animal depictions as manifestations of the Great Mother Goddess but this notion has developed into the person of a nature goddess whose domain encompasses the natural world (Immerwahr 1990: 50; Marinatos 1993: 193–200). As a result, depictions of animals assume a primarily religious symbolism in some accounts (Oulié 1926: 12–13; Czernohaus 1988: 156; Goodison 1989: 49–55). As Vesa-Pekka Herva (2006a, 2006b) has convincingly argued, this symbolic approach is both more prevalent in modern scholarship and in some ways less revealing than a straightforward belief in the 'nature-loving Minoans'. He suggests that the focus on symbolism hampers attempts to study human-environment relations in Bronze Age Crete by reducing animals, plants and landscape features to carriers of human meanings.

Related to the idea that the Minoans loved nature is the idea that Minoan Crete was essentially peaceful and non-violent. This continues to shape archaeological interpretations and remains widespread outside archaeology, even though it has been challenged by various authors who suggest that it arose from Evans's pacifist and escapist fantasies, imposed on the society he was instrumental in revealing (McNeal 1973; Starr 1984; MacGillivray 2000b: 6). Two finds in the area of Knossos, made a year apart in 1979 and 1980, provided evidence of a less peaceful society: children's bones with signs of defleshing and a skeleton on a stone platform interpreted as a human sacrifice (Wall et al. 1986; Sakellarakis and Sapouna-Sakelleraki 1997: 268–311). Further evidence for human sacrifice has subsequently been found at Chania in western Crete (Andreadaki-Vlazaki 2015). A recent reassessment has placed warfare at the heart of Bronze Age Cretan society (Molloy 2012). There has always been evidence to support this view, but it has tended to be downplayed. The 'Harvester vase', for instance, was initially published as depicting a military procession, before being renamed when it was reinterpreted as a religious festival (Savignoni 1903: 83–8; Forsdyke 1954). Before Arthur Evans excavated Knossos, he travelled around Crete and traced what he regarded as a military road with fortified guard houses (Evans and Myres 1895; Brown 2001: 202–7). His use of the term 'Pax Minoica' to describe these signs of the militaristic control of Crete and the surrounding seas by the priest-kings has subsequently been misunderstood and elaborated as an argument that Crete was inherently peaceful (Evans 1930: 6; Momigliano 2020: 65–6).

These interpretative schemes continue to flourish in the analysis of the body of material that has been established as 'Minoan art'. The naturalistic objects highlighted by Evans continue to be regarded as part of a canon of art objects, reinforced by their reproduction in art books and guides to Minoan civilisation (Zervos 1956; Marinatos and Hirmer 1960; Higgins 1967; Hood 1978; Poursat 2008, 2014). These objects have provided inspiration for artists, writers and designers, helping to embed them further in Western culture (Boucher 2014; Momigliano 2020). Some of the most famous examples of Minoan art, such as the 'Taureador fresco' or the 'Snake goddesses', were heavily reconstructed by restorers working for Evans, and so are partly twentieth-century products (Lapatin 2002). Terms invented or adapted by Evans, such as 'horns of consecration' or 'sacral knot', continue to be used as part of an artistic vocabulary which helps to maintain the mutually reinforcing relationship between Minoan art and Minoan religion. Like Evans's restorations, these too are twentieth-century constructs which are in need of re-evaluation.

A more explicitly theoretical approach to Aegean Bronze Age archaeology was established in the 1970s, although it left intact the relationship between nature and Minoan civilisation. Colin Renfrew (1972) introduced systems theory to explain the development of complex society in the Aegean. Drawing on the field of cultural ecology, he suggested that culture intervened between humans and nature; the more complex society became, the less direct contact people in that society had with nature. The 'emergence of civilisation' which gave him the title of his book was also the growing insulation from nature: 'Civilisation is the complex artificial environment of man; it is the insulation created by man, an artefact which mediates between himself and the world of nature' (Renfrew 1972: 13). Art belonged to one of the subsystems and was one of the means of mediation between the two. While animals, including fish, had been depicted from the Early Bronze Age, 'these forms are seen rather schematically until the sudden passionate awareness of nature which grips Crete and the Cyclades at the very end of the middle bronze age' (Renfrew 1972: 438). Here nature is defined in opposition to civilisation, with artists returning to nature once it was safely under control.

Despite the differences in terminology, Yannis Hamilakis has suggested that the same interpretative scheme underlies Renfrew's application of systems theory and Evans's evolutionary ideas expressed in the tripartite Minoan pottery classification: 'As a result of the persistence of cultural evolutionist concepts in Minoan archaeology (combined with Europeanist, romantic myths) Minoan societies are most often viewed as a unified, homogeneous and undifferentiated whole, following a mostly unilinear, progressive development, throughout their history and geographical extent' (Hamilakis 2002b: 11). This evolutionary scheme has tended to obscure the differences between Cretan Bronze Age sites, with archaeologists also downplaying evidence for conflict within or between them.

The twenty-first century has seen a more radical reassessment of Minoan Crete. The term 'Minoan' itself, which Evans championed, has been challenged because of the way it homogenises the archaeology of Bronze Age Crete, with cultural developments mainly originating from Knossos. Whether or not pottery specialists subscribe to its evolutionary implications, there is a growing debate about whether the pottery chronology developed for Knossos can usefully be applied to the rest of the island (Brogan and Hallager 2011; Macdonald and Knappett 2013). The assumption behind the Minoan pottery chronology, allowing archaeological levels to be dated, is that pottery styles change evenly across Crete. The increasing recognition of regional production of pottery styles which are not in step with developments at Knossos threatens to undermine the unified nature of Minoan pottery studies. Knossos was the largest settlement on Crete for most of the Bronze Age and appears at times to have been the centre of political control of most of the island. As a result the undoubted similarities in pottery across Crete at various periods can be seen instead as indicating political relationships between sites (Knappett 1999a, 2002b).

The term 'palace' has also come under scrutiny in recent publications, while the number of this type of building has been steadily increasing as new palatial buildings, with or without central courts, have been excavated (Adams 2017: 70–140; Christakis 2020). Nicholas Platon, who excavated the palace at Zakros, used shared architectural developments as the basis for the widely used palatial chronology, which sits alongside the pottery chronology (Platon 1956). As can be seen from Table 1.1, the palatial phases do not sit neatly within broad pottery phases and so they help move away from the latter's evolutionary associations. Although palaces have only been found at a small number of sites in central and eastern Crete, their development and relations are central to an understanding of the Cretan Bronze Age. According to Platon's scheme, the widespread destructions and rebuildings of palaces at the end of the Old Palace or Protopalatial period marked the transition to the New Palace or Neopalatial period. Following the end of the Neopalatial period, only Knossos continued to operate in the same way, in what has become known as the 'Final Palatial' period (Rehak and Younger 1998). The term 'Postpalatial' is now reserved for the period in which there was no longer any written administration in Crete, but debates about the final destruction of Knossos and discoveries of Linear B documents at Chania make it difficult to date the transition from Palatial to Postpalatial Crete (Preston 2008).

There have been a number of suggestions for more neutral, descriptive alternatives to the term 'palace' which focus on the architectural layout of these buildings, such as 'court compound' (Driessen 2002, 2004: 80). Ilse Schoep (2002b), who prefers the term 'court-centred building', points out that there are a variety of different types of monumental buildings in Bronze Age Crete which have administrative functions, but which do not have a central court. One example is the villa at Ayia Triada, where clay documents and sealings were found. The term 'villa', implying the country retreat of those in the palace, is another term whose use has been debated (Hägg 1997). Although most archaeologists would no longer subscribe to the idea that there was a Minoan royal family moving between residences, these terms implicitly shape assumptions about the exercise of power in Bronze Age Crete.

Even the figure of the priest-king has been called into question, both materially and conceptually; reanalysis of the fragments from which the fresco was restored has shown it is not clear whether the crown belongs to a sphinx or the human figure, which itself could actually be part of a boxing or bull-leaping scene (Coulomb 1979; Niemeier 1987, 1988; Shaw 2004). Although there are other depictions of seemingly important humans holding staffs or flanked by animals, the lack of a clear ruler iconography in Bronze Age Crete has increasingly been noted (Rehak 1995b; Driessen 2002). Some scholars have concluded that hierarchical power structures were lacking on Crete and have suggested alternatives. Discussions of heterarchies and factions, particularly for Protopalatial Crete, highlight the possibility that power was distributed between different groups, whether co-operating or competing (Hamilakis 2002a; Schoep and Knappett 2004). One response has been to use the vague term 'elite' to describe the presumed ruling group. Like 'palace', with which it is often interchangeable, it has the advantage of being generally agreed upon and understood, although the imprecision of this term has also been highlighted (Legarra Herrero 2016). The same term is used for ruling groups throughout the Bronze Age, eliding the social organisation of Prepalatial communities based around single villages with that of the later palaces and their extensive territories. Another response is to shift focus from abstract notions of power to relationships between humans and non-humans from which hierarchies arise.

The decipherment of Linear B in 1952 has meant that the social structure of the 'Final Palace' period, when Knossos was the only surviving palace on Crete, is better understood. Michael Ventris established that Linear B, used both on Crete and in mainland Greece, was used to write an early form of Greek (Ventris and Chadwick 1956). A large proportion of the tablets at Knossos were shown to deal with the administration of the wool industry, from the control of single flocks of sheep to the management of groups of textile workers. Some insights into political organisation emerge from the tablets; by this time a figure called a 'wanax', a male leader, appears to have been at the top of a hierarchical political structure. This structure emerges from the study of tablets from both Knossos and the palaces of the mainland, particularly Pylos, although those from Knossos probably date from over a century earlier. The precise date of the Knossos tablets, and hence the destruction of the palace, is a contentious subject, complicated by the more recent acceptance that some tablets belong to one or more earlier episodes of destruction (Palmer and Boardman 1963; Driessen 2008). Linear B is recognised to have developed from the as-yet-undeciphered Linear A script, used across Crete until the end of the Neopalatial period, but the two scripts were almost certainly used to represent different languages. For a time after the decipherment of Linear B it was assumed that Evans's ideas about Minoan domination of mainland Greece had been decisively overturned, with the use of Greek at Knossos instead providing evidence for a Mycenaean invasion. Some scholars are now more cautious about assuming that a change of language at Knossos signals such an event, arguing instead that closer contact between Crete and the mainland in this period led to a process of adaptation or hybridisation (Preston 2004; Maran 2011). Linear B, adapted from Linear A to write Greek, is one example of this process.

There has also been a growing dissatisfaction in Aegean Bronze Age scholarship with the terms 'Minoan' and 'Mycenaean', with their implications of homogeneous cultural groups facing one another across the Aegean Sea. One answer has been to turn to the terms 'Minoanisation' and 'Mycenaeanisation' in order to examine the processes of culture change (Gorogianni et al. 2016). Thus Knossos and Crete become 'Mycenaeanised' in the Final Palace period and after, when Chania in western Crete appears to become an increasingly important centre. 'Minoanisation' refers particularly to the appearance of Cretan material culture forms outside Crete in the Neopalatial period, although the Minoanisation of Crete itself should also be considered (Broodbank 2004: 51; Knappett 2016, 2018). Rather than assuming that Crete was a homogeneous entity in the Bronze Age with material culture changing simultaneously across the island, it has become increasingly important to consider the relations between sites, and the growing influence of one site in particular: Knossos.

Over a century after his excavations began, Evans's discoveries at Knossos remain fundamental for understanding Bronze Age Crete. As a recent archaeological survey has demonstrated, it was the largest settlement on Crete for much of the Bronze Age (Whitelaw 2017; Whitelaw et al. 2018; Whitelaw 2019). Its political and cultural centrality coincides with the period Evans identified with the highpoint of naturalism in art. The following chapters will argue that these developments are linked, with changing styles and technologies of depiction connecting the palace with the animals beyond its walls. The bodies of these animals became constituents of the collective centred on Knossos, whose expansion was undoubtedly accompanied by violence against humans and non-humans. As is increasingly recognised, generalisations about 'Minoan Crete' obscure the relations between sites on Crete and their different histories (Christakis 2020: 106). Stereotypes involving 'the Minoans' have tended to imply island-wide unity and downplay the existence of conflict. Before considering an alternative approach to 'Minoan civilisation' it is necessary to turn to one of the concepts on which it is predicated: an antithetic natural world.

CRETAN NATURE

Cretan natural history was established in parallel with the archaeology of the island. It started with early travellers' observations of the local wildlife and fossils (van der Geer et al. 2006: 119, 2010: 44–9). The late nineteenth century saw the arrival of more specialised scientists, including palaeontologists and zoologists, who were responsible for recategorising the remains of Cretan animals such as agrimia as scientific specimens. Among these were Felix Raulin, who listed the mammals of Crete, and Vittorio Simonelli, who discovered the remains of fossilised elephants (Raulin 1869: 252–3; Simonelli 1897, 1908). Another nineteenth-century traveller to Crete, Captain Thomas Spratt, contributed a number of observations on the palaeontology and geology of the island, as well as its archaeology (Spratt 1865; Moore 2013: 85–144). His companion, Colonel Henry Drummond-Hay, provided a list of birds (Drummond-Hay 1865). Spratt (1865: 12) describes him as 'a keen sportsman and a patient and earnest ornithologist' who shot many of the birds he observed for his collection of skins. Spratt and Drummond ascended Mount Ida in 1851 to take triangulations as part of the British Admiralty's survey of the island, with the resulting charts mapping its mountains and plains according to the spatial logic of modernity.[8]

They were followed in 1904 by Dorothea Bate, who collected both present-day and ancient specimens over the course of a four-month expedition, allowing her to provide the first synthetic account of Crete's fauna. Her collection included the fossilised bones and teeth of a number of extinct animals, including deer, elephants and hippo, the skins of birds and mammals and even three live rabbits (Shindler 2005: 145). She summarised her discoveries in scientific papers and a contribution to one of the first guidebooks to Crete, Aubyn Trevor-Battye's (1913) *Camping in Crete*, which can be seen as the forerunner to modern field guides to the wildlife of Crete (Bate 1913). Finds at various locations at either end of the island enabled Bate to build on previous work to describe the animals of the Pleistocene, including a dwarf form of elephant and pygmy hippos (Bate 1905b, 1907). She also predicted that the shrews she had found in fossil deposits would have survived, despite not finding live specimens (Bate 1905a). The Cretan white-toothed shrew (*Crocidura zimmermanni*) was indeed discovered subsequently and is now regarded as the only mammal endemic to Crete, that is, part of the older Pleistocene fauna (Vigne 1999: 309). For this reason it is included in the IUCN Red List of Threatened Species, while, as noted above, the agrimi is not because wild goats are regarded as feral rather than endemic.

These specimens were assembled into the modern understanding of the natural history of Crete, with its strange Pleistocene mammals replaced by more familiar ones in the Holocene geological epoch. Debates continue over whether the disappearance of Pleistocene endemic mammals from Crete and other large Mediterranean islands at the start of the Holocene was due to climate change or human impact (Lax and Strasser 1992; Vigne 1999; Broodbank 2013: 212–18). Palaeolithic tools of at least 130,000 years old have been found in the south of Crete, showing that humans were present on Crete for at least part of the Pleistocene, but as yet there is no direct evidence for the hunting of the endemic Cretan deer, hippos or elephants by humans (Grove and Rackham 2001: 73; Strasser et al. 2011). The earliest evidence for the arrival of domesticated animals and plants on Crete in the early Holocene comes from the Initial Neolithic levels of Knossos, settled in around 7000 BCE (Evans 1964). Here Jarman characterised the introduction of domesticated sheep, goats, pigs and cattle at the start of the Neolithic as marking 'a complete faunal break' between the Pleistocene and Holocene, noting that: 'It is interesting to note that man has acted on Crete not as an agent of zoological impoverishment, as he has done in so many mainland situations, but has considerably enriched the Cretan fauna over the past eight thousand years' (Jarman 1996: 221). Yet the arrival of these domestic animals in the Neolithic did not mark the end of species turnover, since the deliberate and accidental introduction of wild animals continued in the Bronze Age and later. In ecological terms all of these introduced species can be described as 'exotic'; during the course of the Bronze Age other exotic animals, such as monkeys and cats, are likely to have been brought to Crete, as will be discussed in Chapter 7. At the same time, 'zoological impoverishment' was underway on the mainland with the gradual retreat of the lion from Greece, although it was present in the Middle East until the nineteenth century (Cheylan 1991: 231; Thomas 2014).

More recent ecological accounts stress the role of humans in Cretan natural history (Rackham and Moody 1996; Grove and Rackham 2001). Environmental historians and geographers have traced similar anthropogenic processes across the Mediterranean Basin, accompanied by modifications to the landscape including burning and forest clearance (Blondel and Aronson 1999: 45). The picture that emerges from the biogeographers and environmental historians is of a landscape which cannot be understood apart from the actions of humans, and associated animals and plants, over the last 10,000 years (Allen 2001). The environment experienced by inhabitants of the Aegean in the Bronze Age was already inscribed with the actions of others. In the words of Jacques Blondel and James Aronson (1999: 199): 'Apart from sheer, vertical cliffs and some remote mountainous areas, there is probably no square metre of the Mediterranean that has not been directly and repeatedly manipulated, and, one might say, "redesigned" by humans.'

There is an increasing debate over whether the Holocene is now at an end and has been succeeded by the Anthropocene geological epoch. Although geologists have tended to look for a recent marker of human activity such as an increase in radioactive isotopes or plastics as a signal for the Anthropocene, some archaeologists have suggested that the decisive break is the emergence of agriculture early in the Holocene. As Bruce Smith and Melinda Zeder (2013) argue, the domestication of animals and plants and the spread of farming resulted in a remodelling of ecosystems by human societies. Mediterranean islands show this process particularly clearly, with the replacement of endemic animals and change in land use. This does not mean that these islands were somehow 'unspoilt' before the arrival of agriculture; Cretan plants were already adapted to fire and browsing by Pleistocene mammals, but humans brought new browsing animals and new plants with accompanying regimes of land clearance (Grove and Rackham 2001: 166; Rackham and Moody 1996: 115–18). Whether or not the Anthropocene stretches back that far, it is clear that the environment of Crete was fundamentally changed by the arrival of a new group of humans, animals and plants in the Neolithic.

A Cretan 'nature' devoid of humans and domestic animals continues to be staged by modern tourism and conservation initiatives. It is exemplified by the Samaria Gorge Biosphere Reserve which is nevertheless populated by goats introduced by humans in the Neolithic and traversed by thousands of modern-day tourists. The same ironies of conservation can be seen on Cyprus, an island with a similar ecological history to Crete, where wild sheep (mouflon) are treated differently from feral donkeys even though both were introduced to the islands by humans and now live beyond their control (Mrva-Montoya 2015). One approach is to regard non-endemic species of animals as part of the cultural heritage of Mediterranean islands (Masseti 2009). These debates are relevant to modern conservation, resulting in modern animal practices focusing on the protection of particular animal species because of their perceived cultural or natural significance. They are ultimately the product of an imposition of a modern study of natural history on the animals and plants of Crete from the nineteenth century.

The status of agrimia on Crete encapsulates the problems of defining 'nature' as separate from 'culture' in general or 'Minoan civilisation' in particular. Nature and culture have been intertwined on Crete since humans brought a new way of dwelling to the island in the Neolithic, involving a particular group of animals and plants. The effects of the interactions between them and the soils and rocks of Crete continue to unfold. Attempts to classify agrimia, bound up with scientific nomenclature, as either 'wild', 'feral', 'hybridised' or 'domestic' goats oversimplify the interactions between humans and these animals on Crete, and the way that these have changed over time (Harris and Hamilakis 2014). Populations of goats, as well as various other animals, have moved in and out of human control since the Neolithic period, their relations defined at various times by herding, hunting or preservation. Environmental historians and ecologists have traced these changing relationships and their effect on the Cretan plants and soils, moving beyond simplistic distinctions between natural and cultural impacts. This opens the way to reconsidering human-animal relations in Bronze Age Crete.

OUTLINE

This book focuses on the changing relations between humans, animals and things in Bronze Age Crete. It draws on a relational perspective derived from the Actor Network Theory of Bruno Latour, among others, which has become increasingly influential in archaeology (Olsen 2010; Watts 2013; Alberti 2016). Latour (1993: 4, 2005: 75) uses the term 'collective' to describe groups of humans and non-humans, abandoning the distinction between nature and society. It is an inclusive term which encompasses the relations between humans, animals and things; the term 'palace' is amenable to this definition because it is already used to describe both a building and a variety of relationships and material culture associated with it, particularly in the use of the adjective 'palatial'. Rather than abandon an old and familiar term, the following chapters will broaden the meaning of 'palatial'. A 'palatial collective' is the entity comprised of the relationships between humans, animals and things which was centred on the buildings known as 'palaces'. The palatial collective was maintained through many different types of such relationships, some of which can be traced because objects which took part in these relationships, such as Linear B tablets, still survive.[9]

The modern collective spans the globe, marking a dramatic increase of scale in the number of connected entities compared to Bronze Age Crete. Topographers, naturalists and archaeologists increasingly visited Crete in the eighteenth and nineteenth centuries and their practices helped to absorb it into a modern, globalised collective. At that time, it was part of the Ottoman Empire, a collective centred on what is now Istanbul. Before that it was part of a maritime Venetian collective, traces of which can still be seen in the main harbour towns of Crete, particularly Heraklion, Rethymnon and Chania (Rackham and Moody 1996: 3–4). Over the millennia Crete has been part of much larger collectives, such as the Roman Empire, or divided into smaller collectives such as the city states of the Archaic and Classical Greek period (Whitelaw 2018: 214). Over the course of the Bronze Age, the collectives of Crete merged, first into palatial collectives and then in the Final Palace period to a multi-centred collective focused on the mainland, with centres at Knossos and perhaps another at Chania in the west. Following Latour (1993, 2004), this book seeks to go beyond the 'modern constitution' which divides the world into natural and cultural phenomena, and which has been applied uncritically to Bronze Age Crete. Instead it focuses on the collectives which expanded and contracted during the Bronze Age (Shapland 2013). The term 'Minoan' obscures this dynamic picture and so will be used only to describe the vision of Bronze Age Crete developed by Arthur Evans which developed over the course of the twentieth century and is now increasingly the subject of critical scrutiny.

It is not necessary to abandon the modern chronological framework of Bronze Age Crete defined by the rise and fall of the palaces; the distinction between Prepalatial and Neopalatial Crete is one of scale, which will be described here in terms of domestic, local and palatial collectives. Palatial collectives comprised a greater number of different types of bodies and things than local collectives and extended over a greater geographical area. Domestic collectives, throughout the Neolithic and Bronze Age, were individual households: groups of humans, domestic animals, houses and their contents. The household continued to play an important role even as palatial collectives formed around the palaces (Driessen 2010; Relaki and Driessen 2020). Local collectives were formed of groups of interacting domestic collectives centred on discrete areas of agricultural land (Haggis 2002). In the Prepalatial period they were focused on communal tombs; they could also be described as mortuary collectives. The palatial collectives which emerged in the Protopalatial period incorporated many different domestic and local collectives, coinciding with the localised tombs going out of use and a decreasing investment in burial. The courts to the west and at the centre of the palaces instead became locations for activities such as dancing, processions or communal consumption activities. In this way the central courts of the palaces acted as a focal point for the interactions between domestic collectives (Driessen 2004). These interactions were more inclusive and communal in the Protopalatial period, becoming more hierarchical and restricted in the Neopalatial period (Driessen 2018). The Final Palace period saw Knossos becoming the only remaining palatial collective, having absorbed other palatial collectives across most of Crete. This collective was itself closely related to palatial collectives on the mainland, and was perhaps centred there rather than on the palace of Knossos. Over the course of the Bronze Age the collectives of Crete became part of wider networks of contact and exchange, resulting in the changes defined by the palatial chronological system.

Naturalism as defined by Evans is primarily a feature of palatial collectives. The agricultural activities of domestic collectives involving animals continued, but these are almost never shown in palatial visual culture. Instead a variety of animals are depicted which are involved, explicitly or not, in non-domestic animal practices. The 'nature-loving Minoans' are a product of a modern interpretation of the way human-animal relations were depicted by Bronze Age palatial collectives. Chapter 2 sets out the theoretical basis for this argument, starting with the famous observation by art critic John Berger that animals have disappeared from modern life, making it difficult for modern city-dwellers to understand the animal art of the past. Berger's observation had a basis in the anthropological thinking of Claude Lévi-Strauss, whose student, Philippe Descola, provides the system of ontologies explored in this book. Descola's (2005, 2013a) scheme of different types of ontologies helps to formalise the idea that there are different ways for humans to relate to the world based on fundamental assumptions about the similarity and difference of other inhabitants of their world. For Descola 'naturalism' is the ontology of modern Westerners who relate to animals as biologically similar but mentally and culturally different. This ontology results in 'nature' being regarded as a separate domain from 'culture', an idea which is far from universal among human groups. This observation was fundamental for the work of Latour, who applied this insight in his anthropology of the moderns (Latour 1993: 91);[10] like a number of other anthropologists, Descola's experiences in Amazonia, particularly among the Achuar, resulted in a realisation that the nature-culture division was not relevant in many non-Western societies (Descola 1996a).

Descola classified the ontology most frequently found among small-scale Amazonian collectives as animist, in which humans and animals are fundamentally similar despite surface appearances. Shamans act as intermediaries because they are able to see through the bodily difference of animals and relate to them as fellow humans. Bronze Age Crete has been regarded as animist (and shamanic) but this ontology is more usually associated with small-scale hunter-gatherer collectives (Bird-David 2018). Another ontology Descola terms 'analogism', which treats the world as a series of discrete entities, and whose collectives are formed by establishing connections between them on the basis of analogies. This is the ontology of most premodern agricultural collectives, including, it will be suggested, those of Bronze Age Crete. Archaeologists have instead interpreted Bronze Age Crete in terms of their own naturalist ontology, producing a separate domain of nature which is appreciated aesthetically and worshipped. Although Bronze Age Cretan animal depictions sometimes resemble our own naturalist animal art, they arise from a different set of human-animal relations. As Hamilakis (2002b: 18–19) suggests, it is necessary to de-familiarise the Minoan past, in this case by refusing to elide Western naturalism and the naturalistic art of Bronze Age Crete.

Disengaging Western and Bronze Age naturalism necessitates a reconsideration of the role of animal depictions in Bronze Age Crete. For Descola (2010b), the role of art objects differs between ontologies. In Western naturalism, animals belong in the domain of nature but are carried across to the domain of culture through artistic representation. This enforces the separation of art objects, as cultural products, from animals, as members of the natural world. A particular style of depiction is associated with this naturalist ontology; mathematical concepts were employed to create the illusionary space of perspective drawing. There have been a number of reactions against the supposed universality of Western realism in art, some of which are explored in Chapter 2 because they open up the way to considering animal depictions in different ontologies. For James J. Gibson (1979), depictions were only ever an incomplete substitute for the perception of real-world forms. This relational approach renders images as part of the process of interaction between humans and animals, providing information about their properties, which he termed affordances. Affordances are not fixed properties but rather emerge in interaction between different entities. As a result, humans do not simply impose meaning on animals, rendering them symbols of human concerns, but the affordances of animals in a given situation help shape human relations with them. Depictions provided information about animals in Bronze Age Crete; the detailed depiction of animal bodies and behaviour often described as 'naturalistic' helped to reveal their affordances.

Gibson's affordance theory is consistent with a broadly pragmatic account, which shifts the focus from what depictions represent to what they do. Working within a semiotic approach developed by C. S. Peirce, a number of scholars have examined how artworks do not simply represent the world, but act within it. Instead of symbols, with a conventional relationship with what they depict, artworks can be described as indexes, establishing connections with both their maker and what they depict (Sonesson 1994: 313–23; Gell 1998: 13). David Summers (2003: 687) uses the word 'trace' to describe artworks which have a physical connection with what they depict. This term does not just apply to artworks; all seal impressions are traces of the seals which produced them. More generally, he and other scholars have suggested that artworks make absent bodies and things present, so that they are substitutes or 'real metaphors' (Summers 2003; Belting 2005). This book focuses on animal traces and depictions, for which the term 'animal thing', from human-animal studies, will be used (Fudge 2012; Poliquin 2012). Animal things are traces when they retain a connection with the animal, such as bucrania (cattle skulls), or metaphors when they stand in for animals, such as stone rhyta in the shape of cattle heads. In order to distance this account from the modern concept of Minoan art, terms such as depictions and animal things will be used in this book. Although now regarded as an artwork, the bull's-head rhyton from the Little Palace at Knossos acted more like a decapitated head when used in Bronze Age Crete, with its rock-crystal eyes glinting and liquid spewing out of its mouth.

Affordances emerge from an interaction between two entities but they are also shaped by context. A charging bull affords injury to any human he catches in his horns, but in Bronze Age Crete this affordance became part of the dangerous activity of bull-leaping. Such activities will be described as animal practices, which are context-specific and arise from the human-animal relations of particular groups. The concept of animal practices, also from the interdisciplinary field of human-animal studies, complements the affordance concept by focusing on the relations between humans and non-human animals in a particular time and place (Elder et al. 1998). Wall paintings at Knossos helped to shape the affordances of cattle in Bronze Age Crete by showing them as animals to be leapt over; these wall paintings had different affordances from metal rings showing bull-leaping, which could be stamped in clay, preserving an impression of the leap. These depictions, which indexed bull-leaping, and made this animal practice present in various different contexts, can be described as animal things.

This book is intended as a contribution to human-animal studies focusing on the human-animal relations of a prehistoric society, outlined in Chapters 3–7.

Whereas Chapter 2 focuses on the diachronic human-animal relationship of looking, the following chapters outline the animal practices of Bronze Age Crete, starting with 'herding'. The lack of historical records makes it difficult to identify particular animals of Bronze Age Crete, with the notable exception of the cattle named, or perhaps simply described, on Linear B tablets from Knossos discussed in Chapter 3. Instead, animal remains and animal depictions contribute to an understanding of the animal practices of Bronze Age Crete, particularly relating to the palace. The palatial collective consisted of a variety of non-human animals, humans and things. As the Linear B tablets show, it was also defined by an unequal relationship between these elements; both the named oxen, who were loaned out for ploughing, and the sheep, shepherds and textile workers also detailed on the tablets were under the control of humans at the centre of the palatial collective. Despite the modern stereotype of the nature-loving Minoans, the human-animal relations of Bronze Age Crete do not provide a utopian contrast to those of the present. Nevertheless, the importance of cattle and sheep to the establishment and maintenance of the palatial collective is highlighted. Just before the palatial collective of Knossos fell apart, a significant part of it was composed of sheep, wool, human shepherds, dyers and textile workers and the clay things which traced the manufacture of textiles. Hundreds of years before, the palaces emerged partly as a means to co-ordinate the domestic production of textiles for overseas export.

Animal practices involving domestic animals are very rarely shown in Bronze Age Cretan material culture. Herding sheep, goats and cattle was an animal practice shown in some Prepalatial and Protopalatial animal depictions but the emergence of the Cretan Hieroglyphic script and associated animal depictions on sealstones signals a shift in human-animal relations. Animal heads of domesticates on sealstones and on clay documents were related to provisioning of communal consumption events. The deposition of animal figurines at peak sanctuaries, ritual spaces at high points in the landscape, helped to define the territories of local collectives. Two animal practices involving cattle, hunting and bull-leaping, are depicted from the start of the Neopalatial period. These are indicators of the expansion of the palatial collective of Knossos, which came to be the only remaining palace following the widespread destructions at the end of this period. It is argued that depictions of cattle-hunting relate to the raiding of the herds of other collectives by Knossos, most likely as part of a wider military campaign. Bull-leaping is an animal practice which came to define the territorial control of Crete by the Knossos palatial collective; cattle would have been rounded up and brought to Knossos or its outposts from across the island for stock-taking; various bull games provided an opportunity for the young men of the collective to show their physical prowess. The horn, hides and meat of butchered cattle were important markers of these activities and they are widely depicted. There are no depictions of cattle sacrifice until later because it was the living cattle which defined the territorial extent of Knossos; the depiction of cattle heads and the deposition of cattle skulls commemorated the consumption, and perhaps also the hunting, of these important animals.

Chapter 4, 'Butchering', examines animal remains both in terms of traditional zooarchaeological approaches which identify them to genus or species and as animal things. Although some zooarchaeologists have discussed animal practices such as ploughing (Isaakidou 2008), traditional zooarchaeological reports consist of species lists and counts of specimens from which the diet of the human members of the collective can be inferred. When confronted with animal bones which do not appear to be economically useful, either less-meaty body parts like skulls or whole skeletons, archaeologists have tended to regard these as ritual deposits resulting from the animal practice of sacrifice. This practice was important in later Greece but is not depicted on Crete until the Final Palace period. By regarding animal skulls as trophies rather than symbols of sacrifice, the focus shifts to the significance of the animal remains. The importance of animal heads in earlier palatial material culture, including naturalistic animal-head rhyta, points instead to displayed heads being an index of consumption events. Whereas animal heads would have soon rotted to become skulls, naturalistic animal-head rhyta, used for pouring wine, preserved the animal head as it appeared in life; these animal things could be used repeatedly in communal gatherings. In Bronze Age Crete, animal heads were more likely to have been animal things related to the provision and consumption of animals. Such consumption events were central to the emergence and maintenance of the palatial collectives.

Whereas bull-leaping was depicted at the centre of the palatial collective on the walls of the Knossos palace, the animal things most closely related to the animal practice of hunting were the seals worn and used by individual human members of the collective. Chapter 5 focuses on these miniature animal depictions which have frequently been seen by archaeologists in terms of Western naturalism, encapsulated by the term 'animal studies'. Among the most finely crafted seals are gold rings, such as the so-called Burgon ring. Instead of exercises in artistic skill and observation, they are argued to be tokens of membership of the palatial collective. When impressed on clay sealings, sometimes used to carry messages across the island, these rings showed how wild animal bodies, among many other entities, had become part of the collective in the Neopalatial period. The closely observed detail, often described as naturalistic, acted to demonstrate knowledge of the hunted animal and its behaviour. The hunting of animals has happened throughout the prehistory and history of Crete but only the hunting of certain animals is shown on sealstones. Hares, whose bones survive from Bronze Age contexts, are never depicted but wild goats are. It is suggested that the wall paintings in palatial buildings, particularly the Ayia Triada 'Park fresco' and other depictions crafted in palatial workshops, such as metal rings, show the ideological basis, and also the reason, for the absorption of goat-hunting in particular. Upland commodities, particularly crocuses, and also goat horns, were important to the palatial collectives and the animal practice of hunting had defined interactions with the uplands since the Neolithic period. Depictions of elaborately dressed women wearing dyed textiles interacting with wild goats, and sometimes boar, established the involvement of the palatial collectives with this animal practice, and by extension the mountain terrain. It is not necessary to enter into the debate about whether these women are deities or priestesses, or the content of the Minoan belief system, in order to understand the way in which the palatial collective established control over the uplands of Crete.

Turning to fishing in Chapter 6, one of the most famous aspects of Minoan artistic production is the 'Marine' style, showing a variety of marine creatures, particularly those with writhing tentacles. Its naturalism or otherwise has been debated since the late nineteenth century since it captures the movement of underwater bodies without always depicting them accurately, at least by modern standards. Arthur Evans suggested that the Marine style of Late Minoan IB pottery could be traced back to the imitation of marine forms in earlier pottery, which was sometimes directly decorated using shells or sponges, a process Evans termed 'nature-moulding' or 'nature-printing'. Importantly this draws attention to the indexicality of these depictions, the direct contact with marine forms. These underwater forms were brought into the palatial collective using these animal things in order to demonstrate the palatial collective's extension over the sea. Fishing and shellfish-gathering, like hunting, has always been a subsistence activity on Crete. The Marine style instead focused on underwater forms (such as triton shells and octopuses) encountered by diving, as well as marine animals (such as argonauts, dolphins and flying fish) which were encountered while sailing across the sea. An analogy with hunting and upland Crete can be made; like crocuses, purple shells were an important source of dye for the textile industry and so the material culture of the palatial collective established control over the underwater world by incorporating these distinctive marine bodies. Textiles were an important commodity produced by the palatial collectives for overseas trade, in return for which they received raw materials such as metals. The deposit known as the Temple Repositories at Knossos includes both painted shells and faience marine forms, as well as depictions of crocuses, agrimia and women wearing elaborate textiles, including the so-called 'Snake goddesses'; rather than interpret this deposit solely in terms of religious beliefs, it is possible to see how it connects the parts of the world which the palatial collective brought together at its centre.

The overseas world, increasingly connected to Crete by trade links over the course of the Bronze Age, was the source of new animal bodies collected in Chapter 7. Many of these first arrived as depictions on small portable objects; these bodies were incorporated into local and then palatial collectives using various types of material culture. Some bodies, like the cat, horse and monkey, became physically part of the palatial collective alongside their depiction when they arrived from overseas. Few physical traces of their presence in the form of animal remains have survived from the Bronze Age – none so far, in the case of monkeys – but depictions help show when they arrived and also the place they were given in the collective. Monkeys, for instance, became associated with the areas of the landscape the palatial collective extended across, shown on frescoes at Knossos both picking crocuses and stealing birds' eggs in a rocky landscape. Other animals became part of the animal practice of hunting as the palatial collective of Knossos started to extend its connections to the mainland where lions and deer could be hunted; their pelts and remains were brought back to Crete as animal things, and there is evidence that live deer were brought to Crete to be hunted and consumed in the Neopalatial period. The interaction between Crete, the mainland and the Cyclades in this period resulted in what has been termed the 'Aegean animal style', which showed animal attack scenes. Through the movement of luxury objects and fresco painters this was absorbed into the 'International style' of the Late Bronze Age Eastern Mediterranean as Crete became part of a wider Aegean palatial collective, interacting with others overseas.

From the point of view of a naturalist ontology, the extent of the natural world can be defined as the animals, plants and other life forms which inhabit the planet. This excludes monsters such as griffins which do not exist and so, as products of the human imagination, belong to the domain of culture. As will be suggested in Chapter 7, regarding the griffin as a supernatural creature which is therefore different from existent animals such as the horse and lion is not a productive way to understand Bronze Age Crete. Like lions, griffins are shown hunting deer on Minoan sealstones; like horses (and wild goats), griffins are shown pulling chariots. These depictions show how the griffin was incorporated into the palatial collective by analogy with other exotic forms. Lions appeared on Cretan sealstones in the Prepalatial period but over the course of the Palatial period the increasingly detailed depictions of lions suggest that encounters with them had resulted in greater knowledge of their affordances. Depictions of griffins follow the same path, but their bodies always remained out of reach. Whereas horses could actually be seen on Crete in the Final Palace period pulling the chariots of the important human members of the palatial collective, griffins could be seen on the walls of the Throne Room at Knossos. Existing at the furthest edge of the palatial collective, griffins became the ideal analogy for the extent of the collective centred on the human who sat on the throne.

Instead of a ruler-centred iconography in Bronze Age Crete there was an animal-centred iconography. As will be argued in this book, animal practices were an important part of the definition of palatial collectives, resulting in a wide variety of animal things. In Chapter 8, about the phenomenon Evans termed 'the naturalistic spirit', the changing style of animal depictions will be linked with the changing importance of animal practices. The peak of naturalism in style, around the start of the Neopalatial period, was associated with the growing influence of the palatial collective of Knossos over most of the island of Crete. Animal things were a way of capturing and disseminating the animal practices of this collective, which took place in different parts of the landscape: the cattle-grazing lowlands, the uplands inhabited by wild goats and the underwater world of octopuses and squid. Animal practices such as agriculture or coastal fishing were almost never shown. In order to maintain itself, particularly through participation in overseas trade, the Knossos palatial collective needed to extend into the productive parts of the landscape, particularly those involved in textile manufacture. It was not the authority of individual human rulers, wielding power across Crete, that secured these resources; instead these palatial animal practices helped to integrate these areas of Crete into the palatial collective at Knossos. The right to hunt or to round up cattle was reinforced by finely crafted animal things, but also almost certainly backed up by violence. As the palatial collective of Knossos was itself absorbed by those of the mainland, these animal things became more stylised and conventional, also crossing the sea to become part of the iconography of the mainland palatial collectives. By this time they had become symbols of the palace, barely related to the animal practices they arose from. The period of 'brilliant naturalism' noted by Arthur Evans was short-lived but cast a long shadow; this phenomenon is central to the understanding of the relationships between humans, animals and things in Bronze Age Crete.

TWO

LOOKING

INTRODUCTION

> That look between animal and man, which may have played a crucial role in the development of human society, and with which, in any case all men had lived until less than a century ago, has been extinguished.
>
> (Berger 1980: 26)

The strikingly realistic stone bull's-head rhyton from the Little Palace at Knossos can today be seen in the Heraklion Archaeological Museum. Discovered in 1908, it was reconstructed from fragments and the horns were restored, finished in gold. When found, its right eye, the painted pupil and iris capped by a rock-crystal lens with a red jasper rim, was still intact (Figure 2.1). As Arthur Evans observed, 'the crystal lens of the eye both illuminates and magnifies the pupil and imparts to the whole an almost startling impression of fiery life' (Evans 1928: 530). In its current setting the bull returns the gaze of the viewer, a look which was extinguished 3,500 years ago when the rhyton was smashed and buried in a stairwell. Although it is now displayed as a work of art in a brightly lit vitrine, the hole in the muzzle indicates its original purpose. When filled through a larger hole in the top of the neck, it would have spewed liquid from its mouth. It is an object, one of many from Bronze Age Crete, which raises the question of whether this is a work of art, as it was also described by Evans (1928: 530), or something else. Enclosed in its case, subject to the gaze of hundreds of thousands of tourists a year, its capacity for interaction has been reduced to looking back through the glass.

2.1. Bull's-head rhyton from the Little Palace at Knossos, LMI. Height of head: 20 cm (AMH Λ1368). © Heraklion Archaeological Museum – Hellenic Ministry of Culture and Sports – 'Hellenic Organization of Cultural Resources Development (HOCRED)'

Like the Little Palace bull's-head rhyton in the Heraklion Archaeological Museum, viewers of Damien Hirst's artwork *A Thousand Years* also encounter a lifelike cattle head in a glass vitrine. In its original incarnation in 1990 the head was a real, rotting cow's head which fed the flies whose lifecycle was terminated by the Insect-O-Cutor in the same vitrine. In its current form the head is made of fibreglass, with real hair, teeth and horns, a simulacrum of a real animal head, but the flies still inhabit the space, fed by sugar water instead (Baker 2000: 84–7). The artwork was inspired, according to Hirst, by doubts about whether his previous works were real or not.[1] In their different ways, both simulacra of cattle heads disrupt the art historical narratives about naturalism with which such objects have traditionally been described. Rather than representations, both act like presences of disembodied heads, although the possibility of physical interaction with the visitor is denied by the glass. Like other postmodern artists, Hirst wanted to challenge the art establishment, so provides a useful starting point for going beyond modernist assumptions about art. This is not to say that there is a direct analogy; Hirst's artwork carries a heavy symbolic load about the relation between life and death. But the act of placing animal remains, or simulacra of them, in the gallery space posed important questions about the relationship between animals and their representation in art.

Hirst's work, and that of other contemporary artists such as Angela Singer who utilise taxidermied animals, is a reaction against more traditional representations of animals in art. Singer's unsettling images, unlike Hirst's, have an animal rights agenda in challenging the viewer to consider the life of the animal behind the artwork (Poliquin 2012: 153–4; Baker 2013: 165–75). Her 'Dead Eyed' series takes vintage taxidermied animals and uses jewellery to replace their eyes and cover their hides (Figure 2.2). Such artworks challenge what Steve Baker (2001: 174) has called 'disnification', the trivialising representation of animals in contemporary culture in which '[t]he animal is the sign of all that is taken not-very-seriously in culture; the sign of that which doesn't really matter'. The concept can be extended to include many instances of Minoan art in contemporary Crete. The charging bull on the North Entrance Bastion at Knossos can now be consumed in the form of a snow globe or fridge magnet

2.2. *Unexplained Recovery* (Dead Eyed series), 2010. Angela Singer. Vintage recycled taxidermy stag, vintage jewels. 810 × 512 × 421 mm. © Angela Singer

from a nearby tourist stall. In the process it stands alongside other disnified representations of charging bulls in contemporary culture such as the bronze bull symbolising the stock market in New York or those adorning the logos of fast cars or energy drinks. As Baker suggests, however, twentieth-century art includes both disnification and a reaction against it, resulting in artists exploring other means to represent animals. Such artworks challenge the viewer to take animals seriously again and offer a different starting point for considering Bronze Age Cretan animal depictions.

In 1971 the art critic John Berger wrote a piece in the magazine *New Society* called 'Animal world'. It was a response to a book by Francis Klingender called *Animals in Art and Thought to the End of the Middle Ages*. 'The realistic depiction of animals, with which art itself began, seems to have changed very little from civilisation to civilisation', noted Berger (1971: 1042), drawing a contrast between the 'special relationship' humans had with animals in the past and the impoverished relations of the present. This brief review became the basis for his essay 'Why look at animals?', which has become well-known for the idea of the 'disappearance' of animals from modern life (Berger 1980). In it he proposed that animals began to become increasingly marginal to daily life in Europe and North America in the nineteenth century, reappearing in institutions such as zoos. As he suggested: 'Everywhere animals disappear. In zoos they constitute the living monument to their own disappearance' (Berger 1980: 24). This notion of disappearance was encapsulated in the gaze; humans can look at animals in the zoo but animals do not look back because they are inured to the presence of visitors. Berger's essay has continued to provoke debate despite his sometimes questionable assertions; cats and dogs have multiplied in cities but Berger did not regard these as authentic animals (Baker 2001: 11–15; Burt 2005; Marvin 2005). Yet his thesis of disappearance has been widely accepted, despite – or perhaps because of – its overly simplistic identification of such a disjuncture. The importance of Berger's essay here is that it placed animals and animal depictions at the centre of a dialogue between art history and anthropology in a way that has continued to be influential in the interdisciplinary field of human-animal studies. It opens up a way to consider the realistic animal depictions of Bronze Age Crete and their reception in the modern world.

By the time the Minoans came into view in the twentieth century, animals had already begun to disappear. As Berger realised when he was looking at images of animals in Klingender's book, it is difficult to see these animals in the same way as those who created the images. Klingender, in his account of Minoan art, was reliant on a characteristically twentieth-century art historical narrative derived from the work of Arthur Evans and Henriette Groenewegen-Frankfort. In comparison with the violence of the Assyrian lion hunts, Klingender viewed Minoan art as 'graceful' and 'life-affirming': 'However varied the fauna in Cretan art, reality is transmuted in a scintillating evanescent vision of life where effortless movement is the all-pervading principle' (Klingender 1971: 62). This principle had been described by Groenewegen-Frankfort in her work *Arrest and Movement* (1951) as 'absolute mobility', epitomised by octopuses with whirling tentacles or quadrupeds shown in the pose known as the 'flying gallop'. For Groenewegen-Frankfort (1951: 200), 'the flying gallop in animals means speed but not flight; it is hardly ever related to a hunting event; in fact the absence of hunting scenes in Crete as compared with Mycenae is very striking indeed'. As will be argued in Chapter 5, hunting was widely depicted in Bronze Age Crete, although the human hunters were rarely shown. By subjecting animal depictions to an art historical analysis, focusing on the depiction of space and time, Groenewegen-Frankfort helped reinforce the idea of nature-loving Minoans by seeing in Bronze Age Cretan art 'the most complete acceptance of the grace of life the world has ever known' (Groenewegen-Frankfort 1951: 216). Berger's essay is significant because it opens the way to an alternative approach to this type of analysis by instead grounding animal depictions in human-animal relations.

The study of human-animal relations has grown rapidly in the twenty-first century, leading to the characterisation of an 'animal turn' in the social sciences. The interdisciplinary field of human-animal studies has become established but is heavily weighted towards animals in the present or recent past (Franklin 1999; DeMello 2012; Waldau 2013). This is partly a result of interactions with the animal rights movement but also because of the wealth of evidence for human-animal relations in the present. Although some scholars have sought to examine human-animal relations through historical sources (Thomas 1983; Ritvo 1987; Fudge 2002b), there has been a tendency to treat ancient imagery less critically. The effect of this is often to reproduce the human-animal relations of the present, imposing the modern culture-nature dichotomy on past societies or regarding animals as symbols of something else. Berger sought to look at animals in past societies in a different way, using anthropology as a means to do so, although he imposed his own dichotomy, of animals as present or absent. Although there is a rich tradition of using anthropological concepts in archaeology, the archaeological study of animals has often been left to those who study animal remains, the zooarchaeologists. Zooarchaeology has traditionally adopted an economic approach to animals rooted in the natural sciences, but a more holistic and critical approach to animal remains has started to emerge (Oma and Hedeager 2010; Russell 2012; Boyd 2017). Here the focus is on the other major source of evidence for human-animal relations in Bronze Age Crete, animal depictions. Berger's challenge was to look at these through premodern eyes; the approach developed here will go beyond the idea of 'animals in art' as a set of passive representations to be appreciated by the active human viewer.

ONTOLOGIES

For Berger, when animals returned the gaze it revealed that they were both similar and dissimilar, sentient but not endowed with language. From this comes the dualism at the heart of human-animal relations: that animals are 'both like and unlike man' (Berger 1980: 4). According to Berger (1980: 9), only with the philosophy of René Descartes did the animal's significance start to be displaced. Yet looking is not straightforward; Berger himself was looking at animals through the eyes of anthropologists such as Roy Willis and Claude Lévi-Strauss. He regarded anthropology as providing an alternative to traditional art historical narratives. It is only by understanding the peculiarly Western point of view from which we look at both animals and Bronze Age Cretan material culture that it is possible to develop different ways of understanding past human-animal relations.

Berger was able to draw upon a long history in anthropology of studying human interactions with animals, and his essay acts as a pivot between the anthropological approaches he discusses and the work of recent cultural theorists who draw inspiration from his essay. Claude Lévi-Strauss too remains a central figure in the study of human-animal relations. His 1963 book *Totemism* evaluated the different explanations for the significance of animals in the ethnographical literature before proposing a new explanation. He suggested that totemism had come to stand for a confusion between humans and nature, marking a lower stage of cultural evolution. Against anthropologists such as James Frazer, Alfred Radcliffe-Brown and Bronisław Malinowski, he argued that totemism was simply an example of a mode of thought. Rather than marking a confusion between humans and animals, the use of animals and plants in genealogies, or to label social groups, was an example of a metaphorical way of thinking. As he elaborated in *The Savage Mind*, animal and plant species provide the premodern individual with 'the most direct manifestation he can perceive of the ultimate discontinuity of reality' (Lévi-Strauss 1966: 137). These recognisably distinct phenomena provide a parallel, metaphorical, set of entities with which to understand social reality. It is not people's interactions with animals and plants which give them their meaning, but simply their differences. As he famously argued:

> The animals in totemism cease to be solely and principally creatures which are feared, admired or envied: their perceptible reality permits the embodiment of ideas and relations conceived by speculative thought on the basis of empirical observations. We can understand, too, that natural species are chosen not because they are 'good to eat' but because they are 'good to think'. (Lévi-Strauss 1963: 89)

In *The Savage Mind*, Lévi-Strauss broadened this observation to outline a theory of concrete thought, in which the significance of animals is again related to their position in the system of differences rather than their intrinsic qualities, or as a result of people's interactions with them. Totemism becomes a part of the process of classification, and the same animal can have a different significance in different cultures. At the centre of Lévi-Strauss's theories are the workings of the human mind, and the animal becomes a useful, but essentially arbitrary, sign, a 'totemic operator' (Lévi-Strauss 1966: 151). His work has been hugely influential in the study of animals, opening the way for a number of anthropologists to describe how animals are 'good to think'.

Roy Willis, for instance, in *Man and Beast*, applied structuralist analysis on a range of African societies and explored why animals are so frequently chosen as cultural symbols:

> The distinctive peculiarity of animals is that, being at once close to man and strange to him, both akin to him and unalterably not-man, they are able to alternate, as objects of human thought, between the contiguity of the metonymic mode and the distanced, analogical mode of the metaphor. (Willis 1975: 128)

Berger quotes *Man and Beast* in his essay, so his debt to Willis, as well as Lévi-Strauss, is clear. But Berger's epigrammatic prose, in contrast to Willis's structuralist terminology, has helped to give these ideas wider circulation. As Berger put it:

> Animals are born, are sentient and are mortal. In these things they resemble man. In their superficial anatomy – less in their deep anatomy – in their habits, in their time, in their physical capacities, they differ from man. They are both like and unlike. (Berger 1980: 2)

His distillation of structuralist ideas has as a result remained influential. One widely cited human-animal studies text begins: 'Animals present a challenge to humans. They are both similar to and different from us' (Fudge 2002a: 7). The relations between humans and animals are shaped by these similarities and differences.

More recently, Philippe Descola, a student of Lévi-Strauss, has used the same premise, that humans assess others in terms of similarity and difference, in order to define a system of ontologies. An ontology, as the term has come to be used in the social sciences, is the fundamental way in which humans understand the world and the entities within it. Rather than universal mental structures, he suggests, these ontologies are derived from experience within a particular social setting (Descola 2013a: 103). Arguing that humans perceive others (human or non-human) in terms of their own 'physicality' and 'interiority' – roughly body and mind – Descola (2013a: 121) has argued that there are four possible 'modes of identification'. In this scheme 'totemism' rests on the sharing of both physicality and interiority, 'animism' on the perception of a shared interiority but different physicality, and 'analogism' on the sharing of neither. Naturalism is predicated on the relatedness of all life forms, which is fundamental to biology (shared physicality), and the use of features such as culture to set humans apart from other animals (distinct interiority). This mode of identification he sees as the modern Western one, which establishes nature as a domain separate from culture while also recognising the biological continuity between life forms.

Philippe Descola's starting point was his work on human-environment relations in an Amazonian horticultural society, the Jivaro (Descola 1996a, 1996b). Like many anthropologists, he abandoned notions of a dichotomy between society and nature as a result of his interaction with societies for whom there is no such distinction (Descola and Pálsson 1996: 7). Similarly, Eduardo Viveiros de Castro (1998, 2004) has suggested that Amazonian human-animal relationships cannot be seen in what he terms our own 'multiculturalist' terms because they offer an alternative 'multi-naturalist' way of seeing the world. In Amazonian multi-naturalism, humanity is the underlying state of all animals and people, so all see the world in human terms, but have different bodies from the point of view of one another. What tapirs see as a house, people see as a watering hole being used by tapirs. This is not simply the workings of the imagination, human or tapir, because: 'Animals see in the *same* way as we do *different* things because their bodies are different from ours' (Viveiros de Castro 1998: 478). Although the mind or spirit does not differ from species to species, the body, and hence the perspective, does. Shamans are important because they act as intermediaries, travelling between bodily perspectives. This attaches particular importance to bodily difference, that is, 'affects, dispositions or capacities which render the body of every species unique' (Viveiros de Castro 1998: 478). Animism becomes a way of seeing the world from the perspective of a particular body which does not start from the assumption that humans and animals are different in mind/spirit. Unlike Descola, however, the 'perspectivism' of Viveiros de Castro is primarily deployed as a case study to undermine Western ontologies (Latour 2009).

The question this raises, which is of relevance to Bronze Age Crete, is how to understand ontologies or perspectives which are not underpinned by the same division of nature and culture or assumptions about mind versus body (Bird-David 1999; Ingold 2000; Descola 2009: 148). Geoffrey Lloyd (2004, 2007, 2010) attempts to find a middle ground, arguing that these are reconcilable but different 'styles of inquiry'. He cites Vivieros de Castro's example of a jaguar being regarded within an Amazonian ontology as essentially a human who likes manioc beer (or blood, as it appears to people). This, Lloyd (2007: 147) argues, shows that jaguars are 'multidimensional phenomena', since zoologists recognise them as belonging to the *Panthera* genus and, crucially, anthropologists like Vivieros de Castro are able to see them, to some extent, from both points of view. This multidimensionality, he suggests, means that seeing jaguars as humans who like manioc beer and jaguars as carnivorous animals is commensurable; he further argues that there is a common level of perception for all humans, the 'psychic unity of mankind', even if culture provides different ways of experiencing the world. In this he is not arguing for a culturally relativist position because variety in perception arises from both the phenomena themselves and from the different styles of inquiry applied to them. Similarly, Descola advocates an approach he terms 'relative universalism', which starts with:

> the relations of continuity and discontinuity, identity and difference, resemblance and dissimilarity that humans everywhere establish between existing beings, using the tools that they have inherited from their particular phylogenesis: a body, an intentionality, an aptitude for discerning differential gaps, an ability to weave with any human or nonhuman relations of attachment or antagonism, domination or dependence, exchange or appropriation, subjectivization or objectivization.
>
> (Descola 2013a: 305)

For Lloyd and Descola, ontologies and their associated classification schemes are relational; they arise from interactions between humans and the entities they encounter in the world.

Bronze Age Cretan depictions of humans interacting with trees, rocks and animals have contributed to a recognition of an animist ontology (Herva 2005, 2006a, 2006b; Crooks et al. 2016; Tully 2018; Binnberg 2019a, 2019b). Closely related are identifications of shamanism in these interactions, and also ecstatic dancing (Morris and Peatfield 2004; Tully and Crooks 2015). Herva has suggested that in Bronze Age Crete depictions guided people's perception of certain stones (baetyls) and trees in the real world, revealing them as animate.

> What is being proposed here, in essence, is that the Minoans co-inhabited their everyday environment with manifold potentially living, conscious and sentient beings. Some were closer to humans in form and behaviour than others, but there is no reason to assume that the perceived agency of things was attributed to some 'supernatural' beings or forces external to things themselves – or that person-like non-human things were considered categorically distinct from the humans themselves.
>
> (Herva 2006b: 232)

Herva's fundamental point is that we should not seek to interpret Bronze Age depictions in terms of our own ontology, in which non-humans are not generally considered to be persons. As Descola (2010a: 339, 2013a: 233) recognises, in practice people might use more than one mode of identification in any given culture, and ontologies change over time. As a result it is possible that rocks and trees were viewed within an animistic ontology in Bronze Age

Crete. Nevertheless Descola argues that there is a primary mode of identification in any one culture, which provides the dominant ontology within which relations between humans and non-humans (including animals and plants, but also spirits or what we might perceive as inanimate objects) are enacted.

Other factors need to be taken into account before attempting to apply either a Western ontology or radical alternatives to Bronze Age Crete or other past societies. As various authors have suggested, the appearance of agriculture marked the conceptual transformation of human-animal relationships (Serpell 1986: 5; Hodder 1990; Cauvin 2000; Cassidy and Mullin 2007; Mason 2007). One of the few to consider the implications of this for studying the historical dimension of human-animal relations is Ingold (1994: 18).

> Adopting an indigenous perspective, my contention is that the transition in human-animal relations that in western literature is described as the domestication of creatures that were once wild, should rather be described as a transition from trust to domination.

Contrasting hunters, who 'know' the animals they hunt, with pastoralists, who look after them, Ingold's history highlights the different kinds of relations between humans and animals, depending on domestication. Clearly the shift to farming did have a profound effect on human-environment relations, but Ingold's generalisations are of questionable use since his indigenous perspective is that of the hunter-gatherer, reasserting the same Western/non-Western duality. Instead, one can reverse this formula and suggest that the different sorts of encounters arising between humans and domesticated animals require more trust than the domination of hunter over prey (Knight 2005: 4; Oma 2010). Another important shift is that domesticated animals are frequently treated as property, again resulting in different kinds of relationships (Tapper 1989; Russell 2002; Orton 2010). Yet the idea that anthropological perspectives can be used to understand the human-animal relations of the past is worth pursuing, while acknowledging the different kinds of relationships that herders have with animals (whether domesticates or wild animals or somewhere in between) compared to hunter-gatherers, or modern Western industrial farmers.

Descola has focused on the way in which modes of identification are articulated by characteristic 'modes of relation' such as predation or protection, but suggests that other dimensions of experience such as spatiality, temporality and figuration are also important (Descola 2013a: 114). This is particularly the case in farming societies, where two typical modes of relation, protection and transmission, introduce a spatial and temporal dimension respectively into the way in which humans interact with animals (Descola 2013a: 321–35); whereas hunter-gatherer societies are involved in a form of fluid exchange with the environment, in farming societies land and animals are divided up and passed down the generations. The combination of a mode of identification with characteristic modes of relation results in what Descola, borrowing from Latour (1993: 107), terms 'a collective', 'a way of assembling humans and non-humans in a network of specific relations' (Descola 2006: 147). In a naturalist ontology this equates with human society, with a separate natural domain, but other ontologies result in different collectives.

The mode of identification which Descola applies to premodern Europe and other stratified societies he terms analogism, which he defines as follows.

> [A] mode of identification that divides up the whole collection of existing beings into a multiplicity of essences, forms and substances separated by small distinctions and sometimes arranged on a graduated scale so that it becomes possible to recompose the system of initial contrasts into a dense network of analogies that link together the intrinsic properties of the entities that are distinguished in it. (Descola 2013a: 201)

Analogism becomes a way of resolving difference and managing divisions by suggesting connections between entities. As such it occurs in societies in which the number of entities is in danger of multiplying beyond control, leading to the imposition of hierarchies: 'the analogic collective is always divided into interdependent constitutive units which are structured according to a logic of segmentary nesting' (Descola 2006: 152). The Mediaeval European great chain of being is one example of analogism, in which order is imposed on a multitude of distinct entities. The modes of relation typically found in analogic (as well as naturalist) collectives are production, protection and transmission. These are themselves hierarchical, defining the role of humans with respect to the non-humans they protect, extract products from or give to others (Descola 2013a: 321). Animal domestication in particular, unthinkable in animist societies, makes possible these particular modes of relations (Descola 2013a: 386). Since Bronze Age Crete was a premodern agricultural society it is this type of ontology which is most applicable.

Descola's scheme can be seen as an elaboration of the 'like us'/'not like us' ambivalence which recurs again and again in discussions of human-animal relations, or what Willis (1990: 13) describes as 'the self and other dyad'. From the basic identification of similarity and difference in body and mind arise four different ontologies which structure human-animal relations in different ways. As Brian Morris (2000: 6) points out, Descola's definition of 'naturalism' is an oversimplification of modern Western human-environment relations, and suggests that his Malawian ethnography shows that there are diverse human-environment relations in any one society. Usefully bringing in a historical perspective, Lloyd (2007: 144) makes the point that historical China and Classical Greece are both analogist collectives in Descola's terms, but are very different in terms of their conceptions of human-environment relations.

Nevertheless, Descola's scheme opens the way to considering ontologies in different collectives without assuming a modernist ontology or imposing a non-agricultural ontology; he has also responded to these criticisms by stressing that these ontologies are inferences which guide perception of certain qualities of the environment, rather than isolated worldviews (Descola 2010a: 338). His scheme, as he intends, offers a useful heuristic for this book, not least because existing accounts of Bronze Age Crete can be clearly identified as 'naturalist'.

STYLES OF INQUIRY

A naturalist mode of identification is frequently applied to depictions of animals and animal remains from Bronze Age Crete. This is exemplified by the use of the Linnaean taxonomic system, which describes each animal using a binomial term consisting of genus and species. Higher levels, such as family or kingdom, describe the evolutionary relationship between organisms. Zooarchaeologists tend to identify animal remains to genus, or sometimes family, since it can be difficult to allocate fragments of bones and shells to species. In the process, animal remains become specimens, which are identified, recorded and compared using widely agreed criteria. A number of studies have sought to identify depictions of animals from the Bronze Age Aegean, including monkeys, birds and molluscs, using zoological classifications (Ruuskanen 1992; Masseti 1997; Karali 1999; Greenlaw 2011; Pareja et al. 2020). Such studies tend to assume that these animal depictions record the details that zoologists use to differentiate different species, by analogy with the illustrations in field guides. If the ontology of Bronze Age Crete differs from that of the modern West, it is important to question this assumption in two respects. The first is whether animals were classified in the same way in Bronze Age Crete and, if not, whether these classifications can be mapped on to the Linnaean system (Legendart 2020). The second, discussed in subsequent sections, is whether style of depiction relates to ontology.

These questions relate directly to the interpretation of the Neopalatial 'Partridge frieze' found in a building known as the Caravanserai to the south of the palace at Knossos (Figure 2.3). Evans (1928: 103–14) named this building the Caravanserai because he assumed it was a place where travellers to Knossos could refresh themselves, using running water piped from a spring. The elaborate nature of the plumbing makes it more likely that water was the focus of activities there, whether for bathing or even dyeing textiles (Schofield 1996; Burke 2010: 37–9). The Partridge frieze was above the level of the doorways in a room known as the Stepped Pavilion, and the fragments that survive show both partridges amid rocky ground and hoopoes in bushes. The identification of the hoopoe (*Upupa epops*) is not disputed because of its

2.3. The Partridge frieze from the Caravanserai at Knossos, LMI. Detail, reconstruction drawing by E. Gilliéron, *fils* (Evans Fresco Drawing A/2 a). © Ashmolean Museum, University of Oxford

colouring and distinctive crest, meaning that it cannot be confused with any bird known to visit Crete (Masseti 1997: 357). There are arguments, however, for different species of partridge: *Alectoris chukar*, the chukar partridge (Evans 1928: 110; Pollard 1977: 149) and *Alectoris graeca*, the rock partridge (Masseti 1997: 357–8). Modern bird handbooks suggest that given the similarity in appearance it is easiest to distinguish these species by location and call (Beaman and Madge 1998: 259; Sterry 2004: 60). Both species are found in the Aegean, but the species currently regarded as resident on Crete is *Alectoris chukar*, subspecies *cypriotes*, which was most likely introduced by humans in antiquity (Blondel and Aronson 1995: 80; Madge and McGowan 2002: 186).

The distinction between *A. chukar* and *A. graeca* is not simply based on readily observable differences but is partly the result of different concepts of what a species is. They were once regarded as the same species, particularly because they are capable of interbreeding, although now genetic analysis has helped confirm the distinction between them (Randi et al. 1992). Whereas interbreeding is an important criterion for defining a biological species (Mayr 1942), DNA analysis allows species to be defined phylogenetically, in terms of evolutionary relationships. The same technique has shown that *A. chukar* in Crete now consists of two distinct genotypes as a result of captive breeding of an Asian lineage of the same species for release for hunting (Barbanera et al. 2009a, 2009b). Given this diversity among partridges, Randi et al. (2003: 2213) note: 'A strict biological species concept [...] is not appropriate for *Alectoris*, because even formally recognized species can interbreed and produce fertile offspring in nature and in captivity.' This illustrates that two modern disciplines, birdwatching and evolutionary biology, result in two different ways of understanding the partridges of Greece. For birdwatchers they are two different species to be ticked off on a list, whereas some biologists would hesitate to differentiate them. The identification of partridges is not an objective judgement but differs according to the viewpoint of the identifier, itself related to the means of identification.

There are numerous ways of identifying and classifying animals attested around the world. The discipline of 'folk biology' has explored the ways in which humans perceive discontinuities between different types of animals and plants. In a classic paper, Brent Berlin et al. (1973) argued that universal underlying similarities could be found in the way different societies classify and group animals and plants. The most phenomenally salient taxa, termed 'generics' (e.g., cat or dog), are the basis of any classification scheme, and are typically arranged in a hierarchy. Above generics in the hierarchy are usually 'life forms' (e.g., quadruped, fish) and then ultimately 'unique beginners' (e.g., animal, plant); generics are themselves sub-divided into 'specifics' and 'varietals' which vary between cultures, such as breed names of cats and dogs. Scott Atran (1990) has suggested that this 'common sense' way of dividing up the world provides a common basis for both modern biology and other culturally specific religious or symbolic uses of animals and plants. As he suggests: 'Applied to a local biota, universal taxonomic principles (including presumptions of underlying natures) tend to yield basic-level groupings that correspond to biological species, at least for the phenomenally salient vertebrates and flowering plants' (Atran 1990: 56). Such classification is far simpler when faced with the animals and plants of a particular time and place; he expands the term 'generic' to 'generic-species' to reflect the fact that usually there is only one species of any given genus per habitat (Atran 1998: 550).[2]

Returning to the Partridge frieze, *Upupa* is a generic-species because it is the only extant species in its genus. *Alectoris* is also a generic-species in present-day Crete, although both *A. graeca* and *A. chukar* are present in the Aegean area, with introduced members of another species, *A. rufa* (Birdlife International 2021). The problems of hybridisation and recognition, as well as the question of whether there were different populations of human-introduced partridges in Crete during the Bronze Age, occur at the species level. At the generic level it is possible to accept that there are similarities between zoological identification of the birds in the partridge frieze and those found in Bronze Age Crete because differences between genera are here equivalent to recognisable morphological differences (Binnberg 2017). The same can be said of zooarchaeological identifications of partridges, since bones identified as *Alectoris* (but not *Upupa*) have been found at several Bronze Age Cretan sites. The largest number of partridges were found at

Kommos, with bones from at least seven individuals found, mostly from Neopalatial or later levels (Reese 1995a: 196–7). Other contexts with partridge remains are the Archive Room of the Protopalatial palace at Petras and a vessel from the Neopalatial Building A at Mochlos, with three bones each (Soles 2003: 18; Mylona 2010: 228). This provides one way of reconciling a modern classification system with Bronze Age animal depictions.

As Descola (2013b: 45–54) points out, however, the study of folk biology is itself an imposition of a naturalist ontology. This is because it is assumed that plants and animals can be separated into a natural domain to be studied by folk biologists, using scientific classification as a starting point. An analogist ontology, by contrast, might require a different way of approaching the fresco. Indeed, Arthur Evans (1928: 114) provided a different interpretation: that these were game birds destined for the table, comparing the frieze with Dutch still-life paintings.[3] The category 'game' crosscuts the Linnaean classification scheme to refer to any animal that can be hunted, and predates modern zoological classifications. In doing so it creates analogies between different types of animals based on their role as prey.[4] On present-day Crete those who catch birds indiscriminately using guns, nets or birdlime often come into conflict with conservationists and birdwatchers, but this activity provides a different way of categorising the birds of Crete. Partridges are non-migratory so can be hunted all year round, but most bird-hunting (fowling) is tied to seasonal patterns of migration. This provides another way of understanding the significance of the hoopoe on the fresco, as a harbinger of spring (Shaw 2005: 102–3). Regarding birds as animals to be hunted and eaten rather than simply identified changes the perspective, but this does not fully answer the question of whether these are incompatible ways of understanding birds. Indeed, as Morris argues, 'utilitarian' animal classification systems tend to group animals in the same way as Western zoologists but also take into account categories such as edibility or harmfulness (Morris 1998: 144–8).

Lloyd's concept of the 'multidimensionality of the phenomena' is important for reconciling these different perspectives; there are many different ways of understanding partridges which are not mutually exclusive. He too doubts the possibility of universal taxonomies because modern biology itself does not agree on what a species is (Lloyd 2004: 101). At the same time he argues against cultural relativism because rather than animals being classified in arbitrary ways, different societies tend to converge on similar classifications based on similar relationships between humans and animals, even if there might be variation in what is grouped together with what Westerners regard as animals, and variations within societies (Lloyd 2004: 103). Instead of asking whether classifications are universal or relative, he instead suggests that: 'In judging each classification we encounter, we have first to ask what it is a classification *for*' (Lloyd 2010: 206). Classifications are thus not generated solely by the human mind or particular cultures but are different 'styles of inquiry' whose results are produced through interactions with the animals a human group encounters.

As a result categorisations are partly generated by, and enacted through, particular types of human-animal relations. The potential hybridisation of partridges, both between different species and between wild and farmed animals, emerges as a threat as a result of conservation biology. As with agrimia, conservation efforts are hampered by the 'genetic pollution' of native *A. chukar* found wild on Crete with farmed birds of the same species but of different lineages (Barbanera et al. 2009a, 2009b). Hybridisation also poses a problem for birdwatchers since their checklists are for wild species; there are a number of birds on Crete including chickens and peacocks which are not included on checklists because they are in some sense introduced or captive (Xirouchakis 2015). Both *A. graeca* and *A. chukar* are regarded as legitimate to spot by some birdwatchers but hybrids released for hunting are regarded as 'an obstacle to identification', as well as a threat to 'the purity of local populations' (Madge and McGowan 2002: 186, 192). For those interested in partridge-hunting, hybridisation is irrelevant but the number killed is not, which is why farmed birds are released for the purpose of hunting. Similarly, the eco-tourists who catch a glimpse of an agrimi in the Samaria Gorge arguably care little whether they have seen an animal which was originally introduced by humans (and hybridised with local goats); it doesn't matter whether they have seen *Capra hircus cretensis* or *Capra aegragus cretica*. Although the animal practices of conservationists and birdwatchers are concerned with the purity of species, this is not the case for tourists and hunters.

The approach taken to the identification of animal depictions in this book accepts the idea of 'generic' animals such as 'cat' and 'dog' based on recognisable bodily differences; since lions and griffins are also depicted as distinct and stable bodies, they are treated the same way. Subsequent chapters examine the relations between humans and animals that shaped which differences were significant and out of which classification could arise; as will be discussed in Chapter 7, the importation of different types of dogs in the Neopalatial period appears to have resulted in the recognition and detailed depiction of particular breeds. At the same time, the difficulty in recognising even generic-species of birds or fish, particularly on seals, suggests that the depiction of fowling and fishing was more significant than the closer identification of these animals. In classifying the partridges of the Partridge frieze as both *Alectoris chukar* and game, Arthur Evans was not contradicting himself but using different classificatory schemes derived from both zoology and hunting. Given that the practice of birdwatching emerges from a naturalist ontology, it is unlikely that the frieze acted as a zoological identification guide. Yet the naturalistic style of depiction of such Neopalatial frescoes has drawn numerous scholars to read them in this way.

STYLES OF DEPICTION

The term 'naturalism' has acquired a number of meanings, slipping between science and art history (Williams 1983: 216–19). It is used in its philosophical sense by Descola, referring to natural causality separate from gods or people, to describe the typical modern ontology. Here nature and culture are seen as separate domains, the latter defined by the unique capacities of the human mind. Although this is clearly appropriate to describe the assumptions underlying the concept of the nature-loving Minoans, it is also important to explicate Evans's use of the term. Locating the highpoint of naturalism at the start of the Neopalatial period, his descriptions make it clear that he was referring both to the subject matter and also to the style of depiction (Evans 1921b: 319, 1935: xvii). The first aspect is apparent in his description of a fresco of monkeys, birds and rocks from the House of the Frescoes at Knossos: 'The subject chosen for illustration may best be described as wild nature. Man is excluded, but animal forms such as monkeys and blue birds appear here and there, amidst a wilderness of grotesque rocks overgrown with flowers and creepers' (Evans 1928: 446–7) (Figure 2.4). 'Wild nature' as artistic inspiration is a product of nineteenth-century Romanticism (Descola 2013a: 55; Shapland 2010b). The idea that Cretan Bronze Age animal depictions are inspired by nature is still prevalent. Sara Immerwahr (1990: 45), for instance, described the 'Monkeys and birds fresco' in the following terms: 'Although not botanically accurate in all details, the individual species are recognizable, and one can appreciate the artist's delight in recording his beautiful natural surroundings.'

As Immerwahr's description indicates, in the modern world a naturalistic style of depiction is also tied up with the idea of realism. The art critic John Ruskin (1853:

2.4. The Monkeys and birds fresco from the House of the Frescoes at Knossos, LMI. Detail, central panel, from reconstruction drawing by Mark Cameron (Mark Cameron Personal Papers, Archives, British School at Athens, CAM 1). Reproduced with the permission of the British School at Athens

181) defined naturalism as 'the love of natural objects for their own sake, and the effort to represent them frankly, unconstrained by artistical laws'. Evans's use of the term owes much to this conception of naturalism, as can be seen in his description of a fish and squid on a sealing from the Knossos Hieroglyphic Deposit as displaying 'truest sympathy with natural forms and surroundings' (Evans 1935: 491). At around the time these sealings were made, depictions of animals and plants appear in a far more realistic style than previous periods, and in various different media.[5] His use of the term 'naturalism' has long been criticised because these depictions are not closely observed, and so are not accurate. Groenewegen-Frankfort detected an unnatural element in the Monkeys and birds fresco.

> To speak of naturalism, for instance – that is, of a conscious interest in, and respect for, the appearance of the phenomenal world – may be tempting in the face of forms so buoyant with life, yet it means ignoring the fact that several of these forms cannot have been observed at all; blossoms and leaves of different plants have been high-handedly combined, birds' plumage altered. (Groenewegen-Frankfort 1951: 196)

Peter Warren (2000: 365) too suggests that Minoan images are often closely observed but no more than 'near naturalistic' because they do not exactly copy nature. Although accuracy is important to Ruskin's definition of naturalism, for Evans naturalism encapsulated a broader sense of a love of the natural world and its truthful, rather than factually accurate, representation. Pointing out the biological inaccuracies in animal depictions has not challenged this broader notion of Minoan naturalism.

From the point of view of universal art history, the accuracy of any image can be judged using terms such as naturalism. Anthropologists have instead tended to regard naturalism as one style, linked to a particular culture (Layton 1977). Descola (2010b) used an exhibition called *La Fabrique des images* to argue that the production of images is instead bound up with different ontologies. His argument is that the relations between humans and non-humans generated by these ontologies will be apparent in the ways in which different cultures depict the world. A mask which opens to reveal a human face behind a bird's head, for instance, epitomises the animist perspective, in which bodily differences conceal a common humanity (Descola 2010b: 25–32). A similar argument leads Ingold to reject the term 'art' altogether, suggesting instead that the way animals are depicted is part of the playing out of relations between humans and animals. 'The activities of hunters and gatherers that lead to the production of what we in the West call "art" should, I argue, be understood not as ways of representing the world of immediate experience on a higher, more "symbolic" plane, but of probing more deeply into it and discovering the significance that lies therein' (Ingold 2000: 112). Herva too suggests Bronze Age Cretan depictions of plants and rocks guided people's perception of them as animate.

> Minoan nature scenes were instruments for perceiving the rich texture of their world. Nature scenes thus directly guided bodily engagement with the environment that people co-habited with manifold non-human beings. (Herva 2006b: 236)

For Descola (2010b: 73; 2013a: 57–60), the art of the naturalist ontology, that of the modern West, instead emphasises the separation of humans from the natural world. Rather than being imbued with meaning, the world is disenchanted and can be described objectively; the invention of linear perspective provided a means to do this. Groenewegen-Frankfort (1951: 3–6) described this as a 'functional' relationship, which turns a painted surface into 'a mere window'. When an animal, for example, is rendered in perspective, it gains 'illusionary corporeality', placing it 'in a definite spatial relation to the observer'. With the addition of a temporal dimension, 'organic corporeality' is achieved. This is the standard of mimesis, or imitation, against which she judged Cretan and Egyptian art, noting the absence of a functional relationship with the viewer. Instead she saw in Cretan frescoes a freedom of expression which imbues animals and plants with a sense of life and movement. As noted above, Klingender's (1971: 62) description of the 'graceful, life-affirming art of Minoan Crete' owes a clear debt to her work.

Composition is another important element of naturalism. Groenewegen-Frankfort observed that a Minoan fresco showing plants, trees, rocks and watercourses is not a landscape, since 'the essence of landscape, the rendering of natural phenomena in space, is lacking completely' (Groenewegen-Frankfort 1951: 196). At the same time she detected 'a sense for the organic' which, she suggested, prevents Cretan art from being purely decorative, making the scenes 'authentic' (Groenewegen-Frankfort 1951: 197). Animals in flying gallop display 'absolute mobility', a freedom of movement which does not relate to the rest of the scene, but is self-fulfilled. This led her to suggest that there is an ambiguity in the depiction of animals, which are surrounded by the components of landscape, but do not fully interact with them. 'Such separable entities as occur in the form of single beasts, plants and stones – though plants frequently seem to merge into one another – are caught in a web of a living world that has indefinite orientation and indefinite multiple relations' (Groenewegen-Frankfort 1951: 201). Although she suggested the Minoan frescoes were 'nonfunctional', and ultimately religious, her attention to corporeality, and the 'web of a living world', opens the way to understanding Bronze Age Cretan art on its own terms. The effect can be described as lifelike, even if the means of depiction are very different from Western naturalism.

Descola (2010b: 165) describes the analogist mode of figuration as 'un monde enchevêtré', 'an entangled world'. Here the role of an image is as a connector, showing how disparate entities can be brought together or how different spatial and temporal scales relate to one another. Depictions of hybrids – chimeras, for instance – show how one larger entity is composed of smaller elements. Maps of the world made by analogist societies tend to illustrate the principle that connects various places, such as the itinerary of an important person. This focuses attention on the role of depictions, showing how they help to reveal the order of the world. Rather than asking whether Minoan frescoes reflect reality accurately, the question becomes how and why they are bringing various animals and plants into connection with the buildings that housed them and the people who looked at them. The principle of relatedness that Groenewegen-Frankfort identified is a useful way of looking at these depictions because it helps remove them from interpretations rooted in a naturalist ontology.

There are also examples of non-naturalistic depictions in Bronze Age Crete that can be explained in these terms. Among them is a group of sealings from Zakros, attributed to a particular seal carver known as the 'Zakro Master' (Hogarth 1902; Weingarten 1983). These seals include a kaleidoscopic set of combinations of different entities, such as boar-lion-butterfly hybrids, or human body parts combined with parts of birds, goats and deer (Figure 2.5a–d). These foreshadow later human-animal hybrids on seals known collectively as 'minotaurs' which involve goats, cattle and other quadrupeds and date to the Final Palace period (Krzyszkowska 2005: 207–8) (Figure 2.5e–f). Lyvia Morgan (1995a: 144–5) suggests that these could be abbreviated depictions of hunting or bull-leaping. Such hybridity uses conjoined bodies to indicate human-animal interactions (Simandiraki-Grimshaw 2010). Whereas naturalistic seals tend to capture a particular moment, such as the moment of a hunt or bull-leap, these hybrid seals offer a synoptic view of the same interrelations. Hybrid seals are difficult for modern viewers to read because they seem to show how entities are connected, rather than how they appear in the world.[6] In some cases the forms are perhaps deliberately ambiguous, and can be read in different ways, revealing the connections between different entities (McGowan 2018). The mode of figuration Descola associates with an analogist ontology operates in this way, combining entities to show the relations between them.

Carl Knappett (2020: 108) sees the Zakros monsters as part of a much wider phenomenon of composites in the Aegean Bronze Age. Whereas hybrids or monsters combine animal and human forms, another means of establishing connections, between different materials, is termed skeuomorphy. In Bronze Age Crete pottery was often used to imitate other materials, particularly metal. The shiny black slip, ridges and stamps of Protopalatial pottery, for instance, could imitate tarnished silver, while other decorative patterns appear to imitate woven textiles (MacGillivray 1998: 58–61). As will be discussed in Chapter 6, Arthur Evans used the term 'nature-moulding' to refer to skeuomorphs of marine forms, particularly shells, stamped into or applied onto ceramics.

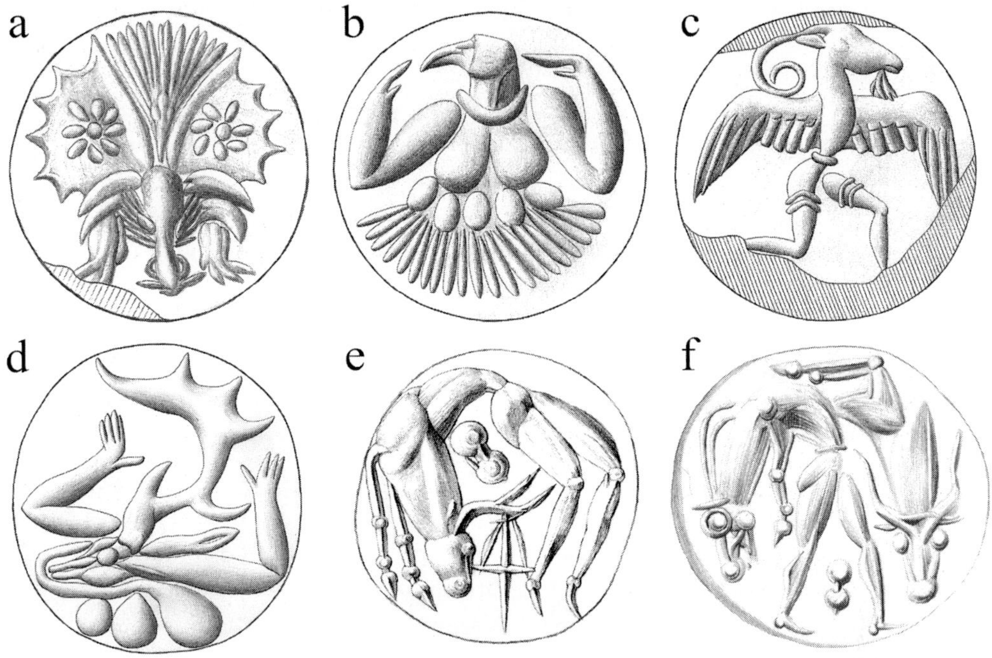

2.5. Drawings of seals showing composite forms. Zakros sealings, LMI: (a) boar-lion-butterfly hybrid (CMS II 7 158); (b) bird-human hybrid (CMS II 7 132); (c) goat-bird-human hybrid (CMS II 7 140); (d) deer-human hybrid (CMS II 7 170). 'Minotaur' seals, LMII-IIIA: (e) cattle-human hybrid, Psychro (CMS VI 298); (f) goat-human-cattle hybrid (CMS VII 123). © CMS Heidelberg

Pottery was also used to reproduce triton shells in their entirety, and later stone versions were also produced. Knappett (2020: 108–29) emphasises the creative combinations of different forms and materials in Bronze Age Aegean skeuomorphy, meaning that it was more than just the substitution of a more expensive material for a cheaper one. This provides another example of an analogist logic at work, using a malleable material like clay to show the potential connections between ceramics and other types of vessels or forms of decoration, or a technology like stone-carving to provide a marine form with different uses and properties. Knappett (2020: 126) also highlights the skeuomorphy of accurately depicted veined marble surfaces in Neopalatial frescoes; rather than seeing this in representational, illusionistic terms, it fits into a wider pattern of referencing one material with another. Whereas some of the frescoes depicting animals in Bronze Age Crete appear to be similar to the depictions of a naturalist ontology, it is worth exploring their significance within a different mode of figuration. As Groenewegen-Frankfort argued, naturalistic frescoes, in the absence of perspective to define a 'functional' relationship with the viewer, showed animals and plants as a set of related entities. In order to understand the relationship of these frescoes to the viewer it is necessary to turn to debates about the function of naturalism in Western art.

FROM NATURALISM TO INTERACTION

The relationship between naturalism and reality has long been a topic of contention in art history. In an influential account, Ernst Gombrich (1982: 12) argued that naturalistic painting 'enables us to recognise the familiar world in the configurations of paint upon the canvas', but that reality is far too rich to reproduce. As a result, the brain relies on a set of learned symbols, the schema, to aid the process of recall which underpins representation. Different traditions have different codes, and so different art styles represent the same thing in different ways. Yet naturalism is not entirely relative, he suggested, because illusionistic art tricks the brain by becoming indistinguishable from the perception of reality (Gombrich 1977: 24). Naturalistic art emerges at certain points in history in order to cast the viewer as an eyewitness, for instance, to a biblical event in Renaissance art. The discovery and gradual perfection of perspective fulfilled this function (Gombrich 1982: 21). From this point of view, Classical Greek painting achieved mimesis through the continued modification of its schemata in order to represent narrative poetry effectively. A different situation existed in Pharaonic Egypt or the Bronze Age Aegean, whose 'visual inventory' of the world was just as complete, but presumably did not require art to fulfil the same function; Gombrich did not regard Minoan art as naturalistic in his terms. Nevertheless, 'in these early cultures the schemata of animals and plants were often refined to an extraordinary degree' (Gombrich 1977: 122). Like Groenewegen-Frankfort, Gombrich saw Cretan art as failing to achieve the mimesis of perspective art, but whereas Groenewegen-Frankfort highlighted the lifelike effect, Gombrich saw a schema standing between viewer and animal.

Gombrich's account of the schema intervening between the real world and the image generated a debate about whether he assigned culture too great a role in perception, or not enough. From a semiotic point of view, naturalism or realism was the result of the conventions of depiction (Goodman 1976: 34–9). Rejecting the idea that viewing a painting is primarily concerned with perception, Norman Bryson (1983: 51) argued that the recognition of a painting's subject is a social act because it takes place within 'the interindividual territory of the sign'. His conception of the sign was one in which the relation between signified and signifier is arbitrary; there is no reality beyond the sign. Instead the image is a 'site of production', bound up in social practice where meaning is created rather than reflected (Bryson 1983: 55). The problem with Gombrich's idea that the schema is gradually perfected was that it suggests that the viewer has access to a reality outside of society, that is, 'nature', against which to judge the degree of mimesis. Instead, as W. J. T. Mitchell (1986: 90) pointed out:

> The 'nature' implicit in Gombrich's theory of the image is, it should be clear, far from universal, but is a particular historical formation, an

> ideology associated with the rise of modern science and the emergence of capitalist economies in Western Europe in the last four hundred years.

Even if Gombrich argued that art can never exactly imitate nature, the latter's existence is necessary for his theory of recognition and his privileging of Greek and Renaissance naturalism. It is also necessary for the judgements about Minoan frescoes described above, as failing to represent nature accurately. It is, however, possible for modern viewers to recognise partridges and hoopoes in the Partridge frieze to some degree, even if the fresco painters did not share the same scientific classification system. Whereas Gombrich tried to account for the possibility of recognising such birds in the art of a different culture by considering the interaction between schema and reality, the semiotic perspective advocated by Goodman, Bryson and Mitchell reduced these birds to symbols within a representational system.

The second reaction against the idea of naturalism as illusion came from the ecological psychology of J. J. Gibson, who came to some of the same conclusions as the semioticians that a picture created according to the conventions of perspective could never fully represent reality.[7] Perspective pictures need to be observed from a fixed point, through a peephole, in order to recreate a scene, and so deceive the eye. Yet, whereas Nelson Goodman (1976: 34–5) concluded that if pictures do not deceive the viewer, except under these artificial conditions, they must be conventional signs, Gibson concluded that they must only be an incomplete substitute for the experience of the world. Distinguishing between the 'artificial perspective' of pictures, with a fixed viewpoint, and the 'natural perspective' of the moving observer, he suggested that: 'Pictures are artificial displays of information frozen in time' (Gibson 1979: 71). For Gibson's biographer, Edward Reed (1988a: 259), this opened the way to a new way of conceiving pictures: 'Once it becomes clear that images are neither copies of things nor conventional constructions, but ways of manipulating optical information, it becomes clear how and why they serve a variety of functions.' Whereas Groenewegen-Frankfort described paintings as functional only when they established a spatial relationship with the viewer, Gibson was concerned with the function of images in the world.

For Gibson, the objects depicted in paintings were to be understood not as representations, but as mediated real-world forms, since they displayed some of the same visual information as those objects; they were perceived essentially in the same way (Gibson 1979: 262). He argued that objects already specify information about themselves, and it is in the interaction between observer and environment that this information is perceived. The meaning of objects cannot be reduced to a conventional system of signs since their meaning is not located purely in the observer's head. Instead of 'meaning' he used the term 'affordance'. 'The affordances of the environment are what it offers the animal, what it provides or furnishes, either for good or for ill' (Gibson 1979: 127). The observer can learn how to perceive affordances, in the process of socialisation into a given culture, but the affordances are not themselves culturally determined because they arise out of interaction. 'Affordances are properties taken with reference to the observer. They are neither physical nor phenomenal' (Gibson 1979: 143). Or, as Reed argued: 'A single object may yield different affordances to different people, not because affordances are subjective, but because they are functional, related to the observer as well as the environment' (Reed 1988a: 293). This functional aspect of affordances is both a strength and weakness; it has the value of refocusing attention on what things do as a key aspect of their meaning, but also needs grounding in other accounts of meaning to avoid the charge of reductionism (Knappett 2005b: 85).

Art objects led Gibson to develop his account of 'mediated perception', suggesting that a more complex account is needed in order to place objects in a cultural, rather than just a perceptual, context (Gibson 1979: 262; Knappett 2005b: 49). Affordance meanings, according to Gibson, can be perceived directly when information is available specifying them. They can also be perceived indirectly when someone selects and displays that information in one or another form of representation, using pictures, language and so on. The pictures and words are also meaningful, but in a way that is partially historical and cultural as well as ecological (Reed 1988a: 307). Knappett (2005a: 50–1, 2007) seeks to reconcile Gibson's work with material culture studies suggesting that there is a need to 'deradicalise' Gibson, distinguishing 'transparent' affordances which are independent of cultural meanings from affordances which operate within a cultural context. In other words, whereas Gibson focused on organism-environment interactions, the environment of humans is inextricably physical, but also social and cultural. This sociocultural environment therefore provides the context in which many affordances are perceived (Knappett 2005b: 49; Turner 2005; Kockelman 2006).[8] Knappett (2005b: 110) also suggests that objects have both pragmatic and significative aspects. Affordances fit into the 'vertical' pragmatic relationships between mind, body and matter, whereas signification can be seen in terms of the networks connecting human and non-human entities.

Although the semiotic and the ecological approaches to pictures were developed in opposition, Bryson's (1983: 16) description of painting as a 'material signifying practice' offers a way of reconciling these two approaches if signification is extended beyond the workings of the human mind. For Gibson, a depiction of an animal mediates between the viewer and animal, allowing the perception, to some extent, of the animal's affordances. For Bryson the image is instead a sign which mediates between people, whose shared knowledge allows them to recognise it as a representation of an animal. In terms of the semiotics of C. S. Peirce, grounded in pragmatism, Bryson's account defines the image as a symbol.[9] In the Peircean scheme an icon signifies its object by likeness, and a symbol by convention; the third type of sign-object relation is an index: 'An Index is a sign which refers to the Object that it denotes by virtue of being really affected by that Object' (Peirce 1998: 291). This means that indexes are related to their object through an existential connection, either the result of contact, such as a footprint, or indication, such as a pointing finger (Sonesson 1989: 39). Although images are often described in semiotic terms as iconic, Gibson's account suggests that certain parts of the world are reproduced in pictures, however incompletely. As Göran Sonesson (1994: 317) suggested, from this point of view: 'all pictures are in a sense indexical, since they isolate a portion of a scene which is present more fully in the actual world of perception'. Sonesson (1994: 283) drew on Gibson's approach, but located perception in what he called a 'Lifeworld'; people perceive from a particular point of view which is shaped to some extent by the culture they have been brought up in, echoing Knappett's approach to affordances.[10]

Regarding art objects as indexes, connected with other objects in the world, allows them to be considered in terms of function and relatedness rather than simply degree of resemblance (their iconicity) or conventional meaning (symbolism). Alfred Gell (1998: 13–16) adopted the term 'index' to describe art objects and also borrowed the Peircean concept of 'abduction', a form of reasoning based on the viewer's assumption of a connection between two things in the absence of any proof.[11] For Gell, the viewer abducted an artwork as an index, which meant that they assumed that an artist had made it, and, by extension, that the artist had agency. Importantly, this stresses the social context in which art objects are understood; what a Western art collector might view as a decorative pattern, its maker and contemporary users might have understood as an apotropaic device, allowing the artist to exercise power over the viewer through the index (Gell 1998: 81–4). Gell (1998: 6, 14–15) preferred the term 'abduction' to 'interpretation'; he regarded the latter as tied to reading the symbolism of an artwork. In making this distinction, he was keen to stress that art objects have indexical relationships as well as symbolic meanings. Sonesson (1989: 38–43) too made use of the concept of abduction as a way of understanding indexical relationships between things, since it describes a form of reasoning based on probability rather than fact. His interest in prehistoric art led him to focus on the particular problem of restorations of frescoes such as the 'Ladies in blue' from Knossos, which are a kind of informed guesswork: 'What makes the reconstruction of Minoan frescoes so much more difficult than a jig-saw puzzle, is not only the fact that so many pieces are lacking in the former, but also that the abductions derived from the regularities of our Lifeworld do not necessarily fit' (Sonesson 1989: 41). These approaches challenge the modern viewer to understand depictions not as art, but as objects located in a particular time and place and embedded in a set of connections with other aspects of that Lifeworld.

Peircean semiotics can be used as a toolbox, providing different ways of thinking about art objects. Since there is a growing body of literature exploring the implications of Peirce's theories for material culture, there is already a set of pragmatic and semiotic approaches to draw upon (Parmentier 1997; Knappett 2002a; Keane 2005; Preucel 2006; Watts 2008). Another art historian to draw on the pragmatic tradition is David Summers (2003: 687), who uses Peirce's index as the basis for his definition of trace as 'an indication of former presence and contact', which he applies not only to the connection between producer and artwork (e.g., the painter's brushstroke), for which Gell used the term index, but also artworks which preserve a physical connection with their subject. For iconic traces he uses the term effigies, which, he argues, 'take their primary authority from causal or indexical relation to what they reproduce, or continue to make present' (Summers 2003: 284). The example he uses is a death mask, which is an object shaped by contact with what is no longer present. All traces are defined by contiguity, the relationship which defines metonymy and indexicality.

More broadly, Summers (2003) argues that all art objects should be seen as 'real metaphors', meaning that they are substitutions for something that is not present. A similar argument is advanced by Hans Belting (2005), who, echoing Peirce, proposes a triad of image, medium and body; images are events transmitted by a medium to a living body. He argues that images prototypically make the dead visible, performing the 'presence of an absence'. As he suggests, '*Iconic presence* still maintains a body's absence and turns it into what may be called *visible absence*' (Belting 2005: 312–13). His starting point is akin to Summers's death mask: the plastered Neolithic skulls at Jericho. Both images rely on an indexical connection with the dead but what is important is their pragmatic function; what they do is to make an absent body present. The same can be said of the verisimilitudinous religious sculptures and waxworks of the dead from antiquity onwards: 'The striving for resemblance marks our attempts to make the absent present and the dead alive' (Freedberg 1989: 201).[12] This pragmatic approach redefines the art object as a substitute for something in real space rather than an illusion or symbol of something else (Knappett 2020: 32–6).[13] Rather than being a window to a different space, Summers also emphasises that the observer has real spatial relations with the art object. Whereas Gibson saw depictions as secondary to experience of the world, Summers shows how the spatial and material aspects of depictions are also integral to the viewer's experience.

This provides a different way of approaching naturalism; rather than deceiving the eye in the case of painting, or approximating to reality in the case of sculpture, what naturalistic art does is makes absent bodies or things present more effectively. This is the argument Jeremy Tanner (2001: 270) makes for naturalism in Classical Greek cult statues, that naturalism enhances the possibilities of the image's interaction with the viewer.[14] Both Mediaeval Europe and Classical Greece are what Descola regards as analogistic collectives; they demonstrate that what we now see as naturalistic art can serve as a point of connection between dead and living or human and deity in societies other than our own. Naturalistic depictions in Bronze Age Crete similarly establish a connection between the viewer and what is depicted. It is this indexical level, in which meaning arises through material interaction, which is so often missing in discussions of Bronze Age Cretan material culture. Instead, the focus is often on reconstructing symbolic meanings (Cain 2001). Frescoes, following Evans, are frequently seen in terms of religious experience (Immerwahr 1990: 40; Marinatos 1993: 50–75). It is undeniable that a Bronze Age Cretan fresco has cultural and symbolic meanings which modern observers can only guess at (interpretations involving unseen deities are symptomatic of this uncertainty), but on a more pragmatic level depictions can be understood in terms of their affordances and their indexicalities through the process of abduction.

One of the most well-known frescoes from Knossos, the 'Taureador fresco', shows a red-skinned human, seemingly assisted by two white-skinned humans, jumping over a bull with dappled coat in flying gallop (Figure 2.6). It is described as naturalistic because of the realism of the humans and animal, and because it appears to capture a moment in time. Arthur Evans made a number of symbolic interpretations which are still accepted by many authors today. The first is that the skin colour of the humans could be read as a convention, with red signifying male and white female (Evans 1921a: 251). This is despite them all wearing codpieces, which he regarded as an act of ritual transvestism, signalling that this was a religious sport in honour of the Cretan mother goddess (Evans 1935: 21–2).[15] In the

2.6. The Taureador fresco from the palace at Knossos and sealing depicting bull-leaping from Ayia Triada. Left: reconstruction drawing based on fresco fragments, LMI-II (after Evans et al. 1967: pl. IX). Right: clay sealing (with scale) and drawing, LMI (CMS II 6 44). © CMS Heidelberg.

case of skin colour, where the convention can be traced between depictions, other possibilities have been suggested, such as that the red/white was a convention for lapsed time (Damiani Indelicato 1988), hierarchy based on initiatory status (Marinatos 1989) and more generally age/skill (Marinatos 1994: 93). Finds of Late Bronze Age Aegean-style bull-leaping frescoes in Tell el-Dab'a in Egypt provided more evidence, including one yellow-skinned leaper (Bietak et al. 2007: 68). Drawing on this find, Marinatos has argued that colour was one of a set of indicators, including activity and dress, to indicate the specialised activities involved in bull-leaping (Marinatos 2007: 127). Bull games, as will be argued in Chapter 3, were related to martial training for young men. From this follows the abduction that instead of gender, the skin colour is more likely to signify time spent outdoors, and hence relative status. More broadly, there has been a growing awareness of the multiple indicators of identity in frescoes, such as hairstyle and dress, rather than starting with the assumption that a binary gender is being represented (Morgan 2000; Alberti 2002). This can be seen as a shift in attention from the symbolic to the indexical meanings of the fresco, tracing connections between frescoes and the real-world details of hairstyle and clothing rather than looking for the meaning of arbitrary conventions.

These indexicalities work in a number of directions. The fresco indexes real-world forms such as hairstyles and codpieces. These have the same indexical significance in the depiction as in the Bronze Age palatial Lifeworld, as indicators of identity or status, although their symbolic significance is more difficult to interpret. Depictions could also index other depictions, as well as the Lifeworld. As Senta German (2005: 85) suggests, images on seals 'served as an index to larger representations of social drama at the palaces, in the form of wall paintings, which were themselves representations of actual events, social dramas'. Hence the seals of bull-leaping indexed frescoes which themselves index social activities, allowing the seal users to connect themselves with these activities (German 2005: 85–6). This imagery could also be distributed using seals. The metal (most likely gold) rings which show the moment of the leap, and details of the bull's anatomy, appear to have been used by the administrators at Knossos; they are found stamped on clay packet sealings (which contained folded messages on parchment) at various sites across the island at the end of the Neopalatial period (Betts 1967; Krzyszkowska 2005: 188–92; Weingarten 2010) (Figure 2.6). The clay of these sealings is consistent with an origin in the area of Knossos (Goren and Panagiotopoulos 2009). As a result, depictions of bull-leaping also appear to index the palace of Knossos, as well as the users of those rings (Hallager and Hallager 1995).

These depictions can also be described in terms of affordances, among which the most obvious relate to size and material because these have particular effects on the possibilities of interaction. A ring, for instance, affords an impression when stamped in wet clay, but the size of the resulting depiction constrains the number of viewers at any one time (Platt 2006: 237; Summers 2003: 274). The sealing, however, is portable, and so affords interaction over long distances. By contrast, a fresco on a wall affords viewing for a larger group of people but cannot be carried around by someone and reproduced in different contexts as a seal depiction can. This was particularly the case for relief frescoes of bull games at the North Entrance, which would have been visible to anyone entering the palace of Knossos. Different again was the Taureador fresco and its related series of bull-leaping frescoes, which were less than life-size and located in an enclosed space in the East Wing of the palace at Knossos, where they would only have been visible to a limited number of people at one time. It is likely that access to this area was restricted given the way in which the possibilities of movement and visibility within palaces were controlled using architectural features (Palyvou 1987). These possibilities of access, or restricted ownership of gold rings, can be described as cultural affordances because these interactions were specific to Bronze Age Crete. The (sighted) visitor to Heraklion Archaeological Museum can view the restored Taureador fresco using the same physical apparatus as a human in Bronze Age Crete, but the museum gallery, spotlighting and glass showcase afford different possibilities for interaction with the fresco.

The detail of the fresco, often described in terms of naturalism, also affords interaction, providing information about the bull-leap. It allows the viewer, for instance, to see the bull as a male domesticate, from the depiction of the penis and the dappling of the coat, only seen in domesticated cattle (Zeuner 1963: 229). Other details, such as particular aspects of gesture or clothing of the human leapers, allowed those with cultural knowledge to interact with the fresco; while the relief bull frescoes in the entrance to the Central Court functioned as real metaphors, making cattle present in the palace, the bull-leaping depicted in the Taureador fresco, and its permutations on other frescoes from the same space, appears instead to have functioned as a visual how-to manual for this practice. The restricted location of these frescoes perhaps meant that they were accessible only to those who were, or would be, directly involved in this activity. This helps to explain why these frescoes display information about different types of leaps and even accidents, with one of the frescoes showing a leaper being trampled by the bull. In order to understand some of the details, it was necessary to have seen, or perhaps even to have participated in, bull-leaping; only those with knowledge of bull-leaping would have been able to interact with some of these details of the fresco. Modern viewers, who generally have little experience of cattle and no experience of bull-leaping, have often doubted whether it even happened, but this indicates that a certain degree of cultural knowledge is required for some affordances of the fresco to be revealed. These, in turn, relate to the affordances of bulls in

Neopalatial Crete, which were animals which could be leapt over. The abductive process allows some of this knowledge to be reconstructed at a distance, by trying to understand these frescoes, and their details, in a Bronze Age Cretan context.

This broadly pragmatic approach offers a different way of looking at the animal depictions of Bronze Age Crete. The bull's-head rhyton now enclosed in a glass showcase in Heraklion Archaeological Museum is another example of a real metaphor, making a bull's head present when it was used as a vessel.[16] Its rock-crystal eyes and liquid pouring from its mouth made it a more effective substitute for a real cattle head, which was, as will be suggested in Chapter 4, sometimes the focus of a consumption event in palatial Crete. Cattle skulls, bucrania, were also deposited in Bronze Age Crete, most likely as a result of consumption events. These skulls can be described as traces since they were contiguous with the butchered animal. Both bull's-head rhyta and bucrania are often interpreted as sacrificial symbols but there are no depictions of sacrifice from Bronze Age Crete until after the rhyton was produced. Cattle heads and bucrania are, however, widely depicted on contemporary seals. The connection between the rhyton and a butchered cattle head is an abduction, an inference based on the Neopalatial context of this object which connects it with other phenomena in this period. Abduction becomes a method of tracing connections between things rather than interpreting their symbolic significance.

These animal heads appear to have functioned in the same way as trophies, which are preserved traces of the hunted animal. From the nineteenth century, taxidermy developed as a means to preserve experiences of animals, particularly encounters through hunting. 'In an age before colour photography and moving images, taxidermy was the technology for making animal action visible, replayable, endlessly available for repeated viewing' (Poliquin 2012: 93). Conversely, early photographs of animals before the age of digital cameras and telephoto lenses were regarded as trophies, and described as 'camera-hunting', gaining their value from demonstrating the photographer's skill in capturing the wild animal on photographic plates (Brower 2005). In both cases the representation has a more complex relationship with the animal depicted than can be reduced to a symbolic meaning; instead, the animal becomes an absent presence. These are both types of indexicality, the remaining part of the once-living animal body and the former presence of the animal recorded through photography. Similarly, Erica Fudge (2012) focuses on historic objects made from animal products, what she calls the 'animal-made-object'; products such as dog-skin gloves nevertheless retain an animal significance: 'the persistent presence of the animal in the animal-made-object seems always to defy the objectification which attempts its absenting' (Fudge 2012: 94). A term used by both Fudge and Poliquin, 'animal thing', captures the ambiguous status of these material presences of former animals.

Here the term 'animal thing' encompasses both objects incorporating animal matter such as taxidermy and things which stand for animals. Using Summer's terminology, the first will be described as traces, and the second as metaphors. This opens up different possibilities for the relationship between animals and objects which are relevant to considering Bronze Age animal depictions. Animal things do not exist in isolation, but result from interactions between humans, animals and objects. Hence animal things need to be considered alongside the practices which produce them.

NETWORKS AND COLLECTIVES

One way to theorise the interaction between humans, animals and objects is with Actor Network Theory (ANT), which extends the capacity to act to a variety of non-humans, both animate and inanimate, but also imperceptible entities such as beliefs or laws, and focuses on the relationships between them (Latour 1993, 2005). These interactions between humans and non-humans result in 'collectives', which Latour defines as 'the project of assembling new entities not yet gathered together' (Latour 2005: 75). Like human-animal studies, these approaches help to undermine human exceptionalism and show how societies, or collectives, have many different constituents, all potentially able to act on one another. This raises the question of whether to replace the term 'animal' with the term 'non-human', rejecting the boundary between non-human animals and things. But such a complete rejection of essentialist categories runs the risk of making all entities interchangeable, their identity a product of the networks of which they are a part.

Definitions of the term 'animal' frequently define humanity by exclusion, with animals deficient in certain human traits (Midgley 1978; Ingold 1988; Fudge 2002a: 164; Calarco 2008: 3). In an influential article, Derrida (2002: 400) observed of the term 'animal' that humans 'have given themselves the word in order to corral a large number of living beings within a single concept'. Rather than defining animals in contrast to humans, Reed (1988b) distinguished between animacy and inanimacy; whereas inanimate beings can be acted upon, animate beings interact. These animate interactions, he argued, are shaped by the affordances of the bodies involved. This aligns with a definition of organisms as autopoietic, that is, self-generating, in contrast to objects, which are made by external processes (Maturana and Varela 1980; Wolfe 2003: 36). Defining animacy in terms of affordances offers a useful starting point to a working definition of animals, highlighting what they do rather than what they are. Different animal species, or individual animals, potentially have different affordances depending on their bodies and the bodies they interact with.

A relational perspective considers both the capacities or properties of particular entities and the relations between them; Knappett (2005b: 77) argues that it is necessary to consider how the properties of objects affect their positions in networks. Descola adopted Latour's notion of collectives to describe the assemblages of humans and non-humans within any given society. Yet he too emphasises the way in which the properties of entities shape their place in a set of relations. As he suggests:

> Now a system of relations cannot be understood independently from the elements it connects, provided these elements are taken not as interchangeable individuals or already institutionalised social units, but as entities that are endowed *ab initio* with specific properties that render them able or not to establish certain links between them.
>
> (Descola 2006: 139)

The affordance concept reinforces the idea that these properties need not determine these relationships, but rather arise from the interaction between entities. To say that humans, non-human animals and things generally have different affordances is not to insist on strict boundaries but at the same time allows for different types of relationships between these different kinds of entity (Latour 2005: 71–2). Latour's thinking on networks has developed over time and, while still stressing the importance of interaction, he has defined different 'modes of existence' which are composed of different kinds of entities with different paths of existence. Animals, and a variety of other beings, belong to the mode of reproduction (Latour 2013: 98–102). The interactions between entities from different modes of existence are 'crossings' which help to define the differences between them (Latour 2013: 63). From this perspective, 'nature' and 'society' are the product of confusing different modes of existence, such as the beings of reproduction with the way humans describe them (Latour 2013: 353). He continues to use the term collective instead of 'civilisation' or 'society' 'to emphasize the operation of gathering or composing, while simultaneously stressing the heterogeneity of the beings thus assembled' (Latour 2013: 296). By multiplying the different modes of existence, Latour continues his project to undermine notions of 'nature' and 'society' and recast the world as a web of connections between a variety of entities (Latour 2017).

The usefulness of Latour's inquiry into modes of existence is that it offers a framework for considering the discipline of Minoan archaeology as well as the various buildings and things that archaeologists have excavated. Archaeologists' excavation reports belong to a mode of existence he calls 'reference'. For the mode of existence he calls 'technology', which includes buildings and manufactured things, he uses the metaphor of a labyrinth. 'Technology always entails folds upon folds, implications, complications, explanations. Its canonical representation, thoroughly studied by the sociology of technologies, sketches it in the form of a series, often a very long series, of nested translations, a labyrinth' (Latour 2013: 228). The prototypical Labyrinth is located at Knossos, the building excavated by Arthur Evans. Following Latour's line of reasoning, Evans's excavation diaries and reports, culminating in *The Palace of Minos at Knossos*, are a chain of references which have been confused with the Labyrinth to create a hybrid known as 'Minoan Civilisation'. Instead of adding to the confusion with further hybrids such as 'Minoan Art' or 'Minoan Religion', the project of archaeologists should instead be to retrace the connections between entities in Bronze Age Crete.

This book will focus on one particular set of connections, between humans, animals and objects. In doing so it draws upon the 'new animal geography', which has sought to locate human-animal relations in particular places (Philo and Wilbert 2000; Philo and Wolch 1998; Wolch and Emel 1995). It is a reaction against the traditional zoogeographic perspective which considers animal distributions rather than human-animal interactions. 'The argument is that these relations are differentially constituted in different places, forming a complex geography which can, in principle, be mapped out onto the places, regions and landscapes spread across the earth's surface' (Philo and Wolch 1998: 112). From this comes the concept of 'animal practices', arising from the concern with human-animal relations in specific places. As Chris Philo and Chris Wilbert (2000: 5) suggest:

> If we concentrate solely on how animals are represented, the impression is that animals are merely passive surfaces on to which human groups inscribe imaginings and orderings of all kinds. In our view, it is also vital to give credence to the practices that are folded into the making of representations, and – at the core of the matter – to ask how animals themselves may figure in these practices.

The idea of animal practices has been used to elucidate the way in which racial and other cultural differences are created in modern America (Elder et al. 1998). The animal practices of the dominant group, such as a particular way of hunting deer, become the standard against which the animal practices of other groups are judged and branded as 'poaching' or animal cruelty. They cite the example of Mexican-style rodeos in the United States, where the lassoing of horses is often banned by local authorities, whereas lassoing cattle is a staple of the American-style rodeo (Elder et al. 1998: 77).

In Bronze Age Crete the dominant group centred around the palaces, and so the animal practices such as bull-leaping can be seen in the same way, as human-animal relations which are place-specific and related to the definition of groups. In Neopalatial Crete, bulls afforded leaping when they came into interactions with humans of a certain status in a particular place, and the depiction of this animal practice in frescoes and gold rings reinforced this relationship between bull-leaping and the palatial collective of Knossos. Bull-leaping depictions acted as animal things, making this practice present in given contexts, whether on the walls of the palace or impressed on a clay sealing in the palm of someone's hand. These interactions between humans and animals, animal practices and the animal things which indexed them together traced the collectives which came together and fell apart over the course of the Bronze Age. Only the animal things now remain, often behind glass in museum cases, where they are part of different collectives, composed of archaeologists and museum visitors, selfies and disnified souvenirs, far removed from the animal practices of which they were once a trace.

CONCLUSION

In 'Why look at animals?' John Berger suggested a link between human-animal relations and animal depictions; before the advent of modern capitalism people were more familiar with animals and this resulted in realistic representations of them in art. This is a short step from the theories of Evans and Groenewegen-Frankfort that Minoan animal depictions reflected a love of nature. But the importance of Berger's essay is that he opened this link up to scrutiny, drawing on anthropological theory to suggest that the importance of animals to humans, as revealed in the returned gaze, is that they are both similar and different. The notion of similarity and difference, central to the structuralist tradition that Berger drew upon, is also the foundation of Philippe Descola's theories of the four basic ontologies that structure human relations with the other entities in the world, including images. But for Descola it is the naturalist ontology which regards animals as similar but different to humans, sharing genes but not consciousness. In analogist collectives, of which Bronze Age Crete is one, all the entities in the world are distinct and so the function of images is to help order them in some way. Instead of windows onto the world, aided by perspective drawing, analogist images become connectors, bringing disparate things together. The detail of these naturalistic depictions did not deceive the eye or celebrate nature but made them more effective connectors.

A consideration of the Partridge frieze from the Caravanserai at Knossos helps to address the problem of understanding naturalistic depictions of animals from the point of view of a different ontology. Whereas it has often been read in the same way as a birdwatcher's handbook, it is problematic to assume that its painters shared the same species concept as modern observers. Lloyd's concept of 'styles of inquiry' helps reveal categorisation as relative to different types of human-animal relation, whether birdwatching, evolutionary biology or fowling, even if at some level the morphology of the birds is perceived in the same way across ontologies. Given that the animal practice of fowling was widespread in Bronze Age Crete, it is likely that these birds were categorised as game. The interaction between humans and partridges which allows them to be perceived as food is an example of the affordance concept. From this theoretical perspective, the fresco displays their affordances as edible animals. Partridges were indeed eaten in Bronze Age Crete, and so, in semiotic terms, the fresco becomes an index of these animals and their edibility. Grounding Gibson's affordance concept in a pragmatic tradition also brings the process of abduction to the fore, drawing connections between different depictions in a Bronze Age Cretan Lifeworld rather than searching for symbolic interpretations. This method will be employed in the following chapters, each of which seeks to connect a variety of animal depictions with a particular animal practice such as hunting or herding.

The Taureador fresco too affords different things to a modern observer, brought up with a naturalist ontology and with different experiences of cattle, than it did to a Bronze Age Cretan. But at the same time it is possible to ask what affordances a Bronze Age Cretan fresco revealed, such as that bulls can be leapt over by humans, and what sort of connection this image had with what it depicted. Bull-leaping frescoes, for instance, established a connection between bull-leaping and the palace at Knossos. In turn, seal impressions could index frescoes at the same time as indexing the user of the seal. Such depictions can be seen as the outcome of animal practices, whether bull-leaping, bull-catching or bull-hunting. Their realism served to provide information about these animal practices and it is this relationship between animal practice and its material trace which is encapsulated in the term 'animal thing'. Animal things made absent animal bodies present, whether as metaphors or traces, and extended the animal practices which helped define the network of humans, animals and things which Latour terms a 'collective'. Naturalistic animal depictions were not a reflection of the nature-loving Minoans' worldview but rather the material trace of a palatial collective centred on Knossos. The following chapters will consider the animal things and the animal practices which they indexed in order to trace how this collective came together.

THREE

HERDING

INTRODUCTION

Several of the Knossos tablets which apparently give totals have figures in excess of 2000; one fragment contains the numeral 19,000. This would have been piety indeed.

(Ventris and Chadwick 1956: 198)

Even before the decipherment of Linear B by Michael Ventris in 1952, it was apparent that large numbers of animals were being counted on Linear B tablets (Figure 3.1). Johannes Sundwall (1936) had observed that many of the tablets totalled 100 animals, and came to the logical conclusion that these were records of hecatombs, the sacrifice of 100 cattle known from the Homeric epics. Drawing attention to the large numbers of animals, and also identifying the ideograms on these tablets not as cattle but sheep, Ventris and Chadwick questioned whether tens of thousands of animals were really being sacrificed at a single event. They regarded these sheep, mainly male, as some kind of tribute, but were unable to explain the huge numbers or the comparative lack of ewes (Ventris and Chadwick 1956: 197–8). It was only when John Killen compared the structure of the records to the Mediaeval British wool industry that similarities with Knossos became apparent, and the large number of male sheep was explained, since these wethers, castrated males, would have provided the finest wool. One of the reasons that previous authors, including Arthur Evans, had identified these sheep tablets as records of cattle was that large numbers of depictions of cattle had come to light, many showing the animal practice of bull-leaping, but sheep were extremely scarce. It thus came as

3.1. Linear B tablet recording sheep at Phaistos, LMIIIA. Length: 11.1 cm (BM GR1910,0423.2). © The Trustees of the British Museum

a surprise that over 100,000 sheep were listed in the Linear B archive of the palace of Knossos at the time of its destruction. These tablets were not simply records of economic activity but enabled the scaling up of the animal practices of herding and shearing, and other practices including dyeing and weaving, in order to produce an animal thing, woollen cloth. This valuable commodity was probably the single most important export product of Crete in the Late Bronze Age, meaning that sheep and their associated animal practices and animal things were at the heart of the palatial collective at this time. Cattle were also listed, and even named, on the Linear B tablets, but in far smaller numbers. This chapter will trace the developing relationship between humans, sheep, cattle and other domesticates in Crete, and seek to explain why the resulting animal things appeared when they did.

Cattle have all but disappeared from Crete now, although teams of oxen for ploughing are still just within living memory (Rackham and Moody 1996: 75; Halstead 2014: 38). The mechanisation of farming on Crete has not resulted in the disappearance of domestic animals from everyday life, at least outside the main centres. The countryside of Crete still supports a major agricultural economy. Compared to the 2,000 agrimia estimated to survive on the island, there are over 600,000 domestic goats, a ratio of 300:1. Sheep are even more numerous; 1.5 million were counted in the 1999–2000 agricultural census.[1] By contrast there were 2,207 cattle recorded in the same census. This suggests that maintaining cattle for milk and meat, now that they are no longer needed for traction, is not economically viable on Crete. Instead, donkeys are still occasionally used for carrying loads, particularly tourists, and tilling the soil.[2] Large numbers of poultry (1.4 million) are kept in captivity, with rabbits and pigs to a lesser extent. The census lists 65,000 beehives on Crete, continuing a practice known from the Bronze Age (Panagiotakopulu 2000: 95–103). Unlike other domesticated animals, dogs and cats are abundant in settlements, where they occupy a spectrum from domestic pets to feral scavengers. In the countryside dogs are often employed to guard areas of land, or sheep and goats. The absence of large predators on Crete, apart from feral dogs, means that flocks are relatively safe from predation (although not rustling).

Although it is tempting to use recent history as a source of analogies for the Bronze Age, it is unwise to assume that the countryside of Crete and its animal and human inhabitants were stuck in a timeless past until the jolt of modernisation. The agricultural records of Venetian and Ottoman Crete reveal the agricultural practices of an island which was part of larger maritime collectives; olive-growing has expanded in the last few centuries according to the needs of an export market (Halstead 1987; Rackham and Moody 1996: 82). Crete was also a major exporter of olive oil in the Late Bronze Age, as finds of transport stirrup jars around the Eastern Mediterranean attest (Haskell 2011), but this relates to another time of extensive overseas connections rather than demonstrating long-term continuity. Instead of providing a window into the Bronze Age, accounts of pre-mechanised agricultural activities in Greece can be used more broadly as a way to understand different possibilities of farming the Cretan landscape (Halstead 2014). As Killen showed, Mediaeval English sheep records provided a useful analogy for the Late Bronze Age wool industry because the scale of production was similar. Another analogy for Bronze Age relations with cattle used here comes from a Mediterranean-type ecosystem, California, where in the eighteenth and nineteenth centuries cattle were allowed to roam free in an agricultural practice known as ranching. The relations between humans, animals and the Cretan landscape were contingent rather than fixed in the Bronze Age, as they always have been across the Mediterranean (Cherry 1988; Horden and Purcell 2000; Broodbank 2013).

This chapter will focus on the relations between humans and domesticated animals on Crete using both written texts and depictions in other media. Domesticated animals include cattle, sheep, goats and pigs which were brought to Crete in the Neolithic period by early farmers; although the complexities of the term will be discussed below, here domestication refers to animal populations which have been subject to selective breeding by humans (Russell 2012: 209–11). Many of these domesticated animals were also domestic animals; the term 'domestic' is not straightforward but here refers to populations of animals which participated in agricultural animal practices such as herding or traction as part of domestic collectives. Dogs, which were particularly involved in hunting, and cats, which were most likely imported to Crete in the Middle Bronze Age, will be discussed in later chapters. Of central importance here are two particular networks of human-animal relations in the Bronze Age, involving sheep and cattle. Although they appear in great numbers on the Linear B tablets, sheep are rarely depicted in Bronze Age Cretan iconography, and are all but absent in depictions from the Late Bronze Age. Cattle, on the other hand, are both frequent in the Linear B tablets and widely depicted in media such as sealstones, figurines and frescoes throughout the Bronze Age. But whereas the animals named in the Linear B tablets are plough oxen, there are no depictions of ploughing in the Bronze Age. Instead, cattle, particularly bulls, are shown being leapt over by humans or hunted. As will be suggested, it is likely that these are domesticated animals, but involved in non-domestic animal practices which are closely associated with the palatial collectives. The emergence of the palatial collectives can be linked with changes, particularly in scale, to animal practices inherited from the domestic collectives.

DOMESTIC COLLECTIVES

Neolithic Knossos

The term 'domestication' leads to an attitude which regards animals as either 'wild' or 'domestic' and which blurs the fact that these terms are an arbitrary division of an endlessly varied spectrum. The relationships of 'wild' animals with men are no more constant in time and space than those of 'domestic' animals, and to classify animals in one or the other category does little to define the relationship involved.

(Jarman and Jarman 1968: 258)

Knossos was one of the first settlements on Crete, and has the earliest evidence for farming on the island. It was founded at the start of the Neolithic period by a collective of humans, animals and plants which arrived from overseas, most likely the Anatolian mainland (Broodbank and Strasser 1991: 237). As Michael and Heather Jarman caution, the presence of cattle, sheep, goat, pig and dog bones in the earliest levels does not mean that a particular set of human-animal relations can be readily inferred. A hare's tooth was also found in the lowest level at Knossos, known as the 'Camp'; now wild on Crete, hares were also brought by Neolithic settlers and might have been kept in captivity too. Other 'wild' animals introduced by humans include badgers, possibly introduced for their pelt, and agrimia, whose ancestors either escaped or were deliberately left to establish populations in the mountains as a future food reserve (Bar-Gal et al. 2002). Whether these animals were hunted, kept in captivity or selectively bred by early Neolithic farmers, the animal remains found at Knossos help define the earliest known domestic collective on Crete: a network of humans, plants, animals and objects centred on permanent dwellings whose interactions resulted in the development of a gradually increasing mound.[3]

Domestication can better be seen as an ongoing process, in which certain farming practices lead to closer co-existence between humans and animals than others; by analogy with traditional farming practices in Greece, some pigs could have lived primarily on household leftovers, while others were left to roam and forage in herds (with or without a human herder) with some going feral, living and breeding beyond human control (Halstead and Isaakidou 2011). Cattle, sheep and goats could either have been penned or allowed to roam in a more or less supervised way; the absence of large predators on Crete would have favoured free-ranging animals, with populations of feral cattle and goats

(Halstead and Isaakidou 2017). Over time the browsing of goats and cattle, along with the woodland clearance of humans, would have promoted the development of pastures on which sheep (and cattle) could graze. Although a range of human-animal relations is included within the idea of domestication (Russell 2002), a notion of animals as human property, whose production was under human control and whose ownership could be transmitted between people, would have defined the animal practices of a settled community like Knossos (Orton 2010). These relations of protection, production and transmission are those identified by Descola with an analogist collective. This does not preclude other types of relationship or reduce animals to the status of inanimate objects, but highlights an important difference between human-animal relations in hunter-gatherer versus farming societies.

Although zooarchaeology has traditionally provided information about which animals were killed and eaten, increasingly sophisticated analyses have provided insights into farming practices. Comparing the proportions of different species in an assemblage and the different age/sex profiles of each species can help determine herd composition and the use of animals for meat or dairy, or, by the start of the Bronze Age, wool production for sheep. Valasia Isaakidou (2008: 110) identified a particular animal practice, which she terms 'gardening with cows', in order to explain the increase in cattle at Knossos over the course of the Neolithic. Traction pathologies on cows' pelvises suggest that female cattle rather than oxen (castrated male cattle) were used for ploughing, and probably also dragging heavy loads (Isaakidou 2006: 104–8). This in turn suggested that the land around Knossos was intensively exploited but not to the extent that oxen were kept for this specialist purpose; rather, cows were preferred because they could also provide milk and calves, and meat at the end of their lives. As a result, there is little evidence that these animals were primarily reared for meat in the Neolithic as had been assumed (Broodbank 1992; Isaakidou 2008: 108–9).[4] Rather, the bones were deposited following consumption events even if the cattle were not being reared primarily for this purpose. Cows have been used for traction by many small-scale farmers in Greece in recent history; keeping an expensive pair of plough oxen has often been the preserve of large landowners and, by analogy, the Bronze Age palaces (Halstead 2014: 60).

The zooarchaeology of Neolithic Knossos is well understood because of the high standard of recovery from the 1957–60 excavations of the earlier houses below the Central Court of the Palace conducted by John Evans (1964), and Isaakidou's (2004) recent restudy of these remains. A similar restudy was undertaken for Final Neolithic animal remains from Phaistos, although these comprise a smaller number of bones from a single deposit (Wilkens 1996a: 241–6). Of eight cattle, six were adult, including one 'very old individual', leading to the suggestion that they were not just reared for meat and so could have been traction animals too. Similarly, the presence of older adult sheep and goats among the twenty-one individuals identified indicates that they produced wool rather than just meat, although there were no very young animals, suggestive of intensive dairying. The ten pigs were generally slaughtered before the age of two, meaning that they were reared for consumption; unlike sheep and cattle, pigs do not have other affordances such as wool or milk production which would make farmers keep them after the onset of adulthood. More recent excavations in the same area have uncovered ceramics and animal bones associated with an open-air hearth. The similarities between the animal remains in this deposit (at least four cattle, six sheep, four pigs and an agrimi) with the assemblage published by Wilkens lead the authors to suggest large-scale consumption beyond the level of the household (Todaro and Di Tonto 2008: 181–6).

There are other things relating to domestic animals at Knossos. One of the earliest traces of an animal are two hoofprints preserved in a mudbrick, an index of a particular sheep or goat's presence in the Early Neolithic settlement (Evans 1964: 144–6). A number of cattle figurines were found in later levels, burnished like contemporary pottery. Human and bird figurines were made in the same way, showing how figurines could create connections between different entities and incorporate them in the household (Nanoglou 2009). There were also spindlewhorls and loomweights, used for textile manufacture, although it is not clear whether these were for plant fibres or sheep and goat hair (Evans 1964: 233–5; Burke 2010: 23). Since they were concentrated in the Final Neolithic levels, contemporary with evidence for wool production in the Eastern Mediterranean, it is probable that they indicate the arrival of woolly sheep (Militello 2014: 320).[5] The Neolithic tell of Knossos contains traces of a collective of humans, animals and objects which endured for 4,000 years on the same site. Although archaeologists have noted its strategic location, a few kilometres inland and on the banks of the river Kairatos, surrounded by fertile agricultural land, its importance in the Bronze Age has also been attributed to an awareness of the long history of this site (Soles 1995).

Prepalatial Crete

The Prepalatial period of Crete is characterised by small-scale farming settlements, frequently situated in defensible points in the lowlands, where crops were farmed and animals grazed. Some Prepalatial sites became palatial centres, making it difficult to trace the development of palatial collectives from local collectives because earlier buildings were levelled or incorporated into the palaces. It is clear from pottery deposits that Knossos in the Prepalatial period was already a centre for consumption activities (Day and Wilson 2002). The faunal remains do not, however, show significant shifts in consumption practices from the Neolithic period (Isaakidou 2007b: 19; Halstead and Isaakidou 2011). Zooarchaeological remains have been published from smaller sites which were abandoned in the Prepalatial period such as Debla, an Early Minoan site in western Crete, where the exclusive presence of adult sheep/goat bones, along with other factors such as the altitude and architecture, led to the suggestion that this was a summer pasturage camp. Finds of spindlewhorls indicated wool spinning but there were no remains of animal enclosures (Warren and Tzedhakis 1974: 318, 334).

One extensively excavated Prepalatial settlement, destroyed at the end of the Early Minoan II pottery phase, is Myrtos Fournou Korifi, on a hill by the south coast of Crete (Warren 1972). It provides an example of a local collective composed of five or six domestic collectives, since the settlement was divided into discrete units (Whitelaw 1983). Various evidence for textile production was found, including two groups of loomweights, spindlewhorls, spinning bowls used for wetting fibre and vats which could have been used for processing wool (Warren 1972: 262–3). A high proportion of the bones were from mature sheep or goats, indicating that they were kept for wool rather than bred primarily for meat (Jarman 1972). Textile patterns were also perhaps painted on a figurine known as 'the Goddess of Myrtos', a female figure carrying a jug (Warren 1972: 208–10). Although textile production was seemingly carried out at the level of the household at Myrtos and contemporary Knossos and Vasiliki, Brendan Burke (2010: 29) argues that Cretan textiles, and perhaps the technology of the spinning bowl, were exported to Egypt in this period. Seal impressions from Myrtos and Knossos show that seals were also in use, but it is not clear whether they were used within the settlement to secure goods or were used as part of the process of exchange (Anderson 2016: 61–2).

Prepalatial local collectives were often centred on tombs, which often had associated gathering spaces (Branigan 1993; Legarra Herrero 2014). Many of these tombs were either looted in the past or excavated in the early twentieth century and so there are few published animal remains, either from the tombs or from associated consumption deposits in the courts outside (Isaakidou 2019: 627). Nevertheless they are mentioned in the original publications as being present at significant tomb sites such as Lebena and Archanes (Hamilakis 1998: 119) and have also been found in the re-excavation of Koumasa, despite not having been mentioned in the original publication (Xanthoudides 1924: 3–50; Panagiotopoulos 2018: 52). A number of cattle rhyta were found in the tombs of southern Crete, mingled with the disarticulated bones of the human members of the collective, and perhaps the butchered bones of its animal members. Rhyta in the shape of other animals, including birds, were also found and would have been used for pouring liquids (Koehl 2006: 75–6). Some cattle-shaped rhyta have human figures attached to their horns, appearing to show a form of cattle restraint or capture (Loughlin 2004a). One such rhyton from

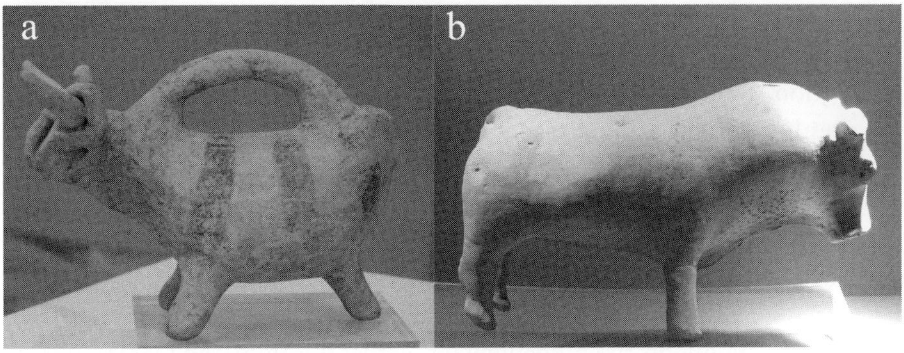

3.2. Prepalatial ceramic cattle rhyta. (a) Rhyton from Koumasa with human figures on horns, EMII–III (AMH Π 4126). Length: 20.4 cm. © Heraklion Archaeological Museum – Hellenic Ministry of Culture and Sports – 'Hellenic Organization of Cultural Resources Development (HOCRED)'. (b) Rhyton from Phourni, Archanes, MMIA. Length 21.1 cm. © Hellenic Ministry of Culture and Sports (N. 3028/2002)

Koumasa has three legs, a handle and a striped body, showing that it retained the appearance of a vessel (Xanthoudides 1924: 40) (Figure 3.2a). Similar, fragmentary examples were found at Myrtos, showing that they were also used in settlements (Warren 1972: 220).

A more realistically modelled bull rhyton was found in a tomb at Archanes, with painted stipples to indicate the bull's coat (Koehl 2006: 73 no. 14) (Figure 3.2b). This illustrates a shift from zoomorphic rhyta which were primarily vessels to those which were primarily animal figures (Miller 1984: 66). The short horns of the Archanes figure are perhaps another significant difference if these are meant to be polled horns, which would have made bulls less dangerous to humans. The Archanes vessel with its emphasis on the physique of the bull hints at the ownership of prime cattle which was an important part of emerging hierarchies. An animal thing like this could have maintained an association between the deceased and cattle ownership. The realism of the depiction made the rhyton a more effective animal substitute, and demonstrated access to skilled crafting, a phenomenon which continued in the Palatial period. The Early Minoan burials at Archanes have sometimes been associated with Knossos because of the proximity of the site, but whether they were related to the Knossos collective or one centred on Archanes, such objects show that bulls were important members of these collectives in the Prepalatial period. These bull rhyta materialise the activity of keeping and restraining cattle in durable animal things, in a form that could be brought into the home or tomb.

PALATIAL COLLECTIVES

The Protopalatial period on Crete, beginning at around the start of the second millennium BCE, is defined by the appearance of buildings conventionally termed 'palaces' at a number of sites across the island. Two palaces, Phaistos and Knossos, were built on sites which, as noted above, had a long history of occupation, and communal gathering, going back to the Neolithic. The palatial sites became the central places of wider regions, resulting in the gradual reorganisation of the agricultural landscape (Haggis 2002; Whitelaw 2012). Palaces enclosed central courtyards where gatherings could take place, almost certainly involving eating and drinking supplied by accompanying storage areas, as well as contributions from those attending. Although various functions have been suggested for the palaces, including as religious centres, Jan Driessen (2002, 2004) highlights the fundamental role that the West and Central Courts played as communal areas, served by the various rooms of the palace. The more accessible West Courts have been seen as the successor to the open areas in front of Prepalatial tombs, which gradually fell out of use. Peak sanctuaries (ritual sites in high places in the landscape), as well as cave sanctuaries such as Psychro, emerged at around the same time (Watrous 1996: 73). These sanctuary sites, with extensive deposits of animal figurines, have been associated with agriculture, but tend to be seen as reliant on the palaces because they appeared at a similar time. The appearance of peak sanctuaries could instead be seen as symptomatic of wider changes which also led to the building of palaces in central places, rather than as offshoots of them (Haggis 1999). At Knossos, Phaistos and Mallia, there was a clear increase in population and settlement size at around the same time as palaces emerged (Whitelaw 2012). These developments also coincided with the use of writing for administration alongside more complicated sealing practices, used to record transactions. Although the palaces clearly were important buildings in these societies, the term palatial collective denotes the increasingly differentiated network of people, animals and objects which connected palaces, peaks and the intervening settlements and farmland. The increased scale of palatial collectives meant that different types of animal management, extending across large areas of Crete, could be practised.

Few traces of animals have survived in the palaces. This is partly because many were excavated early in the twentieth century, when animal bones were not systematically collected. There is also a focus on special deposits, particularly involving animal skulls, which will be considered in the following chapter. The more recently excavated sites, including Zakros and Petras, are still awaiting final publications, although the bones from the 'Hieroglyphic Archive' room at Petras have been published, and are predominantly sheep and goat (Mylona 2010). Moreover, these deposits of bones are traces of butchery and human consumption rather than of the lives of animals. One structure in the north-east part of the palace at Knossos, probably constructed in the Neopalatial period, is likely to have been a row of animal pens, perhaps for pigs (Shaw 1978b; Hood 1985). There is no clear evidence that cattle were brought into the palace for bull-leaping. Instead, the existence of various animals in the palaces and related sites is preserved in the things that stood for them, whether depictions on frescoes or transactions in clay.

Because of its availability and affordances, including malleability when wet and preserving particular forms when dried or fired, clay was the medium for two different types of transactions involving domestic animals in Bronze Age Crete. The first was the moulding and firing of animal figures or figurines, continued from earlier local collectives, but now increasingly deposited at extra-urban sanctuaries. The second was the impression of lumps of wet clay with a seal, or the incision of clay tablets with a stylus, in order to record, among other things, the movement or the transfer of domestic animals. The resulting documents were not fired because they were intended to be temporary records, although they did survive through accidental firing in the periodic destruction events that affected the buildings they were stored in. These buildings tended to be administrative centres, whereas the majority of figurines were deposited at the extra-urban sanctuaries. In both cases the clay object was a substitute for an absent animal, a metaphorical animal thing. This section will examine the palatial collective at both the edge and the centre; peak sanctuaries indicate the role of domestic animals in the expansion of territory while palaces provide evidence that domestic animals were transacted at the centre using administrative practices involving seals and written records. Writing in particular allowed an increase in the scale of animal practices, with large numbers of sheep monitored at Knossos by the time of its final destruction.

Peaks

Peak sanctuaries are upland locations in Crete defined by the deposition of figurines and the remains of consumption events. Some are closely associated with, and visible from, major settlements, such as Juktas above Knossos or Petsophas above Palaikastro. Although the dating of the establishment of the peak sanctuaries is debated, the consensus is that they largely coincide with the emergence of the palaces. The deposition of animal figurines at these sites in particular is agreed to start in the Protopalatial period. At this time animal figurines are deposited at every one of the more than twenty identified peak sanctuaries (Kyriakidis 2005: 149). There are instances of figurines in settlements, including Mallia and Kommos, in the Protopalatial period but these occur in smaller numbers and there are no clear patterns of use (Detournay et al. 1980: 111–13; Shaw 1996: 299–300). Although the peak sanctuary assemblages differ in composition, the animal figurines are almost certainly domesticated quadrupeds for the most part. Some of the Atsipadhes cattle figurines have sexual characteristics, both male and female (Peatfield 2000); this specific detailing suggests an interest in particular animals rather than generic offerings. Sheep and goat figurines, distinguished by their horns, are also found. Few of these assemblages have been published in detail and so identification is problematic, as is dating (Haysom 2015, 2018). In general undifferentiated quadrupeds are usually assumed to be bulls, although Zeimbekis's term 'bovine' is safer for these in the almost complete absence of sexed figurines from the sites of Juktas and Kophinas (Zeimbekis 1998: 244).

Preliminary totals for animal figurines from peak sanctuaries include almost 3,000 fragments from Juktas, of which 70 per cent were bovine and 6 per cent were sheep (Zeimbekis 1998: 77, 244–5). There were at least 3,500 bovines from Vrysinas and 5,000 from Atsipadhes (Tzachili 2012; Peatfield 1992: 64). These sites seem to be typical in that the majority of their figurines were cattle; only at Traostalos, near Zakros, does there appear to be a higher total of a different domesticate, sheep (Pilali-Papasteriou 1985: 150). Piskokephalo is the most pronounced example of the deposition of beetle figurines, seen also at other sites in eastern Crete. Petsophas shows a diverse range of animals, including tortoises, birds and a form usually identified as a weasel, but which could be any one of a number of small predatory mammals (Figure 3.3).[6] Bogdan Rutkowski (1986: 87–91) suggests that nearly all these animals can be related to the domestic sphere in some way; a weasel-like animal could be kept for pest control, and dung beetles would have been associated with sheep dung. Figurines are found in domestic contexts in the Protopalatial period, and throughout the Bronze Age, but peak sanctuaries provide the biggest concentrations. The occurrence of sexed figurines at some peak sanctuaries but not others, and different types of animals at various sites, points to local diversity within a shared practice.

The practice of depositing animal figurines at peak sanctuaries is open to a range of interpretations. Figurines are commonly regarded as sacrifice

3.3. Animal figurines from Petsophas MMI–II. Left to right: bovine, length: 5.5 cm (AN1896–1908.AE.1845); sheep, length: 4.7 cm (AN1896–1908.AE.1841); mustelid, length: 4.6 cm (AN1896–1908.AE.1846); bird, length: 5.2 cm (AN1896–1908.AE.996); dog, length: 5.5 cm (AN1896–1908.AE.1025.2); goat, length: 4.2 cm (AN1896–1908.AE.1842); bovine?, length: 4 cm (AN1896–1908.AE.1025.1). © Ashmolean Museum, University of Oxford

substitutes (Watrous 1996: 49; Kourou and Karetsou 1997: 115). Certainly, cattle would have been expensive animals to sacrifice, explaining the use of substitutes, but at the time they were deposited, in the Protopalatial and Neopalatial period, there appears to have been more of an emphasis on consumption than sacrifice. Conversely, authors do not suggest that the human figurines deposited at peak sanctuaries are sacrifice substitutes; their various gestures are usually understood in terms of adoration, worship or other types of performance (Rutkowski 1991: 52–6; Morris 2001). Animal figurines have also been seen as extensions of herds or flocks. It has been suggested that the offering of animal figurines at sanctuaries was directed at the well-being of the supplicant's animals (Peatfield 1992: 78). For Rutkowski (1986) too, these are substitutes for particular real animals, whose divine protection was secured by these offerings. The different markings on the Petsophas figurines, some black and others piebald, and the sexed Atsipadhes figurines, could show that figurines were individuated. Both humans and animals could have had particular problems that were brought to the peak sanctuary in the form of figurines.

A figurine could also stand for a herd or a flock in general, acting as a token rather than indexing a specific animal. The composition of the figurines, however, does not appear to be representative of livestock in general. It is clear from faunal assemblages that there was an abundance of sheep, goats and pigs in Bronze Age Crete, but cattle figurines were most commonly deposited at peak sanctuaries. Cattle are more expensive to keep, particularly plough oxen, and so might have been given more attention. The almost complete absence of pig figurines indicates that peak sanctuaries were associated more closely with animals herded in the area around settlements than pigs, which were potentially fed with household leftovers or allowed to roam free. This suggests that figurines were deposited in order to maintain the association between animals, particularly cattle, and the land. The emergence of peak sanctuaries in the Protopalatial period can be explained in terms of the filling-up of the landscape, and the increased pressure on grazing space. This increasingly valuable land could be viewed from the high vantage points that the peak sanctuaries afforded their human visitors.

It is likely that the people offering figurines did so for a range of reasons and so these options are not mutually exclusive. What has been neglected is the role of the figurines themselves in these ritual practices. Figurines can be seen as animal things which arise from the everyday relationships between people and domestic animals, whose materiality offers a different set of possibilities for their use. The affordances of fired clay meant that animal forms could be made permanent and deposited at particular sites, connecting them with that place. Figurines act as metaphors for animals, whether as offerings or in supplication for their well-being. As Stratos Nanoglou (2009) has observed of interpretations of Neolithic animal figurines from northern Greece, these objects should not simply be seen as reflections of a pre-existing economic reality. Instead, the acts of making and deposition of these objects in particular contexts are constitutive of the way people understand animals. In the case of Bronze Age peak sanctuaries, animal figurines enabled the everyday human-animal relations of the area around the settlement to be relocated and manipulated in an upland ritual setting. This gave a material presence to domestic animals in a prominent place in the landscape.

Deposition of figurines continued in the Neopalatial period, although many peak sanctuaries fell out of use. Those that remained saw the deposition of more costly items and are often close to palatial centres. At Psychro there is a shift in the middle of the Neopalatial period to the deposition of fine pottery from Knossos (Watrous 2004). It has been suggested that these changes mark a shift from popular to palatial cult practice (Peatfield 1992: 60). At Juktas, the peak sanctuary close to Knossos, and Kophinas, both figurines and larger cattle figures were deposited in the Neopalatial period (Zeimbekis 1998, 2004). At Vrysinas, on the western side of the island, there is evidence for the consumption of domestic animals in the early Neopalatial period, along with the deposition largely of cattle figurines. The meat was cut into portions and stewed in pots; although cattle bones were found, sheep/goats and pigs were more frequently eaten (Mylona 2016). There is also evidence for consumption of cattle at Alonaki, an early Neopalatial building on the slopes of Juktas (Karetsou 2013: 77). The consumption of meat, which would not have been an everyday occurrence for most people, at peak sanctuaries further associates these places with domestic animals. There is little evidence for the consumption of animals in Protopalatial peak sanctuaries, supporting the idea of a shift in peak sanctuary use at the start of the Neopalatial period (Cromarty 2008: 31–41, 79; Haysom 2015: 101). The possible exception is Juktas, which had a Protopalatial ashy layer containing animal bones, but was architecturally elaborated in the Neopalatial period (Karetsou 1981: 145).

At the same time as signs of increasing animal consumption at sanctuary sites, some vessels deposited at Psychro, Juktas and Vrysinas were combined with human and animal figurines in the form of appliqués (Boardman 1961: 57–9; Watrous 1996: 35; Tzachili 2011). This is part of a wider phenomenon of applied ceramic relief, which has also been compared to relief stone vessels and seal impressions (Foster 1982; Tzachili 2011: 192–9; Knappett 2020: 92–7). There are a number of appliqués of wild goats, notably on a Protopalatial jar from Phaistos and on vessels from Psychro, Juktas and Alonaki all of which probably date to the early Neopalatial period (Simandiraki 2006; Karetsou 2013: 82). From around the same period is a large jar from Anemospilia decorated with a bull in relief, painted with piebald markings (Sakellarakis and Sapouna-Sakellarakis 1997: 548–62). In these cases, animal appliqués or relief depictions become part of the fabric of the vessel, and by extension part of consumption practices.

The changing pattern of peak sanctuary activity and use between the Protopalatial and Neopalatial period could reflect changes in the relationship between grazing animals and the landscape. The population growth seen at Knossos and other centres implies that their territories had also increased in size in order to feed growing human populations (Whitelaw 2012, 2018, 2019). This in turn would have favoured the grazing of larger herds further away from the centre, and in upland areas (Halstead 1987). Whereas Protopalatial peak sanctuaries could have marked the grazing areas of local communities, it is likely that the larger Neopalatial settlements were sending animals to be grazed across much wider areas. Cattle-rearing in particular requires extensive land and access to water supplies, and so is likely to have been controlled by the palaces (McInerney 2010: 52–3). This helps to explain the deposition of more elaborate cattle figures at the peak sanctuaries whose use continues in the Neopalatial period. If peak sanctuary deposition is part of the maintenance of territories for grazing animals, the territories appear to be associated primarily with cattle. This is consistent with an emphasis on the breeding and herding of cattle on prime land, leading to large-scale ranching in the Neopalatial period.

The deposition of figurines at cult sites appears to have decreased over the course of the Palatial period, but reappeared once the palaces had been destroyed. One Postpalatial practice was the deposition of clay and bronze figurines of cattle at existing sanctuary sites such as Psychro, but also at palatial sites including Phaistos and Ayia Triada (Gesell 1985: 132; Pilali-Papasteriou 1985). It is unclear whether these objects relate directly to cattle or whether they are more connected with the commemoration of the palatial collectives. These figures did, however, continue to make cattle present in these places, just as the peak sanctuary figurines did.

Palaces

The shift from local to palatial collectives was marked by an increase in scale: both spatially and through an increase in the number of bodies and things integrated into the collective. The origins of writing and sealing pre-dated the palaces but they came together in palatial administrative systems which used wet clay to help maintain an ever more complex network of people, animals and commodities (Schoep 1999). There were two main types of written administration in Protopalatial Crete. Cretan Hieroglyphic was a pictorial script used predominantly in east-central Crete, which was inscribed on sealstones as well as written on clay documents (Figure 3.4). Linear A appears to have developed in southern and central Crete and, as the name suggests, used signs which were more linear than pictorial. This was, almost without

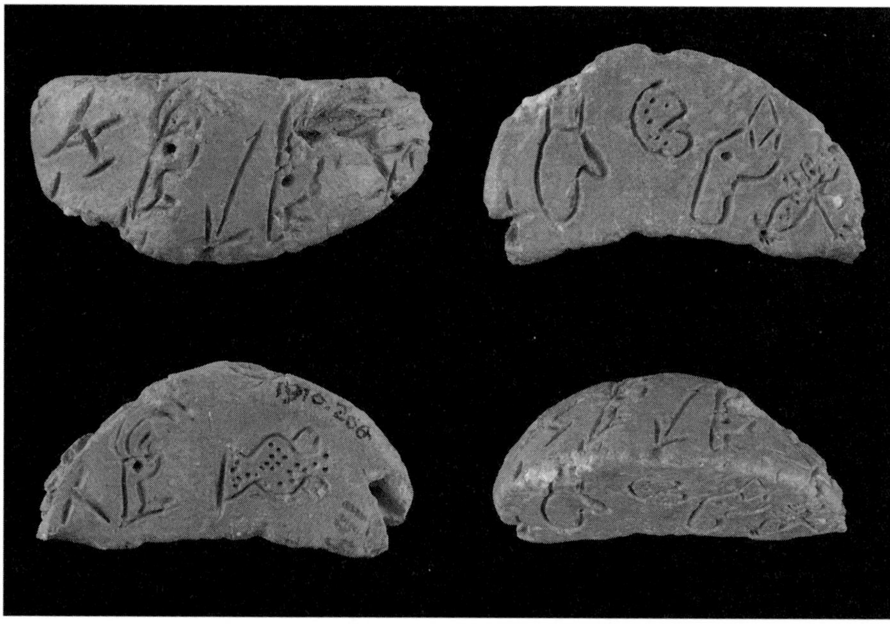

3.4. Clay nodule inscribed with Cretan Hieroglyphic signs including cattle and goat heads, MMII–III. Length: 3.2 cm. Photograph of three sides and overhead view (AN1910.206). © Ashmolean Museum, University of Oxford

exception, not used on seals, and became the predominant script in Neopalatial Crete before being adapted for Mycenaean Greek in the Final Palace period, as a script termed Linear B. Although Cretan Hieroglyphic and Linear A have not yet been deciphered since the language they represent is unknown, the inscriptions can be understood to some extent, particularly through analogies with Linear B tablets. Over time written documents extended control over ever greater numbers of animals, which appear either as contributions to consumption events or, particularly in Linear B, as wool sheep or draught oxen.

The Cretan scripts were identified by Arthur Evans; before he started excavating at Knossos he had identified the script he later called Cretan Hieroglyphic on sealstones and had travelled around Crete to find more examples. The resulting publications suggested that sealstones and writing were intertwined in Bronze Age Crete (Evans 1894, 1897, 1909). But since the decipherment of Linear B, after Evans's death, there has been a tendency to focus on the relationship between script, language and inscribed clay documents. Whereas the first volume of Evans's (1909) *Scripta Minoa* included a variety of images from seals, including animal practices such as fishing, the modern corpus of Cretan Hieroglyphic only includes signs from sealstones if they are also attested on clay tablets (Olivier and Godart 1996). As a result there has been an interpretative split between the 'decorative' elements of sealstones and what are regarded as true script signs. This has obscured the true split between seal imagery and text which occurred at the end of the Protopalatial period when Cretan Hieroglyphic fell out of use. At this time Linear A became the dominant script and seals increasingly showed naturalistic animal depictions rather than script signs. Until this time the heads and bodies of domestic animals were entangled with a variety of other things in a richly visual semiotic system which goes beyond the strict modern definition of the Cretan scripts.

Clay tablets and medallions incised with Cretan Hieroglyphic are found at Mallia, Knossos and Petras from the end of the Protopalatial period (Schoep 2001; Younger 1999). The tablets at Knossos, from the so-called Hieroglyphic Deposit, appear to be late instances of the use of Cretan Hieroglyphic, and Maria Anastasiadou (2016: 27) has suggested that they are the result of incorporating scribes from further east in the administrative system of Knossos. The subsequent use of Linear A across Neopalatial Crete indicates that this was the script adopted by the expanding Knossos palatial collective. Although it is not possible to read people's names or understand words in either script, the exhaustive published corpora of texts written in Cretan Hieroglyphic (CHIC) and Linear A (GORILA) identify when signs are used as ideograms, often when they are accompanied by numbers (Godart and Olivier 1976–85; Olivier and Godart 1996). This means that it is possible to read the counts of animals on Cretan Hieroglyphic and Linear A documents. In addition many of the signs can be traced back from Linear B, helping to identify the animal ideograms securely. Although, by definition, ideograms refer to ideas (and pictograms to objects), the term ideogram tends to be used interchangeably with logogram in Bronze Age epigraphy for signs referring to animals/people/things (Hooker 1980: 40–1). Ideogram is preferred to logogram (a sign representing a word) here because it is a reminder that the relationship between script sign and signifier is not arbitrary; as Carl Knappett (2008) has pointed out, ideograms for pots which have an iconic relationship with those pots retain the 'thingness' of the object.[7] Even the more stylised animal ideograms in Linear B retain the characteristic shape of the animal's horns. It is in this context that Arthur Evans's attempts to trace the pictorial origin and development of signs is important, because it draws attention to the 'thingness' or 'animalness' inherent in the signs.

Protopalatial

The most common type of Protopalatial seal is the three-sided prism (Figure 3.5a). Such seals are generally carved of soft local stone and usually have a different design on each face. Over 600 three-sided prisms are now known. Many designs include animal heads, which are also recognised as signs in the Cretan Hieroglyphic script because they are used on clay tablets. Others show whole animals, which are not generally recognised as script signs. Although some three-quarter views of animals have been described as

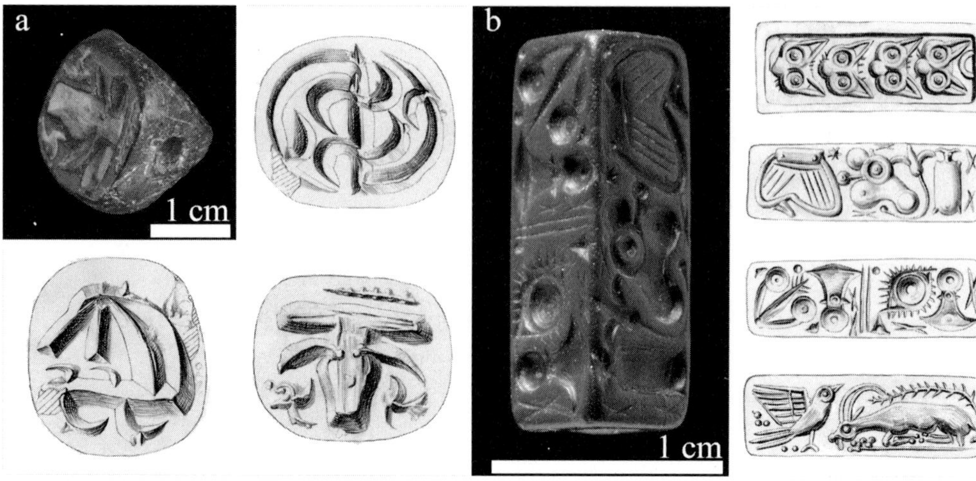

3.5. Soft-stone three-sided prism and hard-stone four-sided prism, MMII. (a) Photograph (AN1910.244). © Ashmolean Museum, University of Oxford. Drawings of impressions (CMS VI 52a–c). © CMS Heidelberg. (b) Photograph (AN1938.793). © Ashmolean Museum, University of Oxford. Seal drawings (CMS VI 100a–d). © CMS Heidelberg

'naturalistic', the majority are described as 'schematic' (Anastasiadou 2011: 276). This is partly because they were made from comparatively soft local stones which could be carved with simple tools. These seals were quick to make and were numerous enough to have been owned by a wider section of the population than in the Prepalatial period. Towards the end of the period, three- and four-sided prisms were also made from hard imported gemstones and carved with a drill, a more sophisticated technology imported from the Eastern Mediterranean (Figure 3.5b). These hard-stone seals typically have script signs on one or more faces, reinforcing the idea that there was a hierarchy of users, with the higher-status users differentiated by imported stones and more complex inscriptions. It is likely that many of these hieroglyphic inscriptions on hard-stone seals were personal names (Ferrara 2015). Both hard- and soft-stone prisms co-existed at the end of the Protopalatial period, as can be seen from the destruction deposit of Quartier Mu at Mallia. There appears to be a range of literacy, with some soft-stone 'pictographic' seals emulating hieroglyphics (Anastasiadou 2011: 355). As Ilse Schoep (1999) suggests, the inhabitants of Quartier Mu at Mallia were using writing and sealstones as a means of differentiation; in this context the soft-stone seals could allow a wide sector of society to participate in the palatial collective without being fully literate.

Despite the wide range of seal designs, a focus on decipherment has led to the separation of hieroglyphic inscriptions on seals from designs which are regarded as ornamental or pictorial. There has also been an emphasis on decoding combinations of elements, either on the same or different seal faces, in the search for meaning (van Effenterre and van Effenterre 1974; Younger 1999; Jasink et al. 2008). Although Evans was also interested in decipherment, he paid close attention to the identification of signs with real-world objects and speculated on the relationship of the seal design to the user. In particular, the designs he regarded as pictographic, he suggested, 'bore a personal relation to their possessor. Several examples of the more primitive class seem in fact to indicate the quality and pursuits of their owner' (Evans 1894: 337). The example he gave was a three-sided prism now in the Ashmolean collection (Figure 3.6a), from which he deduced that 'the owner was evidently a master of flocks and herds'. On this seal a human figure appears on all three sides, and although the design is difficult to read, the human appears to be shown sitting on a stool holding pots, milking a goat or sheep and carrying things on a pole. Evans identified the hanging objects as 'skin-buckets' containing milk; more recently they have been described as hanging pots or loomweights (Burke 2010: 44–7). Another possibility, close to Evans's suggestion, is that these are cheeses. Either way an argument can be made that all three faces relate to animals or animal products. A straightforward connection between three-sided prisms and occupation has been rejected in recent accounts (Krzyszkowska 2005: 95; Anastasiadou 2011: 349) but at least some seals do appear to show

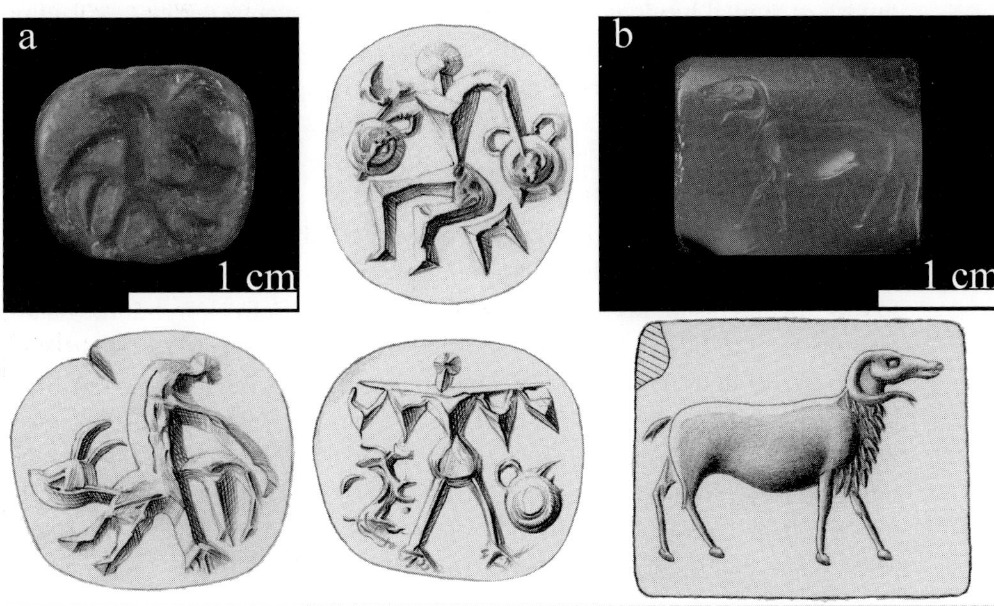

3.6. Seals related to agricultural activities. (a) Three-sided prism with possible milking scene, Praisos, MMII. Photograph (AN1938.750). © Ashmolean Museum, University of Oxford. Seal drawings (CMS VI 44a–c). © CMS Heidelberg. (b) Cushion seal showing sheep, Lyktos MMIII–LMI. Photograph (AN1938.953). © Ashmolean Museum, University of Oxford. Drawing of impression (CMS VI 177). © CMS Heidelberg

activities such as hunting or making pots which could be linked directly with the owner/user of the seal, or which were perhaps akin to family names.[8]

Depictions of clearly domestic animals on seals are rare in Bronze Age Crete, except in the Protopalatial period. Depictions on soft-stone prisms are rarely detailed enough to do more than identify the animal, and it is possible that pigs and goats shown whole in profile on prisms are, as a contrast with the heads also used as script signs, non-domestic animals connected with hunting. A small number of depictions do show goats with spears over their backs, a shorthand for hunting which continues into later periods. By contrast the few seals which show a sheep with a woolly chest almost certainly depict a domestic animal valued for its wool (Krzyszkowska 2012) (Figure 3.6b). These occur at the end of the Protopalatial period, and could indicate the importation of a new breed of woollier sheep from the Near East.[9] Sheep were present on Crete since the Neolithic period, but the arrival of a different breed would help to explain why they only occurred on seals at this time. A unique three-dimensional depiction of sheep herding, a bowl from Palaikastro containing 160 animal figurines and a human herder, also belongs to this period (Bosanquet and Dawkins 1923: pl. VII).

A small number of clay documents appear to record the movement or consumption of animals, using animal heads as ideograms signifying particular animals. As will be suggested in the next chapter, there is evidence for the display of animal heads at communal consumption events, perhaps explaining the significance of the head in this context. One sign appears to be a head without horns, often referred to as a 'calf', although it could be a sheep. It appears more frequently as a syllabogram (see Figure 3.5b), but when associated with a number is regarded as an ideogram.[10] The ideogram appears on two Cretan Hieroglyphic 'bar' tablets from Knossos, which include other commodities alongside the animal; this 'mixed commodity' format in Linear B tablets is associated with provisioning for consumption events. There are so few documents in this script that this reading is far from clear, but the bars at least do seem to be part of transactions involving either sheep or cattle. Other tablets appear to deal with commodities such as wool (Younger 2005). Among the Protopalatial Linear A documents from Phaistos there is one clay roundel with a sheep ideogram inscribed on it and a thumbprint instead of a seal impression; this is clearly a receipt for one sheep, most likely provided to the palace rather than given out by the palace (Hallager 1996b: 189). This document shape will be discussed below since it is characteristic of Linear A and continues into the Neopalatial period.

There are few Protopalatial clay documents and even fewer involving animals. Those that survive deal with small numbers of animals which could conceivably be linked with consumption events organised by the palace. By contrast there are hundreds of Protopalatial seals showing animals, the majority three- or four-sided prisms which are linked with the Cretan Hieroglyphic script and which would have been used to take part in the administration system centred on the palaces. In this respect Cretan Hieroglyphic appears to be a more inclusive script than Linear A, allowing the participation of a large number of seal users. This corresponds to explanations of the palaces as communal buildings, which these seal users were involved in provisioning (Driessen 2002, 2018). The animal head stands for the whole animal but, as will be discussed in Chapter 4, is also the product of the practice of butchering an animal, and sharing out its body. It was established in the Protopalatial period as the predominant way to depict domestic animals.

Neopalatial

There are no Linear A inscriptions on sealstones, but they have been found on a variety of objects including stone libation tables and metal pins, and one signet ring (Schoep 1994). Although this marks a wider use of the script than Linear B, these are all finely crafted objects which are exclusive compared with the appearance of Cretan Hieroglyphic signs on soft-stone seals. Animal ideograms are more frequent on the preserved clay documents of Linear A than Cretan Hieroglyphic. The largest number of documents come from Late Minoan IB destruction deposits, particularly at Ayia Triada, Zakros and Chania. A recent count of documents totals 1,370, of which 328 are tablets (Schoep 2002a: 38). The major difference between Cretan Hieroglyphic and Linear A is not in the syllabograms but the ideograms. The basic animal signs are the same as Linear B, with minor exceptions,[11] but Linear A ideograms are distinguished by an extremely complex system of ligatures used to modify them. The vertical stem of the sign for the animal ideograms is easy to change; for male animals it is cut with two small horizontal lines and for female animals the stem is duplicated, but they are modified in other ways too. Some ligatured signs appear to refer to particular types of animal, allowing fine distinctions to be made between them.

The main Linear A document types with surviving records of animals are tablets and roundels. Tablets are inscribed on the flat surface and record transactions. Roundels, as noted above, are inscribed with an ideogram and have a number of seal impressions around the edge, which almost certainly equate to the number of items (Hallager 1996a: 100–1). The Neopalatial seals used on roundels typically depict the non-domestic animals with which seal users often identified themselves. One roundel from Chania is inscribed with a sheep ideogram and around the edge are nine impressions of a seal depicting two lions (Godart and Olivier 1976–85: KH Wc 2063; Hallager 1996b: 101). This is the trace of the seal whose user was involved in the transaction with a central authority, and the ideogram shows that the roundel stood in for nine sheep. The roundel, then, is the animal thing which materialises the relationship between a palatial official, a shepherd/owner and nine sheep. The affordances of the clay, realised with both the inscription of the ideogram and impressions of the seal, allowed the trace of this transaction to be made durable and retained. Roundels such as this example show the split in the Neopalatial period between seal iconography which focused on non-domestic animals and a script which used signs to indicate particular types of domestic animals.

The largest number of tablets recording animals comes from Ayia Triada, including one which is inscribed with thirteen animal ideograms (Figure 3.7). The number of animals represented is small: nine rams, five ewes, seven she-goats, at least two he-goats and two pigs (Brice 1988; Godart and Olivier 1976–85: PH (?) 31; Schoep 2002a: 121). Some of the signs, for pig and goat, are ligatured, and it is unclear whether it refers to these animals directly or something connected with them. The same can be said of the composite sign 521 related to cattle and associated with 100 on HT 34 from Ayia Triada; it is possible that it refers to 100 plough oxen (Schoep 2002a: 187) or it could be evidence for large-scale cattle-herding being monitored at Ayia Triada, or even for a product such as hides.

Half of the Linear A animal ideograms refer to sheep, with twenty-two signs accounting for 214 animals. Of these, 187 are ewes; tablet ZA 22 (from Zakros) accounts for 102, while two from Ayia Triada refer to twenty-seven and thirty animals. These numbers are exceptional, and apart from another tablet detailing

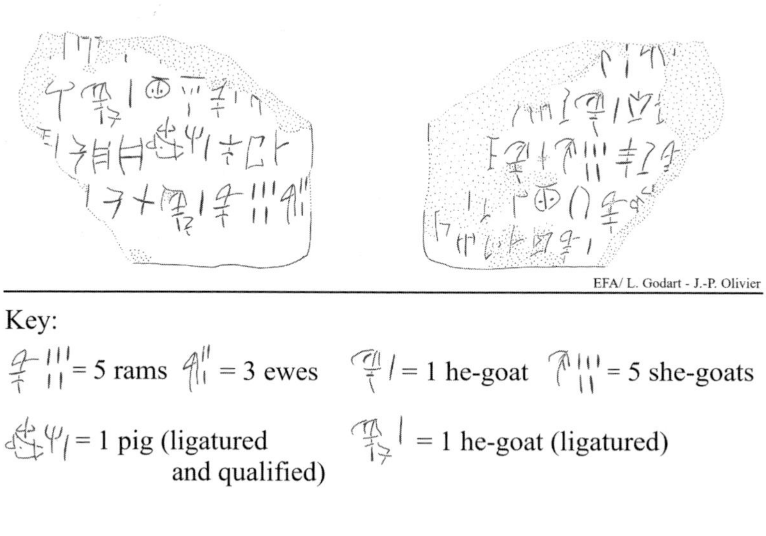

3.7. Drawing of Linear A tablet listing animals, LMI (after Godart and Olivier 1976–85: 318–19, PH (?) 31). Image credit: École française d'Athènes/L. Godart – J.-P. Olivier. Tablet (AMH Π-N1609). © Heraklion Archaeological Museum – Hellenic Ministry of Culture and Sports – 'Hellenic Organization of Cultural Resources Development (HOCRED)'

thirteen ewes and a problematic tablet referring to thirty pigs (HT 118) (Schoep 2002a: 121), the remainder of the ideograms are associated with fewer than ten animals. The number of ewes is difficult to explain because, as discussed below, the Knossos wool industry was based on wethers denoted by the ram ideogram. The problem with understanding ewes in the Linear A tablets as breeding flocks is that the concomitant evidence for wool flocks is missing and there is also a low frequency of textiles recorded in the Linear A tablets (Schoep 2002a: 191). The main problem is, however, the small number of tablets, so it is unclear if they are representative. Despite this, these large quantities of sheep have been seen as indicating an interest other than meat and dairy (Palmer 1995: 148; Schoep 2002a: 121, 186). Yet this is an argument from silence, and there is a rationale for building up general-purpose flocks of ewes (Halstead 1993b: 362–3). If these ewe numbers are typical then they are consistent with small-scale flock management, and possible seasonal or special culling of older sheep. There are hints of the management of textile workers at Ayia Triada (Del Freo 2020), but it is likely that most textiles were produced by domestic collectives, for trade or redistribution by the palatial collective. Given that most of the evidence from these documents is for transactions involving small numbers of animals, including all of the roundels, they appear to be traces of consumption. A comparison with the copious Linear B evidence reinforces this idea.

Final Palace

Linear B studies have contributed directly to an understanding of the animal practices involving domestic animals in Final Palace period Crete, or at least those related to Knossos. John Killen (1962, 1963, 1964) was able to demonstrate that the way the flocks were organised and recorded indicated a palatially administered wool industry. Central to this was the realisation that the sheep ideogram with the male ligature could refer to castrated animals, wethers rather than rams; the Mediaeval English wool industry provided a useful analogy. Work by Jean-Pierre Olivier (1967b) on scribal hands showed that the lists of shepherds and flocks on the Knossos Da–Dg and Dn series of sheep tablets[12] were all written by the same person, designated 'scribe 117' (see Figure 3.1); he also argued that these two series mainly involved the same sheep at different stages in the administrative process (Olivier 1967a). The wool from the palace-controlled sheep appears elsewhere in the tablets, disbursed to weavers also under palatial supervision (Killen 1984, 2008b). It is worth pointing out that conclusions about the Mycenaean economy are drawn from analyses of tablets from both Knossos and mainland sites combined, but here the Knossos tablets are the focus. Knossos has yielded the greatest number of Linear B tablets so far: about 3,500 compared to 1,200 from mainland Pylos (Driessen 2008). Wool and textile production is attested at both, but the Knossos wool industry, involving approximately 100,000 sheep, dwarfs that recorded at Pylos. Scribe 117 alone is responsible for 723 tablets (Olivier 1988: 219).

Paul Halstead has looked at the tablets in broader context, examining how the demands of palatial wool production related to the activities of individual shepherds. The palace set targets for wool yields per sheep and expected the shepherd to maintain the flock at the same number. Although Killen emphasised the importance of wethers in wool production, Halstead (1999a) suggested that shepherds were allowed to remove wethers for consumption and replace them with their own ewes; the palace recorded the sex of the animals with a view to redistributing sheep among flocks but, as long as wool targets were met, accepted this substitution. 'The palace was principally concerned to recoup wool and prescribed numbers of sheep from these flocks, with little interest in which animals were returned' (Halstead 1999a: 161). The palaces did keep breeding flocks, for which these ewes might have been destined, but the animals documented in the surviving tablets are insufficient to maintain flock levels without other contributions from shepherds including yearling sheep (Halstead 1998). Halstead (2003) regarded palatial flocks as 'open systems' with an interchange of sheep between shepherds and the palace, according to centrally defined targets, going so far as to see flocks as an accounting device rather than referring to specific animals.

Since the palace and shepherds were interested in different types of animals, Halstead sees the various exchanges as mutually beneficial. Male lambs are not particularly useful to small-scale farmers, except for consumption, but the palace needed a constant supply of them to maintain its flocks of wethers. The contribution of yearlings in return for flock deficits can be seen in the Do series (Halstead 1998: 193–4). Mature fat wethers were prime animals for consumption, but for wool production could be substituted for a younger animal or even an infertile ewe by the shepherds (Halstead 1999a: 153–4). At a broader level Halstead sees the palatial flocks as a mobile reserve of food; it was beneficial for communities to accept this system since the palace could relocate unwanted sheep as famine relief (Halstead 1992: 74, 1993a: 69). This relates to a view of the palace as a redistributive centre, but it was also at the centre of a collective of grazing sheep, hanging loomweights and weaving humans, among others, whose interactions resulted in an important commodity, woollen cloth.

A hierarchy is also apparent in these interactions. Another human actor on some of the sheep tablets is termed a 'collector'; as Halstead (1999a: 163) points out, their flocks were not run any differently from palace flocks. John Bennet's (1985: 239) observation that the proportion of collector sheep increases with the distance from the palace could imply that the palatial collective needed officials to at least monitor or guarantee the participation of shepherds and sheep. Nevertheless, this was primarily a system in which the Linear B tablets themselves extended animal practices and brought them into the palace at

Knossos. The storage of information on the tablets enabled one person based in the palace at Knossos to control, or at least monitor, the movements and productive capabilities of some 100,000 sheep. Clay tablets have particular affordances which facilitate this; erasures and corrections on the tablets are evidence that they could be modified when a shepherd had fulfilled his contribution (Halstead 1998: 191–2).[13] On the sheep tablets the totals were important, rather than individual sheep; shepherds would have known their sheep but scribe 117 was only interested in them as wool-producing units.

The tablets are connected with the affordances of animals which were of importance to a palatial economy: wool-producing and traction. Plough oxen would have been valuable, high-maintenance animals, perhaps loaned out for ploughing with a supply of fodder (Halstead 1999b: 323). The practice of recording boonyms, the descriptive names of oxen, on the Ch tablets from Knossos helps support this; it has been suggested that those in the palace were keen to avoid the substitution of animals (Killen 1993). Although the names, such as 'ugly' or 'piebald', do not betray any sentimentality, they do highlight the importance of oxen to the palace, and their individual identities. Such oxen, combined with a human driver and plough, could cultivate larger areas of land (5 ha per pair) than humans alone (Halstead 1992: 66). Again Halstead sees this as a palatial strategy which would have worked as a collaborative enterprise with the populace who were needed at harvest time, and perhaps owned the land; it is likely that the harvest was shared (Halstead 2001). The combination of the affordances of oxen and Linear B tablets was central to the ability of palatial officials to co-ordinate significant production across Crete, with 800 tons of grain recorded at the location da-wo (perhaps Ayia Triada or Kommos) in the south of the island alone (Halstead 1992: 67).

As Palaima points out, the oxen listed in the Knossos tablets do not appear in a ritual context, whereas those from Pylos in mainland Greece can be linked with sacrifice and feasting (Palaima 1989, 1992). Oxherds are listed at Pylos, but the animals themselves appear on tablets such as Un 138 listing sixteen sheep, thirteen goats, thirteen pigs and three oxen, which is frequently cited as a list of contributions to a feast (Killen 1994: 72; Bendall 2004: 106; Palaima 2004: 224; Weilhartner 2008: 413). Feasting has become a particular focus of attention in the archaeological literature, but the evidence in the Knossos tablets for animal sacrifice and feasting is sparse, particularly when compared to that for the wool flocks. Small numbers of animals do appear on some of the C(2) series of tablets, as well as a possible reference to sacrifice, pointing to consumption (Bendall 2004: 107; Killen 1994: 73–6). It is worth asking why there is not more evidence for animal consumption in the Knossos tablets given that the retired wool-sheep were almost certainly slaughtered and eaten; this could be the result of the seasonal nature of the tablets, so that the hypothetical cull records are missing (Halstead 1992: 68). This underlines that Linear B documents are the material traces of animal practices that were connected with the palatial collective, particularly ploughing and wool production. Other activities, perhaps including slaughter, remained at the level of the domestic collective.

Totals for animals in Linear B have been given by a number of scholars, both for herding (Olivier 1967b, 1988) and possible consumption (Bendall 2007: 203–21). Halstead (2002: 170–1) gives a total of 159,145 animals in herd management texts, of which 97.9 per cent were sheep, but cautions that some animals were probably recorded twice. The oxen in 'herd' texts are those identified as plough animals, often listed in pairs. In the absence of any other recorded uses for goats and pigs in the tablets, all are classed as animals for consumption, along with the sheep not associated with the wool industry; these total only 1,439 animals (Halstead 2002: 171). Although the Linear A evidence is obviously unrepresentative, it is clear that it is far closer in appearance to the proportions of animals in the Linear B consumption tablets than the herding tablets. One Linear B consumption text defined by Halstead (2002), Ce 50, lists 190 male sheep and 144 female sheep, showing that large numbers of sheep, including ewes, could enter the records as destined for consumption. This is an exceptional record, however, and small numbers of mixed animals are most likely to be associated with consumption.

It is possible to compare zooarchaeological assemblages with the Linear B totals for consumption. Figure 3.8 shows the genus distribution in a range of faunal assemblages against the animals listed in the Ayia Triada and Zakros

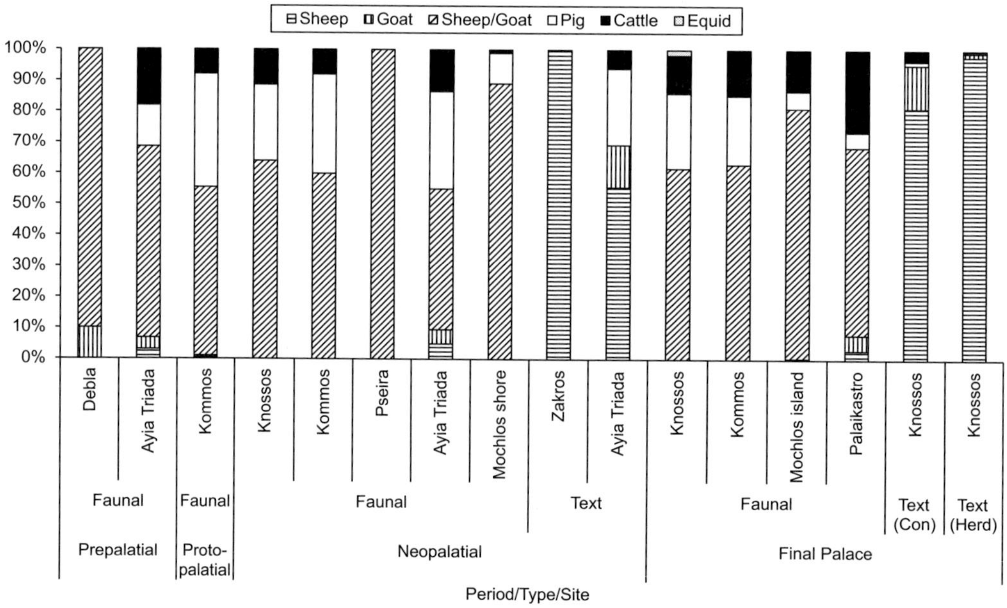

3.8. Chart comparing proportion of domestic animal remains in published faunal assemblages from various sites with animals listed on Linear A/B tablets from selected sites

Linear A tablets, and what Halstead (2002) classifies as consumption and herding Linear B documents from Knossos.[14] The Ayia Triada totals are closer to a number of the faunal assemblages than the Knossos Linear B texts. As Halstead (2003: 258) points out, the comparison is between what he distinguishes as deadstock from faunal remains, animals which have almost certainly been eaten, and livestock listed in the documents. Even the Linear B texts which seem to be concerned with consumption do not show the same proportions as any of the faunal assemblages since they are heavily dominated by sheep and goat. Yet in broad terms the small number of Linear A and B texts with mixed animals are the most comparable to the zooarchaeological evidence for consumption. This can be used to suggest that the Ayia Triada Linear A tablets are listing animals which are mainly destined for consumption. By contrast, even the Linear B consumption texts appear to be listing animals in proportions which are difficult to reconcile with everyday consumption.

The Linear B tablets mark the endpoint of an increase in the scale of animal practices associated with the palatial collective. In the case of sheep, the care for individual animal was delegated to shepherds. For those in the palace animals became units, facilitated by the use of clay tablets and inscribed ideograms to stand for individual animals. Kerry Harris and Yannis Hamilakis (2014: 110) suggest that the palatial economy was a multispecies enterprise, the textile industry one of many 'inter-species collaborative projects'. But it should be acknowledged that this was not an equal enterprise, with sheep being treated as productive units, perhaps analogous to the forced labour or even slavery of the woolworkers (Tapper 1989). By contrast, the animals transacted in Linear A documents were already a unit for consumption. These animals were indicated by pictographic heads, perhaps the part of the animal associated with consumption events. Over time the palatial collectives extended to the productive lives of some domestic animals rather than simply the sharing of their meat. This process can be traced to some extent in the administrative records of the palaces, but in order to explore how cattle were absorbed into the palatial collective of Knossos it is necessary to bring in a wider range of objects.

CATTLE AND THE KNOSSOS PALATIAL COLLECTIVE

Bull Games

The animal practice of bull-leaping has become emblematic of Bronze Age Crete, ever since Evans (1901b: 94–5) found depictions of what he called 'toreadors' or 'cow-boys' and 'cow-girls' on frescoes and seals from the palace at Knossos. It has generated extensive discussion, but recent authors have tended to move away from Evans's suggestion that it was a religious ceremony, instead grounding it in Bronze Age human-cattle relations and their social context (Loughlin 2004a; Zeimbekis 2006; Soar 2009; McInerney 2010: 48–63; Shapland 2012). As Younger (1995) has outlined, bull-leaping was one of a sequence of activities also including capture, wrestling and (debatably) sacrifice. The Prepalatial rhyta discussed above appear to show wrestling or restraint but what defines an animal practice is social context. Bull-leaping depictions emerge in the Neopalatial period, when Knossos was becoming increasingly influential across central and eastern Crete. Depictions of bull-leaping in various media in the Neopalatial and Final Palace periods include gold rings, frescoes and figurines of ivory and bronze. These objects were of palatial manufacture, providing a means to establish and reinforce the relationship between bull games and the palace at Knossos. The various activities involved in bull games were also depicted on seals (Figure 3.9a–d). The appearance of bull-leaping depictions can be understood in terms of a shift from localised herding to a ranching system closely connected with Knossos. Bull-leaping became the animal practice which defined the palatial collective at Knossos.

Modern animal practices have been used as analogies for bull-leaping, including the American rodeo (Evans 1921a: 258). This is an important parallel for Late Bronze Age Cretan bull games because the rodeo took place as part of a seasonal round-up of cattle. Cattle were allowed to roam across the American landscape as part of a system of ranching until cowboys gathered them up and herded them to a central place where they could be checked, branded and some selected for slaughter. This gathering of humans and animals also acted as

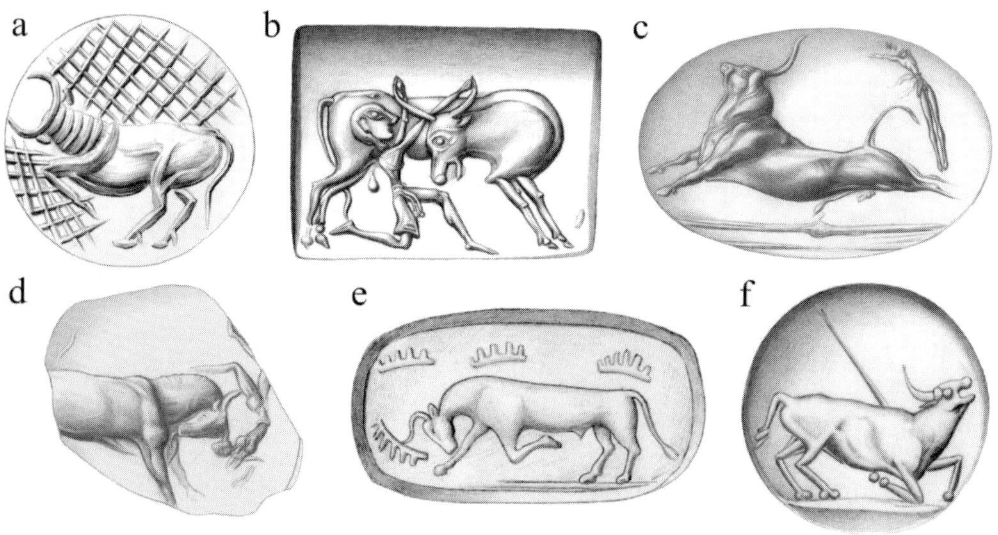

3.9. Drawings of seals showing cattle. (a) Bull caught in net, LMI (CMS VI 408); (b) human grappling with bull, Kalyvia, LM (CMS II 3 105b); (c) bull-leaping scene, Ayia Triada sealing, LMI (CMS II 6 43); (d) bull goring human, Ayia Triada sealing, LMI (CMS II 6 39); (e) bull in landscape, Phaistos sealing, MMII (CMS II 5 268); (f) speared bull, Knossos, MMII–III (CMS II 2 60). © CMS Heidelberg

a chance to demonstrate ranching skills such as 'steer-wrestling' (capturing cattle on horseback) or to show prowess in more dangerous activities such as 'bull-riding'. The result is a performance which both distils and exaggerates the lifestyle and values of the cowboy (Lawrence 1982; Fischer 2015). This makes it an animal practice which defined a sector of American society, and continues to do so. Similar bull games have emerged in various cattle-keeping cultures, including ancient Thessaly and contemporary Sri Lanka (Evans 1921a: 257–8; Zeimbekis 2006: 35). There are also Middle–Late Bronze Age depictions of bull-leaping on Syrian seals, a Hittite relief vase and frescoes at Tell el Dab'a in Egypt (Collon 1994; Sipahi 2001; Bietak et al. 2007). The last of these are clearly related to Aegean depictions, but it is possible that similar bull games were occurring across the Eastern Mediterranean. Rather than looking for the origins of bull-leaping in a particular place, whether Crete or elsewhere, it can be seen as an extension of cattle management. In certain times and places, such as Neopalatial Knossos or the American West, cattle games became an animal practice which helped define particular collectives.

Parallels have also been drawn between Cretan bull-leaping and the Spanish corrida but these are more problematic since the bulls involved in corrida are bred specifically for their 'wildness' and usually are kept away from humans prior to their encounter with the matador. This too is an animal practice with its origins in cattle management, but it has come to be a form of sacrifice which helps define the town in opposition to the countryside (Marvin 1988: 128–36). Whereas the rodeo is a performance arising from cattle management and ownership, the corrida is an act of displacement in which a bull becomes a participant in his own ritualised slaughter. There are other bull-fighting traditions in Spain and Portugal which have superficial similarities with the animal practice depicted in Bronze Age Crete, particularly bull-wrestling (pega de caras) and men jumping over bulls (recortadores). But this form of bull-leaping is an act of avoidance rather than contact, and there remains an opposition between the behaviourally wild bulls and the civilising town arena which is less apparent in the rodeo games.

The deposition of large numbers of cattle figurines at peak sanctuaries across Crete in the Protopalatial period suggests the participation of a significant sector of the population in cattle management or ownership. The large number of three-sided prisms, carved in soft local stone, suggests a participatory rather than an exclusionary system of administration at the palatial collectives of Mallia and Petras; a large section of the population would have owned a seal, which they could have used to participate in administrative practices. At the end of the Protopalatial period there is a shift towards more exclusive, finely carved seals, and in seal imagery related to cattle. A sealing deposit from this time at Phaistos shows that metal rings depicting cattle in a landscape setting were used in palatial administration. At Knossos a discoid sealstone, found in a Middle

Minoan II–III context, shows a bull with a spear in his back (Figure 3.9f). It is notable that the bull on the Knossos sealstone has short horns; this is not an auroch hunted in mainland Greece but a domesticated bull hunted on Crete.

Scenes of cattle-hunting on Crete are often explained in terms of links between hunting and sacrifice: the ritual hunt (Burkert 1983; Marinatos 1986: 42–3). A ritual interpretation fails to explain the coincidence between the territorial expansion of Knossos at the time when scenes of cattle-hunting and then bull-leaping appear. Cattle-hunting could have been both a symptom and a cause of the breakdown of some palatial collectives at the end of the Protopalatial period resulting in the destruction horizons at a number of palatial sites including Petras, Mallia, Phaistos and, to a lesser extent, Knossos. It is also possible that there were stray cattle during this turbulent period which has been ascribed to an earthquake, warfare or both. These cattle could have been hunted by people from Knossos taking advantage of the breakdown of these other palatial collectives, but it is more likely that cattle-raiding was part of warfare since the loss of cattle would have had serious economic as well as social implications. The prevalence of warfare on Crete in the Bronze Age has been greatly underestimated in modern scholarship (Molloy 2012). The depiction of bull-leaping alongside boxing on the so-called 'Boxer rhyton' from Ayia Triada hints that these demonstration of prowess were part of martial training for young men.[15] It is this wider context which undermines Evans's identification of female bull-leapers on the Taureador frescoes on the basis of skin colour alone. At the time of its deposition there were close connections between Ayia Triada and Knossos; it has been suggested that this was an outpost of Knossos in the Mesara following the takeover of the palace at Phaistos (La Rosa 2012: 506). In this context cattle-hunting and bull-leaping can be seen as animal practices which helped define rights over the cattle of Crete and the land they grazed over.

If cattle-hunting was the destruction of the animals belonging to other collectives, bull-leaping was an animal practice which extended the rights of the palatial collective of Knossos to cattle. This helps to explain the emergence of depictions of this activity in the Neopalatial period. Once the other palatial collectives, and their cattle, had been absorbed by Knossos, herds of cattle and sheep could be grazed across the island. A round-up of cattle, in which they were brought to Knossos, and perhaps places like Ayia Triada, for checking and selective castration or slaughter would, by analogy with the rodeo, have provided the context for bull-leaping and other demonstrations of human prowess such as bull-wrestling. As Evans (1921a) suggested, these bull games could have taken place in an open area in the Kairatos valley near Knossos; it has also been suggested that bull-leaping took place in the central courts of the palaces, but there is no clear evidence for this (Graham 1957). The unenclosed space in front of the villa at Ayia Triada seems to be a more amenable location for bull games. Either way, given their affordances in relation to humans, capturing feral bulls would have been a challenging and dangerous task, and one means to catch them would have been using nets. This animal practice is most famously shown on one of the gold 'Vapheio cups' from the mainland whose iconography is linked to Crete, although the place of manufacture is not clear (Davis 1974; Blakolmer 2010: 102–3). A number of Neopalatial sealstones showed netted cattle, sometimes with their heads turned away from the viewer (Figure 3.9a). The netting of cattle would have required a group of people, co-ordinated to put the nets up and then drive the cattle into them, followed by their restraint. One means to do this is by holding the animal's head and wrestling them to the ground. Both bull-leaping and bull-wrestling emerge from the interactions between humans and captured cattle but became performances in the Neopalatial period which demonstrated that these free-ranging cattle belonged to Knossos.

Bull-leaping was depicted on various media in the Neopalatial period, and is particularly associated with Knossos. The relief bull's head which Evans restored at the North Entrance bastion was found with other fragments that suggest it was part of a large relief fresco of either bull capture or bull-leaping visible to everyone who entered the Central Court (Evans 1930: 158–77; Shaw 1995). The imagery on the Vapheio cups could have been derived from the same scene, and it is also likely that contemporary seals related closely to fresco designs, particularly relief frescoes (Blakolmer 2010: 100–3). Frescoes showing bulls were found in a number of other prominent locations in the palace at Knossos, emphasising the connection between bulls and Knossos (Hallager and Hallager 1995). As noted in the previous chapter, depictions of bull-leaping were also distributed using metal rings impressed on clay packet sealings at various sites across the island at the end of the Neopalatial period (Betts 1967; Krzyszkowska 2005: 188–92) (Figure 3.9c). The clay of these sealings is consistent with an origin in the area of Knossos (Goren and Panagiotopoulos 2009). By this time the palatial collective of Knossos, defined by the animal practice of bull-leaping, had expanded to cover most of the island. These clay sealings, enclosing messages, traced the extent of the collective.

The wealth of material culture showing bull-leaping is in stark contrast to the lack of depictions of cattle in agricultural activities; there is one possible milking scene on a sealing from Knossos[16] and no depictions of oxen ploughing known from Bronze Age Crete. By contrast, capturing the moment of the leap was evidently politically significant; objects could be used to make a fleeting activity durable and manipulable. One three-dimensional depiction survives: a bronze bull and leaper said to be from near Rethymnon (Evans 1921a; Shapland 2012). An ivory leaper found at Knossos must also once have been part of a sculptural group showing bull-leaping. Both of these animal things were the result of skilled crafting and restricted materials, further emphasising that bull-leaping, as well as its depictions, was an animal practice of the palatial collective. The frescoes of bull-leaping on the walls of the palace at Knossos too acted as an enduring connection between the palace and cattle, even when the palatial herds were not present.

Cattle depictions continued in the Final Palace period when Knossos was the only remaining palatial site in central Crete. Final Palace frescoes might include the 'Taureador frescoes', which are difficult to date, and those which were preserved in the final destruction of the Palace in the fourteenth century BCE. When the west entrance to the palace was excavated, there was still part of a fresco on the wall showing a bull's hoof above a skeuomorph stone dado (Evans 1928: 674–7; Hood 2005: 66). Earlier layers of the same fresco were found behind, also with traces of cattle, leading Evans to suggest that a bull-leaping fresco greeted visitors to the palace here.[17] This was shown in a published reconstruction, but the preserved hoof, hovering just above the groundline, is at the wrong angle for a bull in flying gallop (Evans 1928: fig. 429) (Figure 3.10a–b). A more likely option, again drawing on seal iconography, is a relatively common design showing two seated cattle, one with the head turned back; one finely carved example was found in a tomb at Ayios Ioannis near Knossos (Younger 1989) (Figure 3.10c). Its frequency led Younger to suggest that it was used by Neopalatial palace administrators, and their successors in the Mycenaean administration at Knossos, since it is also repeated in a stylistically later seal (Younger 1988: xv; 2000: 352–3). The groundline on the sealstone could derive from a dado (Blakolmer 2010). One further argument for this reconstruction is that in this context the cow or bull

3.10. Knossos West Porch fresco. (a) Sketch of standing remains of West Porch (Evans Fresco Drawing M/12). © Ashmolean Museum, University of Oxford. (b) Photograph of fresco fragment of cattle hoof found *in situ* on West Porch wall LMII–IIIA (Sir Arthur Evans Archive). © Ashmolean Museum, University of Oxford. (c) Drawing of seal showing seated cattle LMII–IIIA (CMS II 3 62). © CMS Heidelberg

would be looking into the palace, drawing selected visitors in, just as human members of the 'Procession fresco' do on the adjoining wall (Günkel-Maschek 2012a, 2012b). In any case, cattle continued to be prominently associated with the palace at Knossos until the time of its destruction.

The places named on the Linear B sheep tablets from Knossos help to show that this palatial collective continued to extend across most of the island at the end of the Final Palace period. It is only in this period that the cattle sacrifices are depicted on a small number of sealstones, some preserved on impressions found in the palace at Knossos, and on the 'Ayia Triada sarcophagus'. As will be discussed in the next chapter, this could indicate a shift in emphasis from cattle management to the act of sacrifice in this period. In the Final Palace period, Knossos was part of a wider palatial collective which extended to the mainland, and formalised sacrifice can be seen as a Mycenaean animal practice, whether or not it originated on Crete. Depictions of bulls and bull-leaping continued, but on seals they became less naturalistic, with bulls and people crowded into the seal face. Younger (2000) suggests that stylistic change in seals indicated the arrival of a new Mycenaean administration but this could also have resulted from a change in carving techniques and the decreased importance of accurately replicating the scene. Once the palatial collective centred at Knossos, or perhaps on the mainland, was in undisputed control of Crete, animal practices involving cattle lost some of their social significance. Seal carvers no longer needed to accurately replicate particular scenes with the same degree of realism because they were no longer important as demonstrations of the animal practices which defined the collective. Similarly, scenes of bull-leaping were exported to the mainland, where they become less lifelike (Younger 1976). Arguably bull-leaping was not practised on the mainland but depicted because of its associations with Cretan palatial culture.

Horns, Hides and Meat

Bull-leaping and cattle-hunting were animal practices which emerged from the interactions between Bronze Age Cretans and free-ranging cattle. There are close analogies with the type of ranching that occurred in nineteenth-century California, with feral cattle rounded up (the rodeo) and killed (the matanza) or branded and let loose again (Fischer 2015: 130–6). The purpose of the matanza was to turn the cattle into two valuable commodities, cowhide and tallow (fat). Most of the hide was traded, but it was also used to make clothing and equipment, and leather or rawhide continues to be part of the cowboy costume on display at modern rodeos. One key difference in the method of rounding up is the lasso, since this was closely associated with horse-riding. There is little evidence for horses in Neopalatial Crete, and no evidence for horse-riding until after the Bronze Age. In the absence of horses, cattle had to be trapped by humans on foot, a dangerous activity which would have required an organised group of people, most probably using nets or spears. Although the methods of capture were different, the products derived from cattle were largely the same: meat, hides, horns and probably tallow and sinews. These would have been used to make animal things closely linked with cattle-herding and hunting, but these have not always been recognised.

A variety of carved stone vessels with relief scenes were produced in the Neopalatial period and show scenes linked with fighting, hunting and ritual activities (Warren 1969; Kaiser 1976). They are likely to have been used in drinking ceremonies which helped to reinforce the connections among a group of people who participated in these activities (Logue 2004). Among them is a group of Neopalatial relief stone vessels from Ayia Triada which show young men with hairstyles characteristic of age and status (Koehl 1986b). The largest concentration of these relief vessels was found at Knossos, making it likely that they were produced there and sent to connected sites across the island (Warren 1969: 174). Despite their conventional names, the Boxer rhyton, 'Chieftain cup' and 'Harvester vase', all of which were found in the villa at Ayia Triada, appear to show a series of connections between young men, martial activities and cattle. These vessels, with their narrative scenes, are also closely linked with contemporary sealings which carry the impressions of metal rings. These seal impressions are often more abbreviated, and were used in different contexts, but both the relief vessels and seals were animal things which extended practices such as bull-leaping and cattle-hunting into the palatial centres.

The Chieftain cup is better known for its presentation scene on one side, with two young men standing opposite one another. One holds a sword and a curved object which Evans identified as a 'lustral sprinkler'; it was described in the original publication as a fly-whisk made from horsehair (Paribeni 1903: 324; Evans 1928: 792–3) (Figure 3.11). The other man holds either a staff or a spear. On the other side, one or more people are carrying what have long been recognised as hides (Paribeni 1903: 324; Evans 1928: 743). One piece of warriors' equipment seen in the Late Bronze Age was the shield covered in cowhide; it is likely that these were cowhides destined to be turned into such a shield (Forsdyke 1952: 16). Such figure-of-eight shields were depicted on Neopalatial seals and a variety of different types of objects in the Final Palace period; they appear on frescoes in the palace at Knossos, and clearly show the cowhide patterning (Figure 3.12) (Rehak 1998: 232–4; Marinatos 1999). By analogy with the hides, the object on the other side is likely to be a cattle-tail fly-whisk, as found in cattle-keeping African and Asian societies (Zeuner 1963: 252).[18]

Another product derived from cattle bodies is horn, which does not survive archaeologically. Evidence for horn-working, in the form of cut horn cores, has been found in a Protopalatial deposit close to the palace at Knossos and elsewhere (Evely 1993: 224–5; Isaakidou 2007a). Horn could also have been

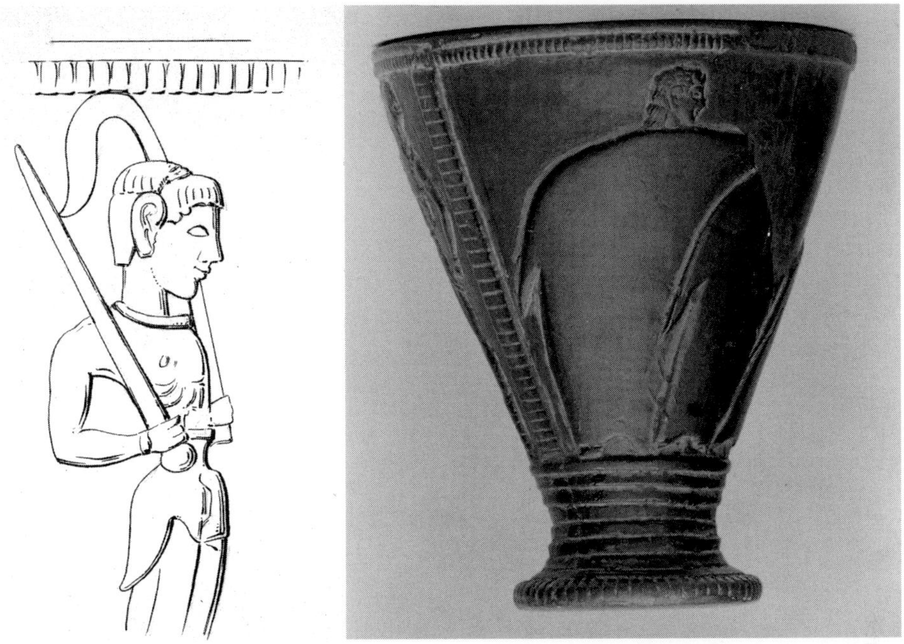

3.11. The Chieftain cup from Ayia Triada, LMI. Height: 11.5 cm. Left: drawing of figure with sword and 'lustral sprinkler'; right: photograph of hide carriers (both Sir Arthur Evans Archive). © Ashmolean Museum, University of Oxford

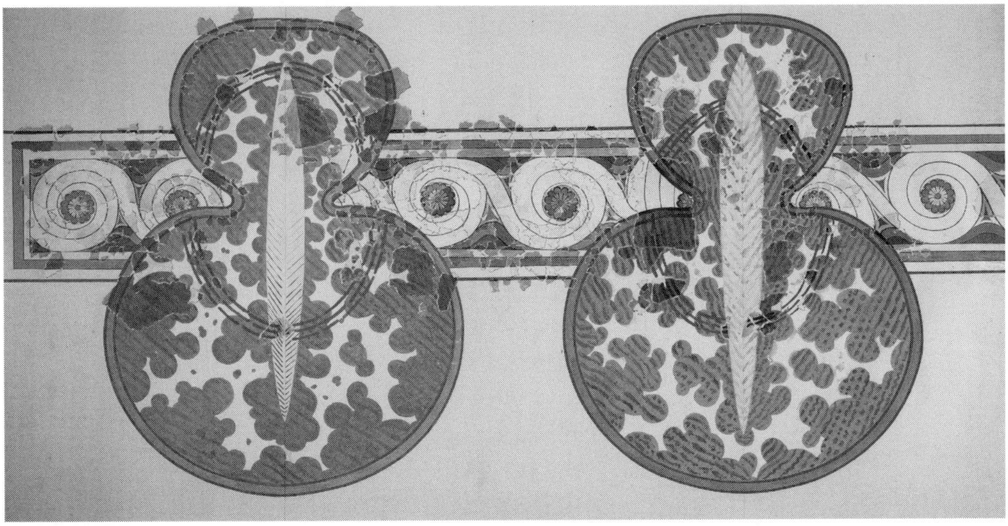

3.12. Drawing of the Shield fresco from Knossos, LMI–II. Watercolour (Evans Fresco Drawing T/18). © Ashmolean Museum, University of Oxford

used to make animal things related to cattle. A characteristic pottery design of the early Neopalatial period is known as tortoiseshell ripple (Betancourt 1985: 113–14; Hatzaki 2007: 163–5; Knappett 2020: 122–4). Although it does

3.13. Tortoiseshell ripple cup compared with modern horn cup. Left: pottery Vapheio cup from Knossos, MMIIIB–LMIA. Height: 7.2 cm (AN1938.456). © Ashmolean Museum, University of Oxford. Right: horn cup, date unknown. Height: 5.8 cm (WCHL: 432). © Museum of Design in Plastics, Arts University Bournemouth, Worshipful Company of Horners collection

resemble tortoiseshell, an abduction can be made that in this context it is more likely to relate to horn patterning.[19] This pattern is associated with a new pottery shape, a shallow straight-sided cup known as a Vapheio cup. The combination of shape and ripple design could be a skeuomorph of a horn cup, or a composite of horn and metal (Figure 3.13). The size and shape of these straight-side cups are consistent with the dimensions of horn cores reported from Bronze Age Crete.[20] The gold cups from Vapheio after which the shape is named are made of gold and are decorated with repoussé scenes of bull-catching; the 'Quiet cup' shows a bull being lured with a decoy cow and the 'Violent cup' shows a bull being caught in a net. Their iconography is linked to Crete, although the place of manufacture is not clear (Davis 1974; Blakolmer 2010: 102–3). Both pottery Vapheio cups and the eponymous gold Vapheio cups could refer, in different ways, to cattle and their horns. Skeuomorphy need not be seen simply as an imitation of one material in another, but rather could have been part of a rich set of connections between different entities which helped to link animal practices such as bull-catching and animal things such as horn cups.

The humans and bulls on the Boxer rhyton have close similarities with those on the Vapheio cups (Davis 1974: 484). Most of the friezes on the Boxer rhyton, however, show another activity, boxing, with young men wearing helmets and gloves, presumably made of leather. The association between bull games and boxing seemingly extends to the Neopalatial peak sanctuary of Kophinas, in the Asterousia mountains on the edge of the Mesara Plain. The deposit contained a large number of bovine figurines, some of which were rhyta, and human figurines, predominantly male where identifiable, which appear to be wearing boxing gloves (Rethemiotakis 2014). Similar finds have been made at the peak sanctuary of Atsipadhes (Haysom 2015: 96). Another Neopalatial site where boxing and bull-leaping appear to be associated is Knossos; the fragmentary relief frescoes found in the East Wing include male figures reconstructed as boxers, alongside another which is holding a bull's horn (Evans 1930: 216).[21] These Neopalatial frescoes brought bull games and martial activities into the centre of the collective.

The Neopalatial period was the time when the animal practices involving bulls and their related animal things were established. If the associations drawn here are accepted, a far greater proportion of seal imagery and other iconography of the Neopalatial period is associated with cattle, or animal things derived from cattle, than has previously been recognised. As noted above, bull-leaping and animal things such as figure-of-eight shields continued to be depicted in the Final Palace period, but by this time Knossos had long established its connection with the cattle of Crete. As Matthew Haysom (2010) argues, the apparently symbolic use of things such as double axes later can be explained in terms of a secondary appropriation as religious symbols rather than a key to their original indexical significance. Neopalatial relief stone vessels including the Chieftain cup and Boxer rhyton were used in drinking ceremonies to reinforce the connections between cattle and the group of young men shown on them. As well as being linked with military training, cattle-hunting and capture were part of the demonstration of the extension of the palatial collective of Knossos over Crete in the Neopalatial period.

These activities can also be linked with the ashy debris from the consumption of meat at various sites in the early Neopalatial period noted above. The relief frescoes of cattle and boxers in the Knossos palace, as well as the tortoiseshell ripple cups, also date to this period, Middle Minoan III, which Evans regarded as the acme of artistic achievement at Knossos. These frescoes, and the associated consumption activities, placed cattle and a group of young men who had the skills to catch them at the heart of the palatial collective. Their interactions, which extended across the Cretan landscape, came to define the extent of this collective, reinforced by portable depictions stamped on clay or carved on stone vessels, and most likely backed up by violence. Traces of these animal practices and the resulting animal things such as horns and hides survive on these fine, naturalistic depictions but their significance for the human-animal relations of Crete in this period has been overlooked. The relation between the animal practice of hunting and territory will be discussed further in Chapter 5.

CONCLUSION

Cattle and sheep were an integral part of the domestic collectives which began at sites such as Knossos in the Neolithic and the palatial collectives which eventually absorbed them. The animal practices of domestic collectives can sometimes be distinguished zooarchaeologically. The bones of individual animals can have pathologies from being yoked to ploughs or carts. Assemblages of bones can show whether sheep and goats in particular were being herded for milking or reared for meat. These examples from the animal side are rare, however, partly because animal bones have not always been recovered in bulk or studied in detail. As Halstead has emphasised, the remains of animals found on archaeological sites tend to be the result of consumption rather than direct evidence for animal management. This is true even for sites such as Debla, whose location suggests that it was a herders' camp in the uplands. Nevertheless, it is clear that sheep, goats and cattle were herded extensively across the island of Crete in the Bronze Age.

The Protopalatial period marked the increasing articulation between the domestic collectives of herders and their animals with the palatial collectives which took over some of these animals' products or bodies. The increase in scale of animal management, as the palatial collectives grew, resulted in the use of material substitutes for animals. Although animal figurines had been produced from the Neolithic period onwards, and animal rhyta were found in Prepalatial tombs and settlements, peak sanctuaries helped to define a relationship between the bodies of animals and areas of land. The animal practice of hunting cattle which appeared on sealstones at the end of the Protopalatial period was part of a process of territorial competition which ultimately led to the takeover of most of Crete by the palatial collective of Knossos.

The end of the Protopalatial period, and with it the disappearance of Cretan Hieroglyphic and the associated multi-sided seals, coincided with the almost complete disappearance of domestic animal practices from seal imagery. They were superseded, in part, by a set of animal practices involving bulls, including hunting, bull-leaping and grappling, that marked a shift in cattle management from local to palatial collectives. It is likely that cattle were brought to the area of Knossos or its outposts seasonally, when bull-leaping took place, but animal things like seals and frescoes helped to establish the link between cattle and Knossos when their bodies were not physically present. The naturalism of these depictions, also seen in rhyta in the shape of cattle or cattle heads, reinforced this absent presence. Depictions of bull-leaping on frescoes in the palace and gold rings whose impressions travelled across Crete traced the extent of the palatial collective. But as the Knossos collective was absorbed into wider Aegean collectives, the focus perhaps shifted from cattle management across a wide territory to cattle sacrifice at the centre. As these activities decreased in importance, the realism of depictions also decreased; what had been an animal practice which demonstrated territorial control became a tradition, echoed in the deposition of bronze bulls at sites like Psychro and Phaistos long after the disappearance of the palace at Knossos.

FOUR

BUTCHERING

INTRODUCTION

> The remains of all the animals found in the upper stratum, the domestic ox, the goat, the horned sheep, the fallow deer, and the hog were parts of offerings to Zeus, while the dog from the lower stratum may belong to an earlier period in the history of the cave, when it was used as a habitation and before it was used as a shrine of the Dictaean Zeus. The perfect condition of the offerings contrasts strongly with the fragmentary remains of the refuse heap.
>
> (Boyd Dawkins 1902: 165)

The 1900 excavation of Psychro Cave resulted in what is regarded as the first zooarchaeological report from Crete (Reese 1994: 191). It was written by William Boyd Dawkins, whose palaeontological background is apparent in the detailed measurements of bones he provided. His concern with defining species, delineating between wild and domestic on the basis of size, foreshadows more zooarchaeological methods. One cattle frontlet (the front part of the skull) allowed him to identify a new variety of domestic cattle he named *Bos domesticus*, var. *creticus*. He also makes an observation on its archaeological context: 'It has been hacked off the rest of the skull and has probably been fixed to one of the sides of the shrine or to the altar, as a Bucranium' (Boyd Dawkins 1902: 162). Despite the brevity of the report it was in many ways exemplary, providing measurements, discussing context and making connections with depictions of animals on sealstones. But at the same time it established these remains of animal heads as specimens, and, as will be discussed below, as sacrificial relics.

Animal skulls have affordances which were useful to people in Bronze Age Crete and continue to be useful to zooarchaeologists; they are durable and retain recognisable animal characteristics after the flesh has rotted away. When found disarticulated from the rest of the skeleton in archaeological contexts with indications of eating and drinking, these skulls can be inferred to be the products of butchery, and hence past consumption events. In this way they often retain an indexical link with past meals. It is this set of associations which allows zooarchaeologists to infer the consumption practices of the past, but the same affordances and indexicalities could have been at play in the display of animal heads and skulls in Bronze Age Crete. The retention and display of animal heads as trophies appears to have been an important means to memorialise the provision of animals for communal meals. Zooarchaeological identifications, particularly when used with an understanding of archaeological context, can help provide evidence for these past practices. By contrast sacrifice can rarely be identified archaeologically; as will be discussed there are methodological problems of recognising either the intentions behind killing animals or the resultant trace. There is no iconographic evidence for this animal practice until the Final Palace period, when whole animals are depicted trussed on tables. Nevertheless animal skulls have become symbols of sacrifice rather than indexes of consumption, specimens rather than trophies.

SPECIMEN LOGIC

The term specimen, used by Boyd Dawkins and subsequent zooarchaeologists, is embedded within a naturalist ontology. Janice Neri traces the emergence of what she terms 'specimen logic' to the early modern period as images of insects became a focus of zoological study. 'Specimen logic turns nature into object by decontextualizing select creatures and items – that is by removing them from their habitats, environments and settings' (Neri 2011: xiii). The practice of zoological identification similarly transforms animal remains into specimens. Although Boyd Dawkins and subsequent zooarchaeologists have to a greater or lesser extent considered animal remains within an archaeological context, their taxonomic identification and recording of measurements inevitably removes them from their original ontological setting. Individual animals become tens or hundreds of specimens identified to genus, often detailed in separate zooarchaeological reports in publications of the excavations. The resulting data have the potential to provide important insights into human-animal relations but the materiality of the remains under study has often been overlooked.

The standards of zooarchaeology have changed over the course of the twentieth century, making it difficult to compare animal remains published from different sites. For Boyd Dawkins the skulls were the most characteristic part for measurements, allowing him to define species. His report is concerned with confirming the presence of particular species, as represented by bones. Until the postwar period, most zooarchaeological reports were similar, providing lists of species for given sites, often with little regard for context. Over time zooarchaeologists have tended to move away from identifying animal remains to species, and *Bos domesticus*, var. *creticus* is no longer a recognised term. Recent zooarchaeological reports tend to record specimens to genus instead (i.e., *Bos*) but also record every specimen that can be identified rather than a few characteristic examples. Zooarchaeological identifications to genus represent a pragmatic decision because individual bones or fragments can rarely be identified to species; in this they are comparable with the folk biological 'generic' discussed in Chapter 2, which is also based on clear morphological difference. This type of identification provides a basis on which to describe past activities – the treatment of cattle or goat bones – but also imposes a 'specimen logic' on these bones, in which animal remains become quantifiable units.

The move towards quantification has been accompanied by standard measures such as NISP (number of identified specimens) or MNI (minimum number of individuals). This allows statistical comparison between contexts and sites, although these are not always directly comparable because of differing recovery strategies. Until the emergence of more sophisticated identification techniques, sheep and goat bones were difficult to differentiate and so are often combined. In every large assemblage sheep/goat bones predominate, normally followed by pig and then cattle (see Figure 3.8). In the absence of unusual deposition contexts, these bones are usually interpreted as reflecting consumption, particularly in settlements. Butchery marks are reported from a number of sites, indicating an increasing level of detail in recent zooarchaeological reports.[1] The reporting of cut marks can help to challenge expectations, showing, for instance, that dogs and badgers were butchered (rather than skinned), and therefore eaten, at Postpalatial Kastro and Vronda (Snyder and Klippel 2000, 2003).[2]

More recent zooarchaeological publications of Bronze Age Cretan sites can as a result be useful tools in examining past animal practices. In these sampling and quantification are considered alongside taphonomy, unravelling the complex relationship between the animals which lived in past environments and the animal remains which were eventually deposited and survived to be excavated and identified (Klein and Cruz-Uribe 1984). Isaakidou's (2004) reanalysis of the zooarchaeological remains from Knossos exemplifies how disposal patterns of animal bones affect the proportions of species, and hence interpretations about human-animal relations. There is also a growing awareness that zooarchaeological interpretations have tended to reflect the economic concerns of the present derived from modern farming practices, in which maximisation of yield is key (Birke and Oma 2013). Because zooarchaeologists often deal with the remains of butchered animals there is also a tendency to focus on meat consumption and kill-off strategies. In response, social zooarchaeology broadens the range of interpretations to examine the variety of relationships between humans and animals in the past, in which maximisation of animal products need not have always been the only goal. Nerissa Russell (2012: 399–400) has called for social zooarchaeologists to pay closer attention to archaeological context in order to move beyond a narrower focus on animal bones as the remains of past meals.

In considering archaeological context, Boyd Dawkins's report can be seen as a forerunner of social zooarchaeology on Crete. The bucranium pictured in his article, as well as being a specimen of *Bos domesticus*, var. *creticus*, is one of the animals which has been 'sacrificed in the Sanctuary of Zeus' in contrast to the 'refuse heap' of the lower part of the cave (Boyd Dawkins 1902: 162). This conceptual division, between rubbish and ritual, has long dominated zooarchaeological interpretations. Whereas the majority of zooarchaeological material is interpreted as refuse, and treated in economic terms, phenomena such as articulated skeletons are often seen as ritual. More sophisticated analyses have adopted a more fine-grained analysis of context alongside the particular body parts represented (Hill 1996; Morris 2008). The concept of structured deposition, employed by Hill in his analysis of British Iron Age pits, provides a way of focusing on the intentional assembly of deposits rather than regarding them as by-products of activities such as rubbish discard or animal sacrifice. The concept has been increasingly used for deposits of animal bones, usually related to consumption activities, in the Bronze Age Aegean (D'Agata 2000; Isaakidou and Halstead 2013; Caloi 2017).

It can be difficult to identify animal sacrifice archaeologically, and in most cases discussed here the animals were butchered and consumed after they were killed. If sacrifice is defined as the ritualised killing of animals, there is some iconographic evidence for this animal practice in the Final Palace period. The animal things which result from sacrifice can be termed offerings, which implies their dedication to deities or other powers. Ultimately it is difficult for zooarchaeologists to identify the intention behind either killing or offering (Russell 2012: 88). More evident is the animal practice of butchery, which can be defined as the practice of slaughtering animals for food, and the subsequent partitioning of the animal carcase for consumption.[3] In Bronze Age Crete, butchery sometimes resulted in animal things which were used to memorialise the eating of particular animals at 'feasting' events. These can be termed trophies, and were usually animal heads. Although sacrifice and butchery are difficult to disentangle archaeologically, and depend on the way they are defined, on Crete the term sacrifice has tended to be used indiscriminately for unusual deposits of animal bones, giving them a ritual significance. This has obscured the intentional use of animal remains, particularly the display of heads. The following discussion builds on recent critical explorations of the term 'sacrifice' in the Bronze Age Aegean (Cromarty 2008; Recht 2011; Halstead and Isaakidou 2017).

SACRIFICE

Finds of skulls in Bronze Age Crete have been readily interpreted as the remains of sacrifices partly because they have been seen in the context of later Greek rituals. A. B. Cook (1925: 926), in his exhaustive study of Zeus in Greek religion, agreed with Boyd Dawkins that the remains at Psychro were 'clearly the *débris* of animal sacrifices'. Psychro, a large and impressive cave on the edge of the Lasithi Plateau in eastern Crete, was of interest in this context as one of the candidates for the Dictaean Cave, the birthplace of Zeus. Offerings of votive objects such as vessels or jewellery continued to be made after the Bronze Age, showing that it remained in use as a cult site (Boardman 1961; Watrous 1996). Activity at the site appears to have begun in the Final Neolithic period, and carried on until the Byzantine period. Deposition of various clay and bronze objects peaked in the Protopalatial to the Geometric period, spanning 1,000 years. But the continuity of activity at the site, and its possible later associations with Zeus, should not be seen as evidence for continuity in practice. The tendency to regard Bronze Age cult practice, before and after the Mycenaeanisation of Knossos, as homogeneous, and related to later Greek religion, is exemplified by Martin Nilsson's (1950) influential work *The Minoan-Mycenaean Religion and Its Survival in Greek Religion*. In a chapter devoted to animal sacrifice he suggested that it was likely that 'the bucranium, i.e. the head of the sacrificed bull, played a part in the Minoan cult as it did in the other cults and in Greek times when it was nailed to a tree in the holy grove or to the temple wall' (Nilsson 1950: 232). Boyd Dawkins's (1902: 162) suggestion that the bucranium was probably fixed to the altar is similarly influenced by later practice since there is no mention of the skulls being associated with the structure David Hogarth identified as an altar. Hogarth (1900: 98–9) places the word 'altar' in inverted commas because it was more like a pile of stones; he describes it as surrounded by offering tables and cups, but not skulls. As a result there is no clear evidence that the Psychro skulls were attached to an altar and so they should not be automatically be seen as analogous to later Greek bucrania.

Unfortunately it is difficult to say where exactly the skulls were found in Psychro; Hogarth (1900: 101) only reports taking away 'specimens of the skulls, horns, and bones found in the Upper Grot'. The Upper Grot was the more accessible area of Psychro Cave, excavated by a number of scholars during the late nineteenth and early twentieth century, and the resulting publications do not give many details about the animal remains. After local villagers had started to find figurines and other objects, Joseph Hazzidakis, president of the Herakleion Syllogos, and Federico Halbherr, an Italian archaeologist, began work there in 1886. They were followed by Arthur Evans and John Myres in 1895–1896, the French archaeologist Joseph Demargne in 1897 and then David Hogarth in 1900. Halbherr reported a black layer containing the bones and horns of cattle, sheep and goats (Halbherr and Orsi 1888: 907). Arthur Evans (1897: 355) describes a similar layer at 1.5 m depth, but explicitly associates it with sacrifice, 'a continuous layer containing what appeared to be a sacrificial deposit of bones, horns and ceramic objects, imbedded in ashes and charcoal. The bones were of deer, oxen, and goats, the horn of an agrimi or wild goat found in this stratum being about a foot and a half in length'. Evans's description of the pottery associated with the bones appears most likely to be Neopalatial, and Boyd Dawkins (1902: 162) associates the skulls with this deposit based on pottery sherds among the bones he studied. Hogarth (1900: 96) describes two strata, corresponding to Boyd Dawkins's report, of a lower layer of yellow clay with 'a little primitive bucchero pottery and many bones', probably Final Neolithic to Early Minoan habitation, and above 'a black mould mixed with ashes, bones and pottery'. All that can be said about the cattle and goat skulls is that they are likely to have been deposited in the Middle–Late Bronze Age, and were associated with a black, ashy layer containing the bones of various animals. The reason this layer was described as sacrificial by Evans is most likely that it fitted later Greek accounts of sacrificial deposits.[4]

The excavators of Psychro were more interested in the bronze offerings, clay relief vessels and, in the case of Evans, an offering table with a Linear A inscription. One fragment had been found by a local villager and Evans's excavation was focused on finding the rest of it (Evans 1897: 350–61); Demargne (1902: 581) in fact found the remainder. Hogarth then employed a large number of local people to clear out the rest of the cave, using dynamite to break up fallen rocks. He blasted his way into the lower cave, where bronze offerings had been deposited in a pool and among the stalagmites. These were retrieved haphazardly, and his report noted the 'grotesque sight' of all the local villagers grubbing for finds in this 'Lower Grot' (Hogarth 1900: 101). He reported that the bronze figurines in this lower part were generally human, whereas in the 'Upper Grot' both he and Evans found a number of clay and bronze animal figurines (Hogarth 1900: 107–8). Some of the relief vessels from this area also showed animals, including wild goats (Boardman 1961: 57–9; Watrous 1996: 35). These figurines and relief vessels provided another means of bringing animals to the cave, as clay and bronze things rather than the animals themselves. But in order to understand the animal remains, particularly skulls, deposited in Psychro Cave, it is necessary to consider other deposits described as sacrificial from Bronze Age Crete.

As at Psychro Cave, Arthur Evans identified a deposit of animal remains as sacrificial at Knossos, in a building just to the east of the palace. The House of the Sacrificed Oxen was named after smashed cattle skulls found in the basement, associated with the remains of what he termed 'portable altars'. Evans interpreted them as follows:

> These sacrificial relics, thus ranged on the floor of the basement Chamber, could have only one signification. The methodical filling in of the ruined building and its final relinquishment as a scene of human habitation had been preceded by a solemn expiatory offering to the Powers below.
>
> (Evans 1928: 302)

Continuing with a quote from the Iliad, 'in bulls doth the Earth-shaker delight', Evans explicitly gave these remains a religious significance. In particular he used them as evidence for the practice of expiatory sacrifice found in the Bible and discussed in an influential work by William Robertson Smith (1889). Regarding them as an act of atonement following an earthquake, Evans suggested the skulls were placed in the basement of the building before it was filled back in with destruction debris, although the stratigraphic evidence is more ambiguous (MacGillivray 1998: 46–9; Mathioudaki 2018: 20–31). Nevertheless, the skulls appear to have been deposited early in the Neopalatial period and then covered with a large dump of broken pottery of a similar date, most likely the result of an earthquake. Evans's report helped to reinforce the equation between sacrifice and animal skulls in Bronze Age Crete.

At the palatial site of Phaistos in the south of the island there is a deposit from the end of the Protopalatial period called the sacrifice pit, 'fossa dei sacrifici', by the excavators because it contained broken pottery and burnt sheep and cattle bones (Pernier and Banti 1935: 207; Carinci and La Rosa 2009: 261–6). These remains have yet to be published in detail but the burning of bones is often regarded as sacrificial because later Greek cult practice involved burning a portion of the carcase; the important aspect of this is not the killing of the animal but the offering of the remains, a practice with parallels in later Greek sacrifice. This is more clearly attested at two Late Bronze Age mainland sites, Ayios Konstantinos in Methana (Hamilakis and Konsolaki 2004) and at Megaron B at Eleusis (Cosmopoulos and Ruscillo 2014), where burnt pig remains have been argued to be evidence for burnt sacrifice. This reconstruction is supported by differences in species and body part representation between contexts. At Pylos, a mainland palatial site, a number of discrete deposits of burnt bone, showing unusual body part representation, have been convincingly argued to be evidence for sacrificial offerings (Isaakidou et al. 2002; Halstead and Isaakidou 2004). These deposits have been connected to Linear B tablets listing animals, including cattle, destined for large-scale consumption; one of the burnt bone deposits found in the archive room has been used to argue that the sacrificial remains were accounted for by the palace administration before disposal (Stocker and Davis 2004). In these cases careful analysis of body part representation, context, and contemporary documents in the case of Pylos, allows the reconstruction of a particular practice.

On Crete it is more difficult to make the same connection between burning and sacrifice; the Linear B tablets from Knossos cannot be linked in the same way with burnt portions. One tablet does refer to goats with a word interpreted as meaning 'sacrifice' (Killen 1994), which Paul Halstead (2002) suggests can be linked with consumption. More generally, there are lists of animals which were to be killed and eaten; this can be described as sacrifice (Godart 1999) or feasting, or both (Palaima 2004). Even the Pylos evidence for burnt sacrifice can be interpreted within a feasting context. Cooking rarely results in the burning of bone; this is either secondary or the result of deliberate practices. At Bronze Age Kommos, a large and well-published site on the south coast of Crete, only five burnt cattle bones (2 per cent of the total for cattle) were found and David Reese (1995b: 188–9) explicitly denies any religious connotations for these remains, which he sees as food refuse. Since burnt bones have been seen as the traces of sacrifice at Iron Age Kommos, associated with altars, it can be assumed that there was no such contextual evidence for sacrifice in the Bronze Age levels (Reese 2000: 474, 491; Shaw 2000: 682–90). Burnt bones of various species are found frequently at Kommos (Reese 1995b: 171) and over fifty are reported from Mochlos (Reese 2004), but no arguments are made to support a sacrificial interpretation. Since the burnt bones at Mochlos are found in the areas with the highest concentrations of animal bones, these do not appear to be evidence for the discrete deposition of burnt bone often associated with this type of sacrifice, but instead look like cooking debris. At Nopigeia-Drapanias, selective burning of bone is associated with marrow extraction (Hamilakis and Harris 2011: 209–10).

By contrast, there are instances of cooking debris being interpreted as related to sacrifice. At the open-air sanctuary site of Kato Syme, the excavators found a substantial Neopalatial enclosure wall around a central podium surrounded by a black layer of ash, animal bones and pottery. 'The abundant faunal remains recovered from the black stratum suggest that ritual activity within the complex centred on animal sacrifice' (Lebessi and Muhly 1990: 324). Angeliki Lebessi and Polymnia Muhly are careful to point out that there is no evidence for burnt sacrifice, but rather the remains of the consumption of animals. Arguing for continuity of cult practice at the site, they suggest that the animal skulls in the Neopalatial burnt layer were the remains of offerings. 'The deity's portion, represented by the victim's head, was deposited in the dying fire, just like the other offerings, while the rest of the victim was consumed by the worshipers' (Lebessi and Muhly 1990: 328). This layer containing ashes, animal bones and pottery seems to be similar in character to the deposits described as sacrificial from Psychro and Phaistos. Vance Watrous (1996: 49) infers from the bones in the ashy layer at Psychro that: 'Animals, including oxen, pig, deer and goats, were sacrificed, cooked and eaten there'. Another deposit containing the remains of butchered animals, including frontlets, with the horns still attached to a portion of the skull, was found at Postpalatial Chania in western Crete (Harris 2014b: 200, 226). Here the excavators tentatively interpreted the higher percentage of deer and agrimi bones in what they called Rubbish Area North as 'waste from offerings in a nearby shrine' (Hallager and Hallager 2000: 194; Hallager 2001). At both Kato Syme and Chania, the boundaries between ritual and rubbish are blurred, with animal remains interpreted as the debris of consumption in a cult setting.

Finds of complete animal skeletons are often described in Crete and elsewhere as indicating sacrifice. A horse skeleton from Final Palatial Tholos Tomb A at Archanes-Phourni was regarded by the excavators as a sacrificial animal. It is possible that the horse was skinned or eaten since there are cut marks on the skeleton, which was carefully arranged but not articulated. The other examples of complete horse burials in the Late Bronze Age Aegean are on the mainland (Crouwel 1981: 32–5; Recht 2011: 103–6). In the same tomb was a bucranium, interpreted as the remains of an animal sacrificed and then eaten in a funerary banquet (Sakellarakis and Sapouna-Sakellaraki 1997: 262). In both cases it is possible, although not certain, that the animal was killed as part of the funerary ritual. In Archanes-Phourni Tholos Tomb B, the remains of a dog, although not articulated, were found, which the excavators argue 'could not be eaten and can therefore only have been sacrificed' (Sakellarakis and Sapouna-Sakellaraki 1997: 263–4). As noted above, there is evidence for butchery of dogs in Postpalatial Crete, but this is one of a small number of dog burials known from Late Bronze Age Crete, which could indicate the sacrifice of companion animals (Preston Day 1984). The deposition of animal bodies, without butchery, appears to indicate a killing with the intention of making an offering. Complete bodies represent a different type of animal thing to dismembered and scattered remains. If the horse participated in the deceased person's warrior lifestyle, the deposition of this animal's bones can also be seen as an act of commemoration, giving the animal a material presence in the tomb along with the dead human. The dog too could have been a hunting companion. The bucranium, even if it was the remains of an animal killed and consumed as part of the burial ritual, can also be seen as the commemoration of the involvement of a valuable animal in this event.

There is growing evidence that human sacrifice was also practised on Crete, although, as with animal sacrifice, this term applies to different phenomena at Neopalatial Knossos and Postpalatial Chania. The two contexts at Knossos are very different; the building at Anemospilia, whose contents were preserved by an earthquake around the start of the Neopalatial period, contained a human skeleton on a platform, with a nearby spear and vessel assumed to be associated with sacrifice (Sakellarakis and Sapouna-Sakelleraki 1997: 268–311).[5] At the

Stratigraphical Museum excavation, finds of children's bones with frequent cut marks are clear signs of butchery and perhaps cannibalism at the end of the Neopalatial period (Warren 1981; Wall et al. 1986). Of particular interest here are the animal bones found mixed with a third example of human sacrifice from Chania, a dismembered young woman (Andreadaki-Vlazaki 2015). The animal bones, mostly from sheep and goats (at least forty-three animals), with four pigs and two cattle, were articulated, with no signs of burning and very few cut marks, showing that the bodies had been cut up but the meat had not been consumed. Dimitra Mylona (2015) refers to this practice, otherwise unparalleled in Bronze Age Crete, as 'a sacrifice not tasted'. There are similarities to the horse and dog burials from Archanes-Phourni, with the completeness of the animal remains indicating that they were offered rather than consumed.

A final category of evidence associated with sacrifice is iconographic, and appears specifically to relate to a practice termed blood sacrifice. Dating to the Final Palace period, there are a number of depictions of animals, particularly cattle, trussed on offering tables (Rehak 1995c). These are shown on a small number of sealstones and, most famously, on the 'Ayia Triada sarcophagus' (Long 1974: 61–8; Burke 2005) (Figure 4.1). The sarcophagus has fresco panels on all four sides, one of which shows a bull whose throat has been slit, with the blood draining into a vessel below the table. There are a number of people around the table who appear to be in a procession towards a structure usually identified as an altar, depicted at the far end of the panel. On the opposite side there is a procession of people carrying offerings, including cattle (perhaps figurines) to a static figure usually interpreted as the deceased. It is not necessary, however, to understand the symbolic aspects of the scenes in order to use the sarcophagus as evidence for the ritualised killing of an animal in this period. Nevertheless it is important to note that what it being shown is the bleeding out of a stunned or dead animal, which is a normal part of butchery; in this context the term sacrifice implies that the blood was regarded as an offering. Underneath the table are two more animals, one of which appears to be a wild goat. Similar depictions of the same scene on sealstones show animal heads rather than whole animals under the offering table, including heads of wild goats (Figure 4.2a). This iconographic evidence provides support for the possibility that the bucranium in the Archanes-Phourni Tholos Tomb A was the result of such a sacrifice, since they are more or less contemporary, and both are related to mortuary ritual. It also raises the question of why this practice was only depicted in the Final Palace period; either it was imported from the mainland as part of the Mycenaeanisation of Crete, or it was practised earlier but not depicted. Either way, there is no archaeological evidence for the blood sacrifice of animals in Bronze Age Crete since this practice leaves no clear traces on animal remains.

4.1. Butchery scene from the Ayia Triada sarcophagus, LMIIIA (AMH Λ396). © Heraklion Archaeological Museum – Hellenic Ministry of Culture and Sports – 'Hellenic Organization of Cultural Resources Development (HOCRED)'

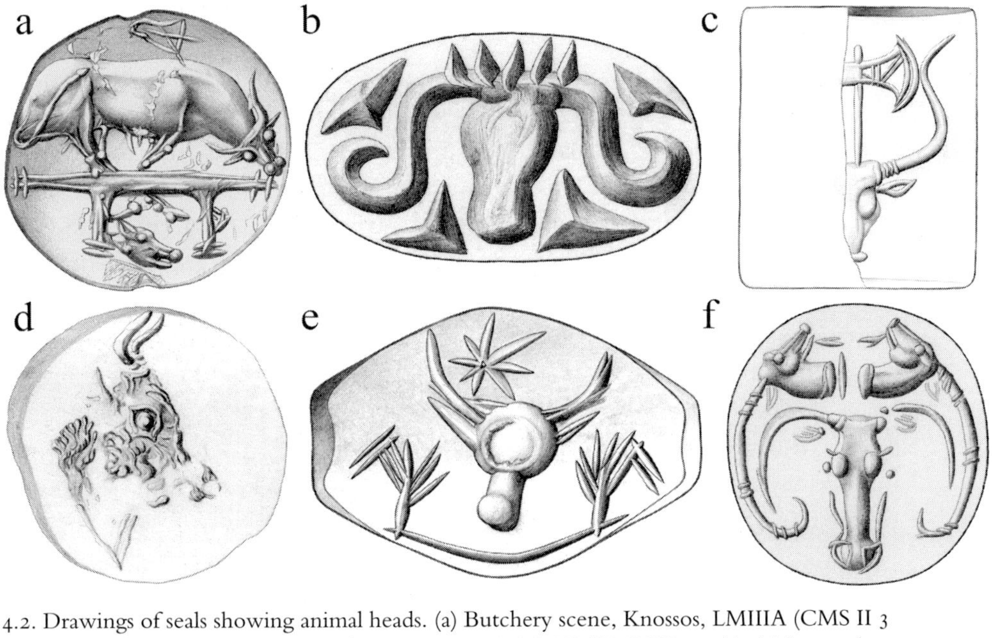

4.2. Drawings of seals showing animal heads. (a) Butchery scene, Knossos, LMIIIA (CMS II 3 338); (b) cattle head from three-sided prism, Kavousi, MMII (CMS VS 3 17b); (c) incomplete cattle head with double axe, Knossos, LMI–II (CMS II 3 11); (d) cattle head, Knossos, MMIII–LMI (CMS IV 168); (e) Talismanic-style cattle head, Psychro, MMIII–LMI (CMS II 3 289); (f) cattle head and two goat heads, Psychro, LMII–IIIA (CMS VI 448). © CMS Heidelberg

The term 'sacrificial' has been used to describe the remains of animals in Bronze Age Crete since the nineteenth century, but it is clear that a variety of practices are involved, which are evident at different times. Some practices, such as burnt sacrifice, or the burial of companion animals, leave distinct traces, but in most cases it is difficult to distinguish sacrifice from butchery archaeologically because animals could have been ritually killed, then butchered and consumed. It can also be argued that all animal killing was ritualised to some extent in Bronze Age Greece because it was carried out in public and was not an everyday occurrence, whereas sacrifice is more difficult to identify zooarchaeologically (Hamilakis and Konsolaki 2004: 145; Isaakidou and Halstead 2013: 88–9). In his examination of sacrificial ritual in Bronze Age Crete, Robert Cromarty (2008: 107) argues that animal remains at Psychro and the sanctuary sites of Kato Syme and Juktas are better described as indicating 'commensal meat rituals' rather than sacrifice since there is no evidence that these were offerings to supernatural powers. As he suggests, the offering of animals to deities is not even explicitly shown in depictions such as the Ayia Triada sarcophagus (Cromarty 2008: 15). Other authors have combined the different types of evidence, both from animal remains and iconography, to argue that sacrifice, particularly involving cattle, was widespread in Bronze Age Crete (Marinatos 1986; Nilsson 1950: 229–35). One way out of this debate is to turn from animal practices to animal things.

TROPHIES

Animal sacrifice and butchery are animal practices which are inferred by archaeologists from excavated deposits of animal remains. Some types of animal sacrifice left distinctive animal things, in the form of whole-animal burials or burnt offerings, but most animal remains found by archaeologists are the products of butchery, whether or not the animal has been sacrificed (Seetah 2019). Among the animal things which result from butchery are joints of meat, or other animal products such as hides and trophies. A trophy can be defined in this context as the part of an animal which is displayed in such a way as to retain the connection with the animal itself; it is an index of that animal. Preserved heads or skulls (particularly with horns or antlers) are obvious trophies since they are the most clearly iconic remains of the dead animal. They also have little meat on them, although the horns of various animals do have a number of uses. Before the modern practices of taxidermy, preserving the head for any length of time was impossible, except by defleshing it and retaining the skull. It follows that animal skulls found in Bronze Age Cretan contexts were either once decapitated heads kept for a short period, for instance, during consumption events, or curated skulls (and modified in the case of frontlets) kept on display for a much longer period. Just like the skulls of people in Neolithic Jericho described by Hans Belting (2005), they made the absent animal visible, a reminder of the animal that had been butchered.

Trophies can be distinguished from discarded heads by archaeological context. Postpalatial Vronda in the hills of Lasithi in eastern Crete has provided a clear and recently excavated example of a primary context for the display of bucrania and horn cores. The southern part of Room B4 in Building B contained a deposit consisting of a circle of cattle bucrania, at least one pair of agrimi horns and boar mandibles. Although it has been suggested that they were from sacrificial victims (Mazarakis Ainian 1997: 295–6), the excavators focus on the treatment of the skulls. 'In this Early LMIIIC context, where horns of consecration and bull imagery are so prevalent and have ritual significance, it seems likely that the cattle skulls and agrimi horn cores were deliberately altered to form objects of ritual or decorative significance' (Preston Day et al. 2009: 43). Although they do not exclude the possibility that the animals were sacrificed, they point out that the deposit is related to these remains being used as wall decorations rather than containing consumption debris which is associated with sacrificial ritual (Preston Day et al. 2009: 62).

In other cases, including the skulls and frontlets discussed above from Psychro and Knossos, skulls were displayed and deposited as trophies as part of consumption events. The House of the Sacrificed Oxen skulls are most likely to have commemorated a consumption event before the building was filled with dumped pottery. There are close parallels for this activity at Protopalatial Phaistos; in Room IL a platform was built around the remains of a 'feasting deposit' which contained numerous cups, three cattle rhyta and a 'partial skull of an agrimi with almost entirely preserved horns' (Caloi 2017: 46). Whereas the House of the Sacrificed Oxen deposit is difficult to reconstruct because of its uncertain stratigraphy, Room IL at Phaistos provides a clear example of structured deposition, in which the debris from a single event have been collected and commemorated in a particular space in the palace; a similar Protopalatial context, Corridor III/7 also contained animal bones (Caloi 2017: 38–9). A partial agrimi skull with horns intact was found at the Neopalatial site of Nopigeia-Drapanias in western Crete together with a deposit of animal bones and pottery consistent with feasting. The skull's horizontal placement on a layer of sand at the bottom of the ditch appears significant (Hamilakis and Harris 2011: 205). A potentially similar Postpalatial example of the structured deposition of a cow skull and goat horns is reported at Thronos Kephala (Isaakidou and Halstead 2013: 91). In all these contexts it appears that animal heads were retained and displayed as indexes of particular animals. Animal heads were significant in themselves because they were the tangible outcome and manifestation of the animal practice of butchery, before the remainder of the animal was eaten. Further evidence for the significance of these animal things comes from their depiction in other materials.

Animal Heads

> A circumstance which is to be treated more fully in another place may be pointed out, namely that the skulls of the animals sacrificed were probably nailed up in the sacred grove and that detached heads of animals are represented on gems, but this will hardly suffice to establish an innate sacral character for the animal-headed *rhyta*, even if it may originally have been closely associated with them. They seem to have been fanciful vessels of luxury which were sometimes also dedicated to the gods or used in the cult.
>
> (Nilsson 1950: 146)

As Nilsson pointed out in his discussion of animal sacrifice, animal heads are frequently depicted in Minoan material culture. Animal heads occur on seals of the Palatial period (Figure 4.2). Rhyta in the shape of whole animals, particularly cattle, were made from the Prepalatial period onwards but head-shaped rhyta only started to be made at the end of the Protopalatial period (Koehl 2006: 32). Earlier examples were made of pottery but some of the finest and most realistic examples, made in the Neopalatial and Final Palace periods, were carved from stone and depicted cattle or lion heads (Figure 4.3). No complete metal examples survive from Bronze Age Crete but a silver and gold bull's-head rhyton and a gold lion's-head rhyton from the Mycenae Shaft Graves show that they were once in circulation (Karo 1930: 77–8, 93). Metal rhyta survive less frequently than

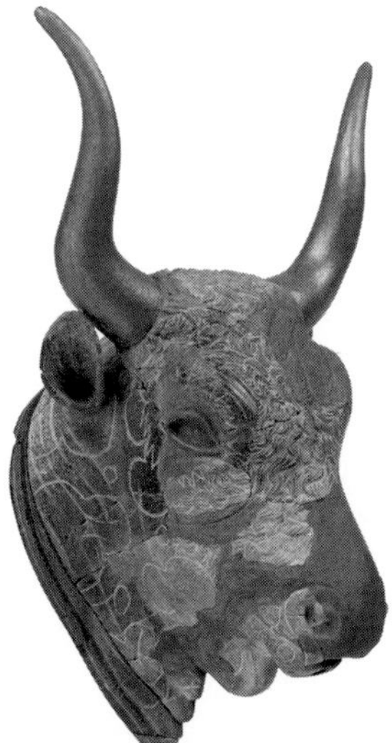

4.3. Bull's-head rhyton from Zakros, LMI. Height of head: 14.8 cm (AMH Λ2713). © Heraklion Archaeological Museum – Hellenic Ministry of Culture and Sports – 'Hellenic Organization of Cultural Resources Development (HOCRED)'

stone examples because they could be recycled, but further evidence for their existence comes from New Kingdom Egyptian tombs, associated with apparently Aegean tribute bearers (Keftiu) (Wachsmann 1987: 56–60). Among the animal heads depicted here are cattle, lion and griffin, seemingly made of gold. Gold foil was also used to cover the horns of some stone cattle-head rhyta, and there is circumstantial evidence for wild goat-head rhyta since there are similar gold foil horn coverings from Juktas and Knossos (Koehl and Karetsou 2011). Although the evidence for griffin-head rhyta could be seen as supporting the idea that they were 'fanciful', the investment in making these rhyta demonstrates the importance of animal heads in Bronze Age Crete.

Rhyta are pouring vessels, and the animal-head rhyta typically have a hole in the back for filling and a hole at the mouth from which the liquid emerges. Emily Miller (1984: 118–31) argues that some of the Neopalatial animal-head rhyta, particularly in stone, are primarily animal sculptures which capture the vitality of an animal. She describes the bull's-head rhyton from Zakros as follows:

> The vibrant naturalism of this work reflects an experience of nature far more intense than perhaps we can understand today. Both artist and observer perceive the subject not as a natural object viewed at a distance but as part of an order in which they too participate. No aspect of nature may be isolated but exists in a complex web of interrelationships.
>
> (Miller 1984: 120)

From this point of view 'naturalism' is a way of drawing the viewer in, and allowing them to relate to the object as if an animate form; the realism of the depiction, Miller argues, means that it doesn't seem like a disembodied head. But it can also be argued that the realism of these rhyta means that they were able to act as substitutes for the recently removed heads of animals. Even though the head is not life-size, the detail of the hair and the form of the sculpture allow it to be perceived as lifelike. The choice of material also contributes; rock-crystal lenses, as found on the Little Palace bull's-head rhyton (see Figure 2.1), have been found elsewhere, suggesting that this was one way to animate such objects (Popham 1984: 239; Rehak 1995c: 456). Although the materials mark these rhyta out as exceptional, their stylistic features capture the presence of a real animal. Whereas Nilsson discusses vessels such as these in terms of sacred versus profane, they can instead be considered in terms of life and death. Such objects resembled newly decapitated heads but could be kept without rotting, and could be re-animated by being filled with liquid. From this perspective animal-head rhyta are an important piece of evidence for the significance of real animal heads rather than fanciful vessels of luxury.

The fine crafting and valuable materials also mark these rhyta out as exceptional. It is likely that interaction with these unique rhyta would have been socially restricted, as Wendy Logue (2004) argues more generally for carved stone vessels. Their impact would have been particularly strong on those who were not familiar with such objects and would have been enhanced if they were only used in occasional socially charged events (Bevan 2007: 11). During these events the rhyta would have been filled with liquid, which poured out of the mouth of the animal. Although rhyta are often regarded as libation vessels, it is likely that they were used to fill individual drinking vessels with an alcoholic beverage. Wine could perhaps have acted as a metaphorical blood substitute given that real blood rapidly coagulates (Marinatos 1986: 31; Laffineur 1987; Koehl 2006: 263). Animal-head rhyta are found at a limited number of sites on Crete, and by far the largest number of fragments comes from Knossos (Rehak 1995c: 438). A recently identified stone fragment from Juktas appears to have come from an exceptional life-size example, whereas most are smaller than life-size (Karetsou and Koehl 2013). They are always found in a fragmentary state, a phenomenon which Paul Rehak (1995c: 450–2) connected with sacrifice by arguing that they were also ritually smashed, becoming simulacra of the sacrificed animal in the process.[6] Such fragmentation appears to be a wider practice in Bronze Age Crete, making it difficult to connect directly with animal sacrifice (Knappett 2020: 177–82).

These animal-head rhyta offer an important insight into the animal practice of butchery which has been subsumed within discussions of sacrifice. Animal heads are ideal trophies both because they are a distinctive part of a particular animal species and because they are not one of the meaty parts of an animal valued for consumption. Whereas animal heads start to rot after the animal's death, these stone or metal examples maintained their vitality until they fell out of use (or were decommissioned by smashing them). Perhaps the heads in the House of the Sacrificed Oxen were buried soon after death, before their features disappeared, leaving only the memory of this event. By contrast the stone animal thing in the same form from the Little Palace allowed such events to be re-enacted when the rhyton was filled with liquid which then drained from the animal's mouth. They also allowed for more careful choreography, without the resistance of a real bull, and repetition, without having to kill such a hugely valuable animal. These animal things could also be carefully controlled, kept out of sight until the occasion of use. They point towards the importance of the animal thing which resulted from dividing and eating an animal's body.

Although the origins of animal-head rhyta have been sought in earlier, less naturalistic animal-head vessels (Tuchelt 1962; Koehl 2006: 353), other media suggest that animal heads were important in the Protopalatial period. Animal heads, as discussed in the previous chapter, were used as signs in the Cretan Hieroglyphic script. When used as ideograms they could signify an animal, most likely brought to the palace for consumption. Although the head of the animal can stand for the whole, a case of metonymy, it is also possible that the actual heads of animals did have a role in demonstrating the success of a hunt, or the delivery of an animal for consumption. Since these depictions and the animal-head rhyta most frequently show the fleshy heads of animals, they point to the importance of animal heads in palatial Crete. It is likely that the bucrania found at Psychro, Archanes-Phourni or the House of the Sacrificed Oxen were instances of animal heads being used to commemorate particular consumption events. It is possible that all were on display for the duration of the event and then buried, just like the agrimi skulls from Phaistos and Nopigeia-Drapanias. Animal-head rhyta could be reused in different events, but their form indexed consumption activities involving the display of heads.

Double Axes

Double axes have sometimes been associated with sacrifice on Bronze Age Crete, particularly because of their association with cattle heads in iconography (Nilsson 1950: 227–31).[7] Because double axes appear in various other contexts, they have usually been seen primarily as a religious symbol instead (Evans 1921: 423–47; Briault 2007; Haysom 2010). This idea is reinforced by the flimsy bronze double axes at sites like Psychro Cave which could have been deposited as votives or displayed on poles; these are the same shape, with curved blades, as the double axes which appear with bucrania in iconography. These would not have been useful as chopping tools because they were too insubstantial, although it is possible that some were used to slice meat or for skinning.[8] There is, however, another type of double axe which is frequently found in Bronze Age Crete. This is more solid and is usually squarer, sometimes has curved blades from use and sharpening and would have been a functional tool for uses including butchery (Lowe Fri 2011: 60–4). Some are inscribed with Linear A characters, helping to date them to the Neopalatial period, and were also sometimes deposited as votives in the same way as the thinner curved versions. Both types occur together at Psychro Cave (Figure 4.4). Maria Lowe Fri (2011: 73–4) suggests that the axe pictured in Figure 4.4a has use-wear traces consistent with chopping bones. Another double axe from Rethymnon is inscribed with a bull's head (Mavryiannaki 1978; Rehak 1995c: pl. Lc). When hafted on a long pole, these more compact double axes would have been an ideal weapon with which to paralyse animals, by severing the spinal cord with a well-aimed blow to the back of the neck.[9] This could have preceded the cutting of the throat, after which the immobile animal would have bled to death (Nilsson 1950: 231; Aldrete 2014: 43–5). After that the double axe could have been used as a cleaver to chop up the carcase.

There is clearer evidence for sacrificial equipment from a Linear B tablet from Pylos, Ta 716. This lists two sets of a bridle and chain for the animal, an

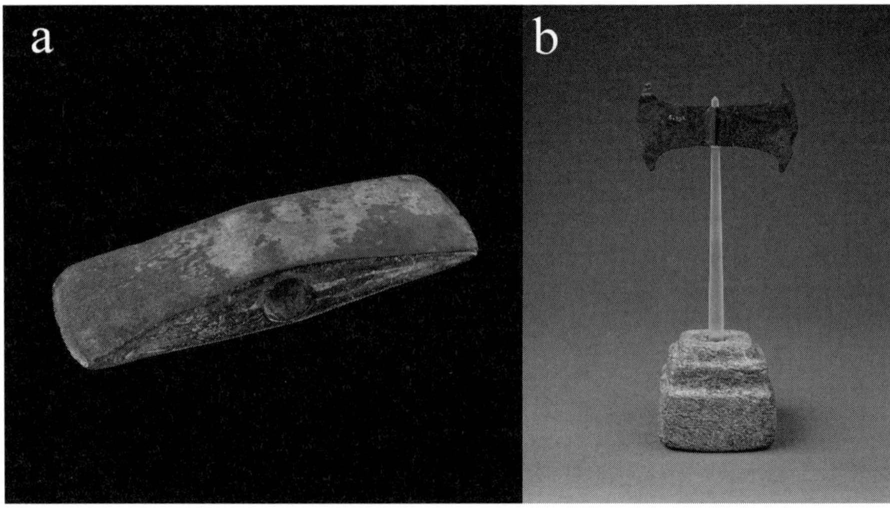

4.4. Double axes and stand from Psychro Cave. (a) Solid double axe, LMI. Length: 15.8 cm (AN1927.1386); (b) flimsy double axe on stone stand, MMIII–LMI. Length: 15.1 cm (AE.593 and AE.594). © Ashmolean Museum, University of Oxford

axe and a knife (Palaima 2004: 236; Palaima and Blackwell 2020). The axe ideogram looks like a double axe but one end is blunter than the other, and it has been described as a 'stunning-axe' (Speciale 1999; Nikouloudis 2001: 21). Although this is based on the shape of a small incised character, the other method used cross-culturally to stun cattle prior to killing is with a blow to the centre of the forehead with a blunt object (Aldrete 2014: 35–6). It is notable that the stylised hair swirl on the forehead of the bull's-head rhyton from the Little Palace is the ideal point for such a blow, where the skull of the animal is thinner. A number of stone mace or hammer heads suitable for stunning animals have been found from Bronze Age Crete, including at Knossos and Zakros (Platon 1971: 145; Marinatos 1986: 22). One stone hammer with rounded end was found in the entrance of a Neopalatial tomb near Knossos but the context had been disturbed and so it is not clear whether the remains of three sheep/goat skulls are directly associated with it (Muhly 1992: 96; Reese 1992). Whether or not the same sacrificial kit existed on Neopalatial Crete as at Pylos, either double axes or blunt instruments such as this could have been used to immobilise animals before their killing and butchery.

Double axes were found in association with fragments of a bull's-head rhyton in another tomb near Knossos, dating from the early Final Palace period. Excavated by Evans and Mackenzie in 1910, the Tomb of the Double Axes had been disturbed in antiquity but among the remaining objects were flimsy votive double axes and the stone ears and inlays from a bull's-head rhyton (Evans 1914: 33–59) (Figure 4.5). A solid-bronze double axe was found at the entrance, although it is not clear if it was originally part of the assemblage

4.5. Tomb of the Double Axes at Isopata, near Knossos. Sketch by Emile Gilliéron, *fils* and plan of tomb with contents (mounted together) (Sir Arthur Evans Archive). © Ashmolean Museum, University of Oxford

within the tomb. Uniquely among tombs of this period, the sarcophagus excavated in the floor was also in the shape of a double axe. No animal bones were reported from the tomb, although they were not routinely reported at this time. Evans (1914: 70–94) compared this find with the Little Palace deposit in which the bull's-head rhyton was found because this also contained a pyramidal stone base usually associated with votive double axes. In both contexts there is an association of substitute animal heads with double axes. The same association occurs at Juktas, and a complete bull rhyton was found at Psychro. At both of these sites, the remains of animals were also found. Although it is difficult to determine when the consumption of animals ceased at sanctuary sites, it appears to be a Neopalatial phenomenon, whereas the association between bull's-head rhyta and votive double axes continued into the Final Palace period. In this way the use of actual animal heads to memorialise consumption could itself have been memorialised through the creation of substitutes of both animal heads and the tools with which the animal was butchered.

It is significant that the moment of killing, or at least bleeding out, of the animal was only depicted in the Final Palace period, and even then only rarely. In the preceding periods, the animal head and the likely tool for butchery were instead the focus. The cult scenes involving double axes are also from the Final Palace period or later, whereas in the Neopalatial period they are generally carried in processions by young men who are likely to be warriors (Haysom

2010: 43–7). As argued above, warfare and cattle-raiding are likely to have gone together at the start of the Neopalatial period, meaning that the double axe could have been used as a tool and weapon in this period, before becoming a cult artefact later. One way to reinforce comradeship is the sharing of food, and it is possible that the increase in evidence for feasting activities at certain Neopalatial sites was tied with an emerging warrior lifestyle involving male fraternities (Koehl 1986b: 104–10; Driessen 2018: 296–7). Whereas the display of cattle heads had come to signify communal consumption in the Protopalatial period, the conjunction of the animal head with a weapon which was perhaps also the instrument of slaughter would have given these events a different significance. It is likely that animals were indeed sacrificed at such events, and that double axes were used to slaughter animals ritually, but there is no direct evidence for this animal practice. Neither the double axe nor the bucranium can be reduced to symbols of sacrifice because they are immediately connected with the animal practice of butchery. The animal practices of sacrifice, which left few material traces, and butchery, which produced the animal heads and bones found in the thick black layer, were instrumental in creating these new configurations of humans, animals and things at sites such as Psychro Cave (Shapland 2021).

CONCLUSION

The first zooarchaeological report from Crete helped to establish the link between animal heads and sacrifice which has become part of the archaeological imagination ever since. It also helped to establish skulls as specimens to be counted and measured. In both cases, attention was shifted from the head to what it represented, the act of sacrifice or the presence of a particular species of cattle. Although it is difficult to date the Psychro animal remains precisely, they appear to be part of a set of practices involving butchery and consumption, most likely involving the display of the animal heads in some way. The use of butchered heads as animal things is attested at other sites on Crete from the Protopalatial period onwards, sometimes associated with structured deposition following a consumption event. Vessels resembling these butchered animal heads were created in other materials, with particularly naturalistic examples created from stone, and perhaps metal, in the Neopalatial period. These animal-head rhyta appear to have acted as substitutes for real animal heads, their association with consumption reinforced by the wine which most likely poured out of their mouths. Animal heads on seals and rhyta in the shape of animal heads can also be seen in the context of conspicuous consumption involving the display of animal things.

The special treatment and depiction of animal heads occurs throughout the Palatial period in Crete and into the Postpalatial period. By contrast, the practices of blood sacrifice, involving animals trussed on tables, the 'sacrifice not tasted' at Chania and the Archanes horse burial, all appear in the Final Palace period. On the mainland there is clear evidence for the kind of burnt animal sacrifice that continued into later periods. Feasting remained an important activity within the Mycenaean palatial collective, which extended across Crete and mainland Greece, but in Bronze Age Crete the animal head, rather than a burnt portion, appears to have been a distinctive animal thing associated with the consumption of animals. The animal practices of sacrifice and butchery are difficult to disentangle archaeologically, but even when deposits are described as sacrificial, the animal head indexed the act of butchery and consumption rather than the act of killing.

FIVE

HUNTING

INTRODUCTION

> [A] group of Agrimia, a species of Ibex, that had been browsing upon the scanty pasture, were standing motionless upon its pinnacle. We had seen several others in the ascent, some forty in all; but they were too wary of any approach of man. They were not to be taken even by a Highland deer-stalker and keen sportsman like my friend and companion Drummond, but bounded away, as soon as we were perceived, over snow and steep, crag and precipice, until they had gained another commanding peak far out of reach of gun and rifle, and there again they watchfully grouped themselves with their ponderous and sabre-shaped horns curved in relief against the western sky.
>
> (Spratt 1865: 12–13)

The reason for Captain Spratt's ascent of Mount Ida, also known as Psiloritis, was not to go hunting but to take topographic measurements. Triangulating other peaks from the highest point in Crete was an important part of his survey of the island. It was through this process that Crete came to be mapped according to the abstract space of modernity.[1] A similar spirit of scientific inquiry led Spratt's companion, Colonel Henry Drummond-Hay, to record the birds he encountered; Drummond-Hay's consolation for not bagging any agrimia on the same ascent of Ida was to find a type of chough, 'the skin of which he was deficient in his fine collection of British birds' (Spratt 1865: 12). Spratt's detailed maps duly appear at the back of *Travels and Researches in Crete* and Drummond-Hay's species list appears as Appendix 5 (Drummond-Hay 1865). The book itself is both an example of the imposition of scientific knowledge on Crete and a fine-grained description of the animals, plants, people and places that Spratt encountered. Although this description is rooted in the attitudes of a British naval officer in the nineteenth century, there is a timeless quality to his account of agrimia outlined against the sky. There is also an important historical reminder that agrimia were far more widespread on Crete before modern weapons became common.

This chapter will explore the way in which the places that animals inhabited in the Cretan landscape were absorbed into the palatial collective through a particular animal practice: hunting. Although agrimia have declined in numbers, and their range has shrunk even since the mid-nineteenth century, their ecological niche hasn't changed since the Bronze Age; they are still encountered in the mountainous areas of Crete and are difficult to hunt because they are agile climbers. Birds too have long been hunted on Crete and occupy various ecological niches, whether the cliffsides, reedbeds or scrubland. Bird remains at archaeological sites suggest that they were eaten, and there is iconographic evidence for the catching of at least some types of birds using nets. Hunting and fowling were animal practices that defined the landscape of Bronze Age Crete.

Hunting is not explicitly depicted in the frescoes from Bronze Age Crete that survive but it is commonly shown on sealstones (Krzyszkowska 2014). Sealstones have rarely been regarded as an important source of hunting iconography because they are more usually discussed in terms of style or symbolic content. Those seals of a style known as 'naturalistic' include the so-called 'animal studies', borrowing a term used in Western art for realistic drawings of animals. Such terminology establishes these depictions as lifelike images of animals whose purpose is self-evident. This chapter instead seeks to explore why a particular type of object, the sealstone, whose size meant that it was closely associated with the individual wearer, was richer in hunting imagery than any other medium. By contrast, more exclusive types of material culture, including frescoes, relief vessels and gold rings, depict the animals which were hunted, but in different contexts. This, it will be suggested, was because notions of ownership and hunting territories came to be extended across Crete, claims which were established partly through the manipulation of animal iconography at palatial centres. These images are more visible now than the hunted animals themselves, although zooarchaeology is helping to correct this. Hunting landscapes too have been neglected but it will be suggested that certain structures known as guard houses could have functioned as hunting lodges following the contest for land which was outlined in Chapter 3, and the absorption of the Cretan landscape into the Knossian palatial collective. The ecological relationship between animals and the places they inhabit results, in Bronze Age Crete and elsewhere, in a relationship between hunting and territory.

ANALOGIES

Hunting in prehistoric Greece was not just a subsistence activity as is often assumed, but helped to define time and space in farming communities (Hamilakis 2003). The importance of hunting has been recognised in histories of the practice in Mediaeval Europe and later European colonialism; it has also been suggested that it would have received more attention if it was not regarded as distasteful in academic circles (Almond 2003: 5; MacKenzie 1988: 1). The Mediaeval English wool industry which John Killen found to be a useful analogy for the workings of the Linear B tablets took place in a landscape which was defined by the hunting rights of the nobility. Deer were imported to Mediaeval England in order to be hunted, just as they were imported to Crete in the Neopalatial period (Isaakidou 2007b; Sykes et al. 2013). Tapestries, illuminated manuscripts and paintings provide a wealth of detail about elite hunting in Mediaeval Northern Europe, although they largely ignored the hunting practices of the poorer sectors of society (Smets and van den Abeele 2007). Similarly, Cretan Bronze Age iconography, particularly sealstones, can be seen as evidence for the importance of hunting as an animal practice rather than a subsistence activity.

Some of the hunting practices of Bronze Age Crete can be inferred from another important source of analogies: the texts of Classical authors. Although Homeric similes describe hunting, it is the later hunting manuals of authors such as Xenophon and Arrian which give a wealth of detail about different animals, methods and equipment (Hull 1964; Phillips and Willcock 1999). These authors hunted a similar range of animals in similar terrain, providing useful insights into Bronze Age hunting techniques. Xenophon, for instance, describes the different types of nets and snares used for trapping different prey species, as well as hunting using spears or bow and arrow. By the time he was writing, in the fourth century BCE, different breeds of dogs were recognised, including a nimble Cretan hound (Hull 1964: 34). As argued in Chapter 7, these could have been descended from hunting dogs imported from Egypt in the Late Bronze Age. Different breeds are apparent in the diversity of sizes of dog skeletons in prehistoric and later Greece (Trantalidou 2006). Dogs were largely employed for tracking and driving animals into nets, as well as holding them at bay before the hunter arrived. There is little evidence for practices such as hare-coursing, in which hares are caught by dogs who are fast enough to outrun them, or hunting on horseback in ancient Greece, although they have since become dominant types of hunting with dogs across Europe.

The concept of sportsmanship which was carefully codified in the Mediaeval period points to an underlying theme in the history of hunting. The term 'sport' is widely applied to forms of hunting which are not solely for the purpose of food gathering, often involving a form of social display.

A corollary of this is that in various periods hunting has become a practice which is political and exclusionary, with restrictions on the animals killed, the people participating and the territory involved. As John MacKenzie (1988) has shown for Africa in the colonial period, the banning of local methods of hunting and the imposition of the idea of licences and hunting reserves was a means of preventing the indigenous population from practising traditional methods of hunting or exerting control over areas of land. The same can be said for the forests and parks of Mediaeval Britain, which redefined the lower classes of society as poachers if they caught animals such as deer (Almond 2003). The phenomenon of poaching is bound up with notions of land ownership and hunting rights. As Barry Lewis (2009: 102) suggests for Britain: 'The ownership of places, in this instance land, that is favoured for hunting either for sport or for the purposes of accumulating wealth for the few, began with the emerging social elite of the Bronze Age and perhaps, in some areas, in the earlier Neolithic.' Similarly, MacKenzie (1988: 13) suggests that the shift from utilitarian hunting to sport hunting occurs at a number of points in history, when land rights are privileged over the right to hunt animals. Another purpose for hunting which reappears in a number of periods is that it acts as training for warfare, and at times becomes indistinguishable from it if hunting expeditions are also used to control territory (Allsen 2006: 209–32).

These provide useful analogies for Bronze Age Crete. Iconography provides an important source of evidence for hunting techniques, although this has often been overlooked. A wide range of animal depictions need to be disentangled from the notion that they are 'animal studies' in order to place them within the animal practice of hunting. If it is accepted that hunting was an important high-status pursuit, it becomes necessary to ask whether the same patterns of exclusion from hunting certain animals and areas of land are apparent in Bronze Age Crete as elsewhere. The decline of peak sanctuaries in the Neopalatial period allowed the Cretan landscape to be reorganised and reconceptualised in terms of non-domestic animals and plants such as wild goats and crocuses. It will be suggested that access to these became restricted at the same time as the Knossos palatial collective extended its reach over large areas of Crete, both the lowland farming areas and the upland hunting and grazing areas. This, in turn, helps to explain some of the specific features of an iconography which is traditionally explained in terms of the nature goddess.

PRESENCES AND ABSENCES

One of the barriers to recognising hunting on Crete has been an inadequate history of zooarchaeological publication. With new recovery techniques, particularly water sieving, this is beginning to change, revealing both utilitarian and elite hunting of non-domestic animals. Until recently birds have been infrequently reported in Cretan zooarchaeological reports, largely due to insufficient recovery of fragile remains (Mylona forthcoming). A number of genera were identified at Kommos, with seabird remains, shearwater and pelican, and one swift bone, found in the same contexts as dove and partridge, which are usually considered as game birds. This suggests that they were all being eaten (Reese 1995a: 204).[2] Partridge bones have also been identified at the more recently excavated sites of Mochlos Artisans' Quarter and Petras (Soles 2003: 18; Mylona 2010: 226). The large-scale hunting of birds, whether songbirds or waterfowl, is attested across the Mediterranean up to the present, and there is no reason to doubt that this occurred in Bronze Age Crete. Partridges and rock doves have long been among the most prized birds, but a wide variety of birds on north–south migration routes over Europe have been targeted by hunters.[3]

The buildings of the Mochlos Artisans' Quarter, on the shore opposite the island of Mochlos, were destroyed at the end of the Neopalatial period. In one of these buildings a tripod cooking pot was discovered with the remains of a meal that was never eaten: hare stew (Soles 2003: 68–9; Reese 2004: 118–19). Hare bones have been identified at a number of other sites in Bronze Age Crete but not in the same quantities as on Mochlos, except at Postpalatial Kastro (Snyder and Klippel 2000: 68).[4] There is no reason to doubt that hares were widely hunted and eaten in Bronze Age Crete, as they are today. A further piece of negative evidence strongly suggests that hares were commonly hunted by a wide section of society: their complete absence in Cretan Bronze Age iconography. Hare-hunting was not an animal practice that connected seal users with the palatial collective because it could not be controlled. This perhaps points to the small-scale hunting of hares with devices such as clubs and snares in the area around settlements rather than the elaborate hunting expeditions with nets and dogs and large numbers of people of Xenophon's time. Large-scale hunting expeditions in the Bronze Age would have focused on larger prey such as agrimia, feral cattle and pigs, and, following their introduction, deer.

Agrimia are difficult to distinguish from domestic goats in zooarchaeological assemblages since they are descended from domesticated goats (Bar-Gal et al. 2002; Horwitz and Bar-Gal 2006). They can be identified by their horns, and sometimes the robusticity and size of their bones. Their remains occur sporadically throughout the Neolithic and Bronze Age of Crete. At Phaistos there are a number of Final Neolithic to Protopalatial contexts with agrimi bones or horns mixed with other consumption debris (Wilkens 1996a: 243, 2003; Masala 2009; Militello 2012: 262; Caloi 2017: 46). Finds of agrimi skulls have been discussed as trophies in Chapter 4, including the Neopalatial site of Nopigeia-Drapanias in western Crete, where a partial agrimi skull with intact horn cores was found at the bottom of a pit containing pottery and other animal remains

(Hamilakis and Harris 2011). Horns have been recovered from a number of Postpalatial sites including Chania, Palaikastro and Vronda (Bosanquet and Dawkins 1923: 87; Klippel and Snyder 1991: 180; Wall-Crowther 2007: 195; Harris 2014b: 200, 226). The display of agrimi heads or skulls as trophies provided an indexical link with both the consumption event and the hunting of the animal.

This chapter will focus on the hunting of feral populations of domesticated animals on Crete during the Bronze Age. Whether animals are considered feral or domestic is often more a question of the ways they are managed by humans than morphological difference. Originally domestic populations of goats, pigs and cattle could have been managed by hunting if they were no longer regarded as being controlled by humans.[5] These feral animals have not usually been identified zooarchaeologically, except for horns from agrimia. It is possible that some cattle skulls with large horn cores found at Bronze Age sites including Knossos and Chania are also from feral populations, or perhaps from a different type of cattle introduced for ranching in the Palatial period (Keller 1909; Zeuner 1963: 232; Nobis 1990, 1993; Reese 2016b; Halstead and Isaakidou 2017: 115–16).[6] As noted in the previous chapter, domestication covers a spectrum of activities; cattle were introduced to Crete by humans but once free-range populations increased, cattle could have been either rounded up or hunted depending on the affordances of individual animals, or notions of ownership. Deer will be discussed in Chapter 7 with other exotic imports in the Bronze Age, but they were also hunted, perhaps initially in enclosed parks, and probably established feral populations in the Late Bronze Age. One way to explore the prevalence of hunting of feral animals on Crete is to turn from bones to seal depictions.

THE ART OF THE CHASE

The extent of hunting iconography in Bronze Age Crete has not been fully recognised. This is particularly the case for sealstones and rings because of traditions of study and interpretation which privilege style over content. Animals on seals are frequently depicted in association with spears, arrows and nets, and these have largely been accepted as showing the animal practice of hunting or fowling. Another group of depictions showing the same animals, but without obvious indications of the hunt, tends to be interpreted differently. If shown with human figures these are often regarded as religious images. Depictions of animals alone or in groups have been regarded as artistic exercises termed 'animal studies' (Betts and Younger 1982: 104–5). Unless a weapon is explicitly shown, the association with hunting has been neglected because many depictions lack the visual cues that modern observers associate with hunting. As Judith Weingarten (1989: 312) has observed, a Cretan seal depiction of an animal in flying gallop which has traditionally been seen as 'joyous exercise' is more likely to show 'desperate flight'. Hunting is one animal practice which is clearly shown in seal depictions from both Crete and the mainland in the Late Bronze Age. If those depictions which are regarded as 'animal studies' are included, it becomes the most commonly depicted animal practice on seals.

The following section presents an analysis of iconography, particularly on sealstones, which instead focuses on information such as pose and other contextual indicators which relate to hunting. But first it is necessary to examine how the interpretation of these images has developed. Arthur Evans rapidly published the seals he collected on his first visit to Crete in 1894. His preferred interpretation for animals on three-sided prisms was that they reflected the 'pursuits of the possessors', including 'the game that they hunted' (Evans 1894: 372). His approach to the amygdaloid and lentoid seals typical of the Late Bronze Age was different; he initially regarded them as amulets or talismans which were not primarily used as sealstones (Evans 1894: 272). Although he was to change his mind about these sealstones when he discovered their impressions preserved in clay sealing deposits at Knossos, this explanation has never entirely gone away. In particular one of the main stylistic groups of seals which emerged at the beginning of the Neopalatial period is labelled 'Talismanic'. They were described by Evans as 'a series of intaglios of rough execution with stereotyped designs of a rustic class' (Evans 1921b: 672–3). These designs include marine animals, particularly fish and cuttlefish, but also objects including jugs and amphorae, and other animals including birds and goats (Figure 5.1a). They were carved on hard stones, frequently brightly coloured orange carnelian or green jasper, which Evans found in use as charms thousands of years after their manufacture (Evans 1894: 276). Many villagers refused to part with these stones because of their efficacy, but did allow Evans to make impressions of the designs instead (Evans 1894: 276). Other travellers noted the same phenomenon; Edith Eccles (1943: 44) noted that: 'the best milk-stones have white markings, blood-stayers red'. They are comparatively common, with over 900 examples known (Krzyszkowska 2005: 133–7). Artemis Onassoglou (1985) gave the Talismanic seals the fullest treatment, constructing typologies to define and identify each particular design rather than asking what they meant, but the label 'Talismanic' remained. This has continued to imply that these seals had a symbolic meaning of some kind.

Evans's first-hand observations perhaps provided a type of conceptual support for the idea that sealstones could have been used as charms or talismans. Victor Kenna (1969) formalised the label 'Talismanic' to describe this particular style, arguing that the design itself was seen as a source of power. He also suggested that seals with naturalistic designs did not function as talismans, and were instead used primarily for sealing (Kenna 1960: 66). Talismanic seals,

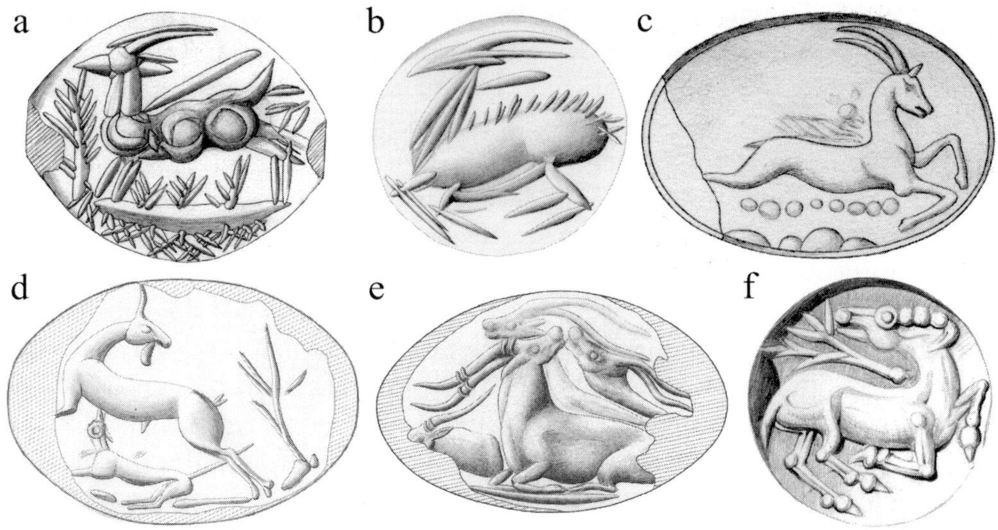

5.1. Drawings of seals showing goats. (a) Talismanic-style goat with spear, Mochlos, MMIII–LMI (CMS II 3 258); (b) Cut-style goat with spear, LMI–II (CMS II 3 342); (c) goat and dog?, Phaistos sealing, MMII (CMS II 5 259); (d) goat and dog, Knossos Hieroglyphic Deposit sealing, MMIII–LMI (CMS II 8 353); (e) four goats, Knossos Temple Repositories sealing, LMI (CMS II 8 492); (f) Spectacle Eye-style goat, Knossos, LMIIIA (CMS II 4 122). © CMS Heidelberg

meanwhile, he saw as unsuitable for making seal impressions; although this idea has been rejected (Betts 1974: 311), it is true that very few sealings preserve impressions of Talismanic seals, suggesting that they were not widely used in administration. One of the questions this raises is whether designs in the Talismanic style had a different significance from largely contemporary naturalistic seals, which can loosely be defined by their smooth contours either from disguising drill marks or carving in soft local stone. Authors who have questioned the separation of the two styles have tended to argue that all seals potentially had a talismanic function (Betts 1974). Edmund Bloedow (1992) has rehabilitated Evans's argument that: 'the rude figure of a running agrimi or wild goat with a shaft in his side [...] may well have brought good sport to the huntsman' (Evans 1935: 541). The problem with this argument is that it arbitrarily isolates some sealstones with depictions of particular animals and attributes a magical function to them while presumably not extending the same argument to all designs. But it is useful to try to see beyond stylistic differences and explain all depictions of goats in the same way. Bloedow (2003) argues that depictions of goats standing with spears over their backs should inform the reading of those without. He suggests the same for running goats, some of which have a spear in their back, so that all standing or running goats are explained as abbreviated hunting scenes. 'Particularly important in determining the significance of the wild goat is the human-agrimi relationship. The most striking factor here is that the agrimi was hunted' (Bloedow 2003: 6).

The Talismanic style is now largely seen as a technique in which the drill marks of the engraver are left undisguised, which, as such, would have been relatively quick to manufacture (Pini 1992; Boardman 2001: 43–4; Murock Hussein 2018). The 'Cut' style can be seen as a development of the Talismanic style (Figure 5.1b). Both use undisguised cut marks on hard-stone seals but the circular drill marks of the Talismanic style are replaced in the Cut style by smoother animal bodies and more frequent straight cuts. In the case of birds in particular these styles are often difficult to distinguish (Onassoglou 1985: 197; Pini 2000a: 212, 2010: 327–8). Jukka-Pekka Ruuskanen (1992: 43–4) argues for a 'naturalistic-talismanic group' of bird seals showing the overlap between the two contemporary styles. The Cut style is useful for dating because it appears to start later in the Neopalatial period; it is attested in a small number of Late Minoan IB sealings, and sealstones have been found in Late Minoan IB graves, but production appears to continue into Late Minoan II (Krzyszkowska 2005: 147). Drill marks also characterise the 'Spectacle Eye' style of the Final Palace period (Figure 5.1f) (Younger 2000). These styles reveal the affordances of hard stones through the crisp marks of the drill at a time when soft-stone seals were still being carved with hand tools. The combination of skilled technique and colourful hard stone produced seals which were instantly recognisable as coming from specialised workshops. Some of the difficulties of definition illustrate that there was a continuum of Neopalatial seal designs from naturalistic to schematic, with the naturalistic examples generally taking longer to produce and involving a higher degree of skill. There is also a difference in usage between the detailed 'naturalistic' seals, which are most frequently used in sealing deposits, and the more schematic Talismanic- and Cut-style sealstones, which are abundant in museum collections but are rarely known from sealing deposits.

Naturalistic seals are generally associated with the Neopalatial period, although similar depictions occur both earlier and later. Although scholars have differed both on the definition of naturalism in Cretan seals and the date of its appearance, there is a shift at the end of the Protopalatial period towards more detailed depictions of animals, often in a landscape (or underwater) setting. A sealing deposit from Room 25 of the palace at Phaistos, known as the 'archivio di cretule', shows a group of seals in action at the end of the Protopalatial period (Weingarten 1994; Krzyszkowska 2005: 104–8; Relaki 2012). The deposit was preserved by the destruction of the palace, which fired the sealings. They had been kept as a record of access to a particular room or containers within that room, showing that seals of different styles were being used simultaneously. With the exception of two very similar seals showing a bull in landscape setting (see Figure 3.9e), the most frequently used seals were more abstract designs, many of which appear to relate to textile patterns, and which go back to the Prepalatial period (Relaki 2012: 317–20). The deposit appears to mark a time of transition from traditional seal designs used by corporate groups to radical experimentation with different shapes of seals and ways of depicting animals (Blakolmer 2020a). These latter seals have been identified with the birth of 'naturalism'.

> The real break between this earlier formal tradition of pictorialised pattern and symbolism, and the new tradition of naturalism, seems to come suddenly with the latest of the seals which impressed the Phaistos sealings at the close of MMII; the motifs from nature used in both traditions are largely the same but the whole ethos has changed. (Betts 1989: 2)

This is exemplified by sealings showing a goat hunted with dogs in rocky landscapes and, for the first time, running in 'flying gallop' (Figure 5.1c).

Another group of sealings from Knossos, the so-called Hieroglyphic Deposit, displays a similar mix of Cretan Hieroglyphic signs and more naturalistic sealings such as that showing a goat in a rocky environment. Another sealing showing what is now recognised as a goat was described by Evans (1921b: 274) as 'a dog seizing a hind, executed in a naturalistic style, with a tree behind' (Figure 5.1d). Unlike the archivio di cretule, it is not clear if this is a single deposit, but the seals are likely to have been accidentally fired by destructions early in the Neopalatial period (Krzyszkowska 2005: 112–16). This is the time in which Evans located the peak of naturalism, which was closely allied with the depiction of animals and landscape: 'By the close of M.M. II, as we learn from the sealings of the Hieroglyphic Deposit at Knossos, the engraved designs on these lentoids were already attaining a considerable degree of naturalistic perfection' (Evans 1921b: 671). Some of these seal impressions were made by sealstones of hard stone carved with a drill, allowing the modelling of the animals' bodies, which is an important element of this naturalistic style of depiction.

Another deposit containing sealings which can be dated to the early Neopalatial period at Knossos is the Temple Repositories. The sealings and other material in the deposit were preserved in a destruction event which damaged Knossos at this time, and a group of material was collected together and buried in two cists under the floor in the west wing of the palace. Again, Evans (1921b: 695–7) described these sealings as 'subjects of natural inspiration', including a goat and kid, marine animals and trees bending in the breeze.[7] A sealing showing four wild goats in various poses is remarkable for its complex composition, capturing the movement of the animals' heads and providing a sense of depth with their overlapping bodies (Figure 5.1e). Here the drill cut adds the detail of the animals' serrated horns. These seal impressions from early Neopalatial Knossos show the way in which the new compositions of the Phaistos seals were developed in a context in which the drill was already being used to create detailed animal bodies on seals.

A large group of naturalistic seal impressions comes from the sites which were, unlike Knossos, destroyed at the end of the Neopalatial period, some of them made using metal rings (Figure 5.2). Again naturalism is an elastic term which includes both the most carefully worked and detailed depictions on hard stones and metal rings as well as the less detailed soft-stone seals. In terms of administrative function, it has been demonstrated that naturalistic seals functioned no differently from those in other styles. In her analysis of the Ayia Triada sealings, Judith Weingarten (1988: 105–6) describes one sealing, showing two goats rutting (Figure 5.2a), as 'a textbook image of the flowering of Minoan naturalism' but dissociates ideas of quality or competence from naturalism, suggesting that seals did not have to be naturalistic to be effective for administrative purposes. Another group of naturalistic seals was preserved in the final destruction of the palace at Knossos; it can be difficult to date some of these seals but some clearly were produced in the Neopalatial period. It was these seal impressions from the final levels at Knossos which Evans focused on when discussing seal iconography in *The Palace of Minos at Knossos*. Here he does accept that some were depictions of hunting and fowling but again it was Victor Kenna who distilled Evans's ideas and lost some of these subtleties. His explanation for these designs was that they 'came from a preference for studies of animal life and nature, and the desire to produce these motifs in a natural likeness' (Kenna 1960: 66). He saw the naturalistic 'bird studies' of Crete as a reaction against the stiff formalism and religious significance of Egyptian bird depictions: 'This is nature in art without the intervention of extraneous values, unknown in the glyptic of the Ancient Near East, in Anatolia, or in Egypt'

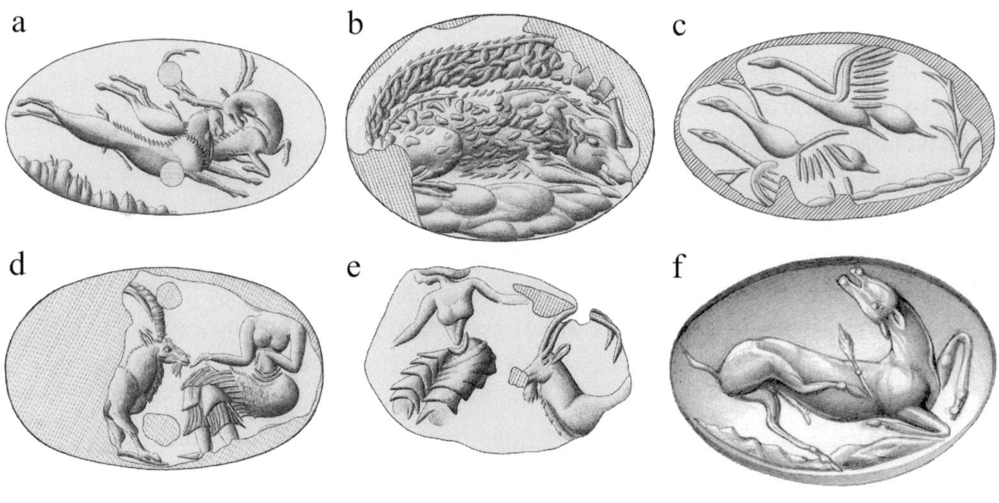

5.2. Drawings of Neopalatial ring impressions showing animals. Ayia Triada sealings, LMI: (a) two agrimia rutting (CMS II 6 70); (b) two boar (CMS II 6 72); (c) four waterbirds flying (CMS II 6 124); (d) woman feeding agrimi (CMS II 6 30); (e) woman and agrimi (CMS II 6 31); (f) calf with arrow in side, Chania sealing, LMI (CMS VS 1A 154). © CMS Heidelberg

(Kenna 1968: 38). Rather than birds being symbols, this was art for art's sake. Not every scholar has adopted this modernist conception of seal depictions uncritically. As Lyvia Morgan (1995b: 171) has suggested, 'These are not studies of creatures in the wild, observed in the dispassionate manner of an ancient naturalist'. Nevertheless, this implicit assumption that animal depictions are artistic exercises lives on in the use of the term 'animal studies' (Betts and Younger 1982: 104–5; Younger and Rehak 2010b: 159). The continued use of this term also demonstrates that these seals carry a depiction of an animal whose effect on the viewer is to suggest first-hand observation. Detailed naturalistic designs, however, are not necessarily passive records of observation but instead could connect the seal user with these animals and associated practices by revealing a detailed knowledge of them.

The idea that the depictions of animals on seals are formal exercises is at the heart of John Younger's (1988, 1993) exhaustive catalogues of seal iconography. He identifies fifty-four main animal pose types which he suggests encapsulate all the variation in the depiction of animals in Aegean glyptic. He regards variation as essentially stylistic, that is, reducible to the conventions of individual artists and workshops.

> These seals with animals make up the bulk of the iconography; they convey almost no information about the Minoan-Mycenaean society and only a little insight into the artist's view of the natural world, limited as it seems to have been to a few animals (most commonly the lion, the bull, and the agrimi) in 54 (most commonly, 36) conventional poses.
>
> (Younger 1988: xi)

Instead it can be argued that details such as pose are part of a shift towards narrative in Late Bronze Age seals (Crowley 1989a; Weingarten 1989). In particular, many of these poses could indicate hunted animals.[8] This is not a new observation, but there is not a widespread acknowledgement that most animals on sealstones, and not just goats, are involved in the animal practice of hunting. Evans (1935: 523) suggested that animals with heads over their shoulders showed the animal being chased. Bloedow (2003: 23) takes this on, seeing animals with front legs bent as a shorthand for wounded animals collapsing. Eleanor Loughlin (2004b: 184) too suggests that the seal depiction of a calf kicking at a projectile in its side shows the 'characteristic cattle response to pain' (Figure 5.2f). Even seemingly conventional details, such as the direction in which the animal faces, could be significant; Bloedow (2003: 24) observes that the majority of standing goats face in the same direction. Both pose and direction can be analysed statistically to determine whether there is any relationship, for instance, between animals shown with spears, pose and the direction they face. The demonstration of such relationships below will be used to question the idea that there is an arbitrary relationship between the seal user and the seal depiction, and to show instead that there is a connection between the seal user and a particular animal practice. The close observation and detailed depiction of animals encapsulated in the term 'animal study' can also be seen as the product of providing information about animal practices such as hunting rather than an artistic interest in the natural world.

HUNTING BY NUMBERS

Discussion of Bronze Age Cretan material culture tends to focus on a limited number of finely crafted objects. As Younger's approach demonstrates, it is useful to consider sealstones as a group in order to look for patterning, whether this is seen in terms of maker, date or animal practices as here. This has become possible with the authoritative publication of sealstones in the *Corpus der minosichen und mykenischen Siegel* (CMS).[9] The following analysis is based on 3,117 seals and sealings published in CMS which depict animals and were found on Crete (Figures 5.3–5.4). Most were also made on Crete, but given the difficulties of identifying off-island manufacture the main selection criterion is that seals were used on Crete during the Bronze Age.[10] Unstratified seals are assumed to have come from Crete either because they are in Cretan museum collections or because they were acquired on Crete by particular individuals and subsequently found their way into overseas museum collections.[11] As such the sample is biased by the history of collection and publication but is assumed to be broadly representative of seals circulating on Crete in the Bronze Age.

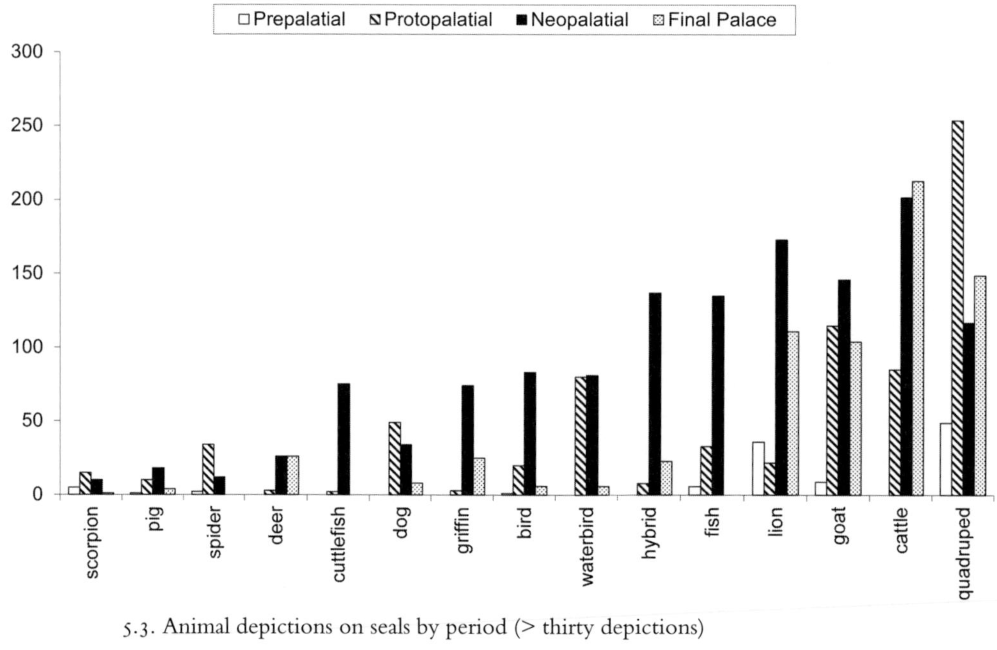

5.3. Animal depictions on seals by period (> thirty depictions)

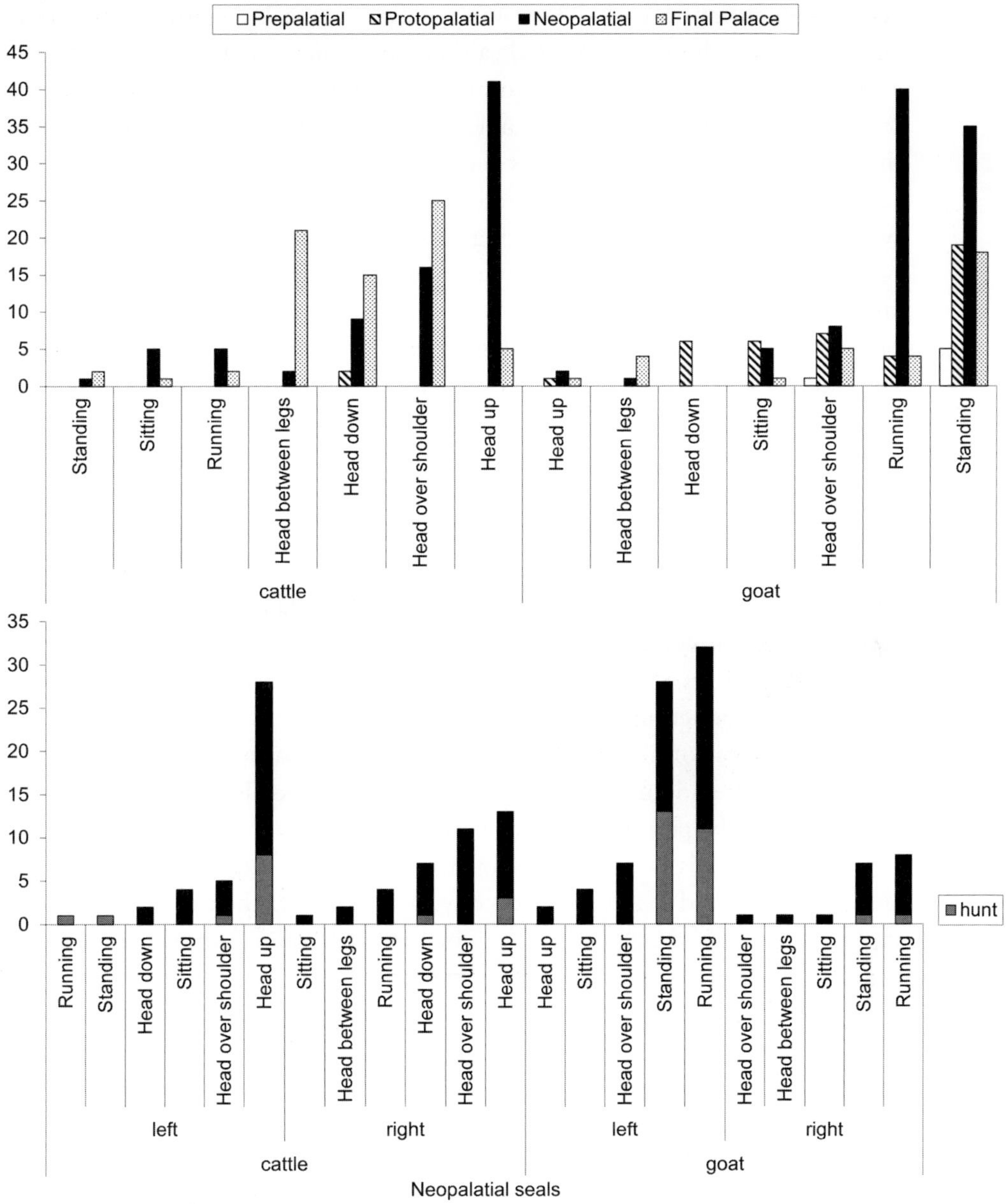

5.4. Analysis of single cattle and goat depictions on seals. Above: by period/pose; below: by direction/pose for Neopalatial seals only (hunt indicated by spear or arrow)

Seals are dated by period although this can be problematic since to a certain extent stylistic development of seals is independent of the pottery chronology and corresponding palatial periods (Betts and Younger 1982: 120; Krzyszkowska 2005: 10–11).

Analysing animal depictions on seals quantitatively provides a different perspective on judgements based on changing style.[12] Figure 5.3 shows the types of animals which appear with the greatest regularity on seals over the course of the Bronze Age, with lions, goats and cattle most frequent. 'Quadruped' refers to animals which are difficult now to identify more closely, usually because the seal or sealing is damaged or incomplete, but these are most likely to be one of these three animals. Inevitably this graph compresses a wide variety of different seal designs, particularly those in which the animal is shown alone, and those in which the animal is shown interacting with other non-human animals, humans or both. Such designs with a narrative content, such as a lion attacking another animal, or a human leaping over a bull, are largely a feature of Neopalatial and Final Palace seals. Late Bronze Age seals tend to be more finely carved and so more detailed, both in the carving of the animal body and the use of additional elements in order to provide context. The introduction of rocks and plants marks a significant change in the way animals are depicted, linking them with a particular landscape (Krzyszkowska 2010). The depiction of spears over animals' backs makes depictions of hunting more explicit. Some of the animal depictions on Protopalatial three-sided prisms appear to be linked to hunting, fowling and fishing but they very rarely include explicit indications of the animal practice involved. Statistical analysis is one way to approach the significance of depictions whose individual significance is now unknown; it provides a context for abduction of meaning.

The depiction of hunting can be explored by isolating different variables: pose, direction and animal type. Figure 5.4 includes all single depictions of goats and cattle by period, grouped by the most frequent pose. The lower graph shows only the seals dated to the Neopalatial period, including the direction the animal is facing and with those shown with a spear over their back or an arrow in their side marked as 'hunt'. As can be seen there are a number of different poses, as Younger noted, but each type of animal appears to have a particular pose most frequently associated with depictions of hunting. For cattle this is with the head raised; goats are shown standing or running with their head forward. By comparison, hunted lions (which will be discussed in Chapter 7) are most frequently shown with their head turned over their shoulder. Some of this patterning is a result of the more simple repeated designs on Talismanic seals, compared with the more elaborate one-off naturalistic designs. Although a small proportion of animals are depicted explicitly in hunting scenes, the same poses are found most frequently in seals without indications of hunting, suggesting that these poses were sufficient to denote a hunted animal. The choice of pose could relate to animal behaviour in the case of cattle; the 'head up' pose, for example, shows cattle bellowing (Loughlin 2000, 2004b: 184). Although modern observers have tended to rely on weapons as a cue for hunting, the pose was sufficient to indicate this animal practice.

A further variable noted above is direction. Figure 5.4 also shows that hunted animals are predominantly shown facing left. Seal designs are conventionally published as drawings of the impression, meaning that sealings and seals are treated in the same way. As Bloedow (1992) suggested, goats generally face left, but cattle show more variation; the 'head up' pose is an exception. As with the association between hunting and pose there are exceptions; these could be temporal or related to different workshops, or even the gender or status of the user. But there is a statistically significant association for each animal type between direction and the particular poses most frequently associated with hunting on Neopalatial seals. This can be compared with an earlier convention for three-sided prisms on which goat heads face right (Anastadiadou 2011: 368). Interestingly this is not the case for goat heads in Cretan Hieroglyphic inscriptions on seals which can face in either direction, as do other signs. Animal ideograms in Linear A face right, whereas these non-domestic animals on seals generally face left. This could explain why the convention of direction for goats is clearest in the Talismanic seals, which started to be produced early in the Neopalatial period, when Linear A was taken up across the island (Schoep 1999: 219). Linear A used conventionalised animal ideograms to document domestic animal practices, leaving seal depictions to describe a variety of non-domestic animal practices in greater detail. The shift in northern Crete from prisms to new seal shapes, allied with the existing technology of hard-stone carving, resulted in new types of depiction. This separation can be seen as opening the way to a different kind of depiction, which locates the animal in time and space. This 'naturalistic' style of depiction can be seen as one which gives enough information about an animal to suggest a particular animal practice with which the seal user was, or wanted to be, associated. The depiction of animal practices on seals, particularly hunting, continued into the Final Palace period.

Rather than being 'animal studies', many seal depictions of animals can be linked with hunting. This phenomenon is particularly apparent in the Neopalatial period, but is not restricted to the naturalistic style. Neopalatial depictions are more detailed than those on three-sided prisms, since they frequently include other elements, such as spears or plants. But impressions were predominantly seen in bureaucratic contexts and were no bigger than a couple of centimetres in diameter. This could explain the use of specific poses for different types of hunted animals, since a spear or the details of horn shape might be difficult to discern. Some materials used for seals obscure the design when seen on the face, so the act of sealing provides the most reliable way to read it. Even if the depiction was only seen by a few people, sealing was an act which constituted the palatial collective by establishing a relationship between the seal user and the palace. It was in these same contexts that writing was also employed; the combination of seals, clay and writing maintained the relationships between the humans and domestic animals of the palatial collective. What remains to be explained is why seals used in these transactions so often depicted hunting.

HUNTING AND THE PALATIAL COLLECTIVE

Although a large number of seals show hunting explicitly, whether or not the spear is included in the scene, some of the most naturalistic depictions of goats, cattle, boar and birds show them in different contexts. This is particularly apparent in the sealing deposits preserved in the island-wide destructions at the end of the Neopalatial period, from Zakros on the east coast of Crete and Ayia Triada by the south coast. Packet sealings at both Zakros and Ayia Triada carry impressions of the same ring showing bull-leaping, suggesting that they were both in direct contact with Knossos, and so part of the Knossos palatial collective. Other metal rings preserved in the Ayia Triada sealing deposit show finely detailed depictions of agrimia and boar which appear to relate to animal fertility. One seal impression showing two male goats rutting (Figure 5.2a) closely parallels the mating scene on the 'Burgon ring' (see Figure 1.1). Seal impressions involving women and goats appear at all three sites and relate to other objects: a relief stone vessel known as the 'Sanctuary rhyton' from the palace at Zakros and the so-called 'Park fresco' from the villa at Ayia Triada, both found in the same destruction levels as preserved the sealings. Another similar set of connections occurs in the Temple Repositories at Knossos, a deposit from earlier in the Neopalatial period. These interlinked depictions help show how the Knossos palatial collective incorporated hunted animals and the landscape they inhabited. In turn this demonstrates how the Knossos palatial collective extended its reach beyond the lowlands of Crete, by absorbing the animal practices, hunting and fowling, that defined upland and marginal areas.

The Sanctuary rhyton from Zakros was found in fragments in the palace and is a chlorite vessel bearing traces of the gold foil which once covered its relief surface (Platon 1971: 161–9). Most of the goats are clustered around a building usually identified as being part of a peak sanctuary. Two birds also occupy the building, but there are no humans shown. Another significant feature is the clump of crocuses growing to the side. The Sanctuary rhyton shapes viewing in a particular way. The goats above the doorway overlap slightly, and the left-hand one's head is frontal, looking back at the viewer, rendering its characteristic curved horns an ellipse. Although there is ambiguity in the relative depth of parts of the building, the scene gives the impression of a single frontal viewpoint (Shaw 1978a). The small scale suggests that only a few people would have been able to view the depiction at any one time; these relief scenes could primarily have been meant for consumption by a privileged few (Logue 2004: 172). In this way it reverses the traditional social relations of the peak sanctuary; a small number of people gained a privileged perception of a peak sanctuary from a single point of view in contrast to the mass participation and multiple objects of communal rituals at the peak sanctuaries of the Protopalatial period. This in turn relates to the increasing exclusivity of the use of peak sanctuaries in the Neopalatial period (Peatfield 1987), when the Sanctuary rhyton was produced. The apparent narrativity of the scene has invited a number of readings, including that of the nearby peak sanctuary of Traostalos or a palatial zoo enclosure (Platon 1971: 167; Bloedow 1990). Whether or not it depicts a particular place, it serves to connect a number of entities: goats, birds, crocuses, a rocky landscape and religious paraphernalia including horns of consecration and a portable altar. This object reveals a much more extensive set of connections than the abbreviated sealings from the same site.

Less well preserved than the Sanctuary rhyton is the burnt and fragmentary Park fresco from Ayia Triada and so scholars rely on modern reconstructions of the scene which covered three walls of Room 14 (Figure 5.5). Some elements are shared with the Sanctuary rhyton: goats, birds, crocuses and a rocky landscape. But here the birds are being stalked by at least one cat and there are also two humans present, women in elaborate dresses. These are usually regarded as religious figures, sometimes as one goddess and one priestess/adorant. The central woman is described in the final publication of the frescoes as the goddess of the world of nature (Militello 1998: 282). 'If we have rightly understood the context of Room 14 as an epiphany scene, the cat fresco would have exalted with its luxuriant landscape the power of the nature goddess whose epiphany was depicted in the adjacent east wall' (Militello 2000: 85). Morgan (2005: 28) echoes this description: 'the world of nature within the sphere of the divine'. Two sealings from Ayia Triada show women in similar poses accompanied by agrimia, calling the reconstructions on which these readings are based into question (Figure 5.2d–e). If these sealings are excerpts from the fresco they show a much more active relationship between women and goats rather than the animals being regarded as passive parts of the goddess's domain. The crocuses in the fresco are also significant because they were an important commodity in Bronze Age Crete, used for dyeing woollen cloth (Day 2011a, 2011b). The fresco revealed a relationship with the territory

5.5. The Park fresco from Ayia Triada Room 14, LMI. Detail, fresco panels, reconstruction drawing by Mark Cameron (Mark Cameron Personal Papers, Archives, British School at Athens, CAM 90). Reproduced with the Permission of the British School at Athens

beyond this room at Ayia Triada, including both the goats and crocuses found in the uplands. The act of feeding goats, shown on the seals, and perhaps this fresco, demonstrated the women's protection of or control over these animals. Although access to the Park fresco could have been restricted, it potentially addressed more viewers than the Sanctuary rhyton, surrounding them in a rocky landscape inhabited by women, cats, goats and birds. Frescoes such as this extended the palatial collective into the uplands of Crete (Tully and Crooks 2019).

Another deposit which combines goats, crocuses and elaborately dressed women is the Temple Repositories, which was introduced above because of its naturalistic sealings. The sealing showing four goats (Figure 5.1e) is closely paralleled in the poses of the four seated goats on the Sanctuary rhyton. Another sealing showed a goat and kid, a scene which also occurs in the Temple Repositories on a larger scale in the form of a faience plaque of an agrimi and kid standing on rocky ground. A similar faience plaque shows a cow and calf (Panagiotaki 1999: 81–7). Other objects made of faience include the famous 'Snake goddesses' with their dyed textile dresses and other dresses showing crocuses. These will be discussed more fully in Chapter 6, but the animal plaques and the bare breasts of the female figures all have a potential connection with fertility. It is not necessary to interpret these figures as goddesses in order to trace the relationship between them, the flowering crocuses and suckling animals. This deposit, and the medium of faience in particular, brought these various things together at the palace of Knossos. Earlier than the Park fresco and Sanctuary rhyton, this deposit shows that this set of connections centred on the palatial collective of Knossos.

Birds occur on both the Park fresco, where they are hunted by a cat, and the Sanctuary rhyton, where they perch on the building. Birds and waterbirds are shown frequently in both the Ayia Triada and Zakros sealings, as well as bird-woman hybrids (Figures 5.2c and 5.6a–d). There is one sealing of a waterbird in the Temple Repositories, and birds, perhaps partridges, were painted on Cycladic vessels which were also contained in the deposit.[13] The relationship between women and birds can be related to the animal practice of fowling; occasionally women are shown holding birds too (Figure 5.6e). One sealing from Ayia Triada shows a bird caught in a net (Figure 5.6b). The same sealing was described by Evans (1935: xvii–xviii, fig. 424) as epitomising the 'naturalistic spirit' of Middle Minoan III, although its restudy by the CMS team revealed the existence of a net. This too was impressed by a metal ring, suggesting it was a palatial animal practice. Although there are no frescoes from Crete showing this scene, a contemporary fresco from the island of Thera, which shares many aspects of Cretan iconography, does show an elaborately dressed woman holding birds in a net (Papageorgiou 2014). It is found in Xeste 3, a building which also contains frescoes showing crocus gatherers and a figure

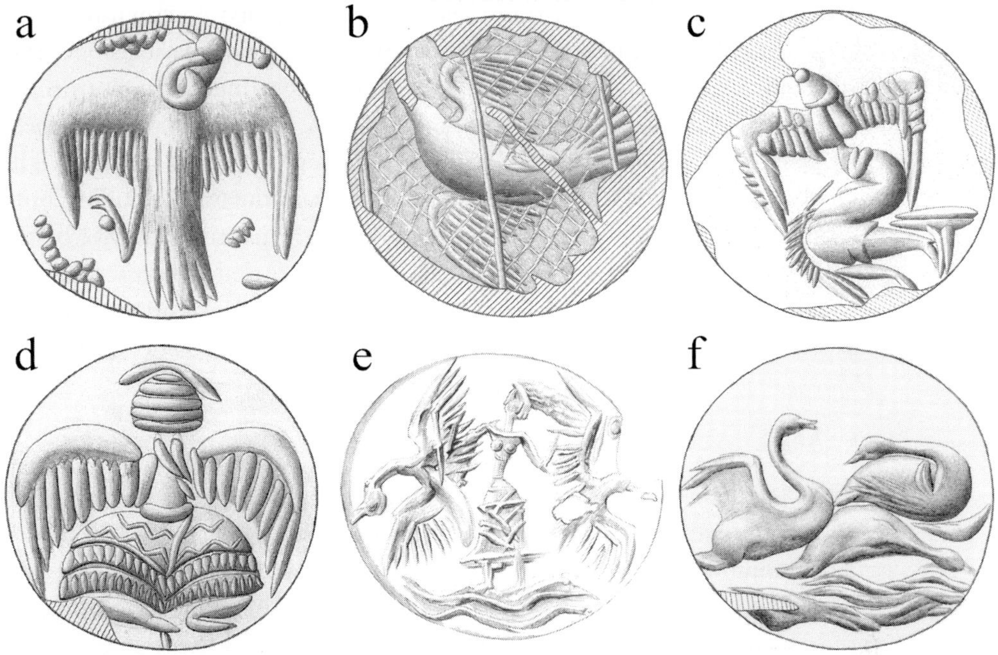

5.6. Drawings of seals showing birds, women and bird-women hybrids. Ayia Triada sealings, LMI: (a) flying bird (CMS II 6 111; (b) bird caught in net (CMS II 6 123); (c) bird-woman hybrid (CMS II 6 106); (d) bird-woman hybrid with ?boar's-tusk helmet, Zakros sealing, LMI (CMS II 7 129a); (e) woman holding two waterbirds, LMI–II (CMS VII 134); (f) three waterbirds, Mirabello, LMI (CMS VI 458).
© CMS Heidelberg

generally identified as a goddess. The context leads Irini Papageorgiou to describe them as live offerings to the Great Nature Goddess (also known as Potnia). Echoing the descriptions of the Ayia Triada fresco she suggests that in the fresco programme of this building 'the Potnia is represented as sovereign of the cosmos, the earthy, watery and airy worlds' (Papageorgiou 2014: 125). As Anne Chapin (2004) points out, it is important to ask what the function of such frescoes was; as well as acting as backdrops to ritual they were important statements of political relations. Instead of showing hunting, these depictions appear to show the metaphorical protection or ownership of hunted animals and the landscape where they were found.

These finely crafted forms of material culture helped to reveal the ritual basis for control over the animal practice of hunting and fowling. Activities such as tree-shaking and rock-hugging, shown on gold rings, were also part of this ritualisation of the landscape, although they could have had earlier roots in an animistic ontology in which rocks and trees were recognised as powerful entities (Crooks et al. 2016; Tully 2018). The depictions on metal rings are significant because they appear to be restricted to the highest levels of the palatial administration. As noted in Chapter 3, bull-leaping is shown on rings, which helped to define the ownership of cattle by the Knossos palatial collective, but hunting is rarely shown. Instead the Ayia Triada ring impressions show agrimia and boar on their own or with women. The naturalism of these depictions helps demonstrate a detailed knowledge of these animals. The Sanctuary rhyton and Park fresco show how the mountains, including both peak sanctuaries and wild goats, had become part of the palatial collective by the end of the Neopalatial period. The palatial collective centred at Knossos claimed to extend across the mountains by creating a network of associations between ritual practice, finely crafted objects and the animals and plants of the uplands.

THE LANDSCAPE OF HUNTING

The link between identity, status and seals showing hunting and fighting has been made for mainland Greece (Laffineur 1992; Cultraro 2004). It has also been argued that hunting was a means of territorial control and a means of training for war in the Argolid (Morris 1990: 150), and 'an important ideological resource for aspiring as well as established Mycenaean elites' (Hamilakis 2003: 244). Yannis Hamilakis also highlights the role of hunting in controlling land. The reason the same interpretation is rarely made for Crete has much to do with the traditional contrast between 'peaceful' Minoans and 'warlike' Mycenaeans. By tracing the development of the animal practice of hunting on Crete, it becomes clear that it fulfilled many of the same functions as on the mainland.

At the level of the local collective, hunting was an animal practice which had the capacity to differentiate people. The affordances of agile mountain-climbing agrimia mean that hunting them would have been seen as an indication of a hunter's skill and prowess from the Neolithic period onwards. There is zooarchaeological evidence for hunting agrimia, as noted above, from the Final Neolithic period at Phaistos. There is also a clay figure, probably an agrimi, from a Prepalatial tomb at Porti with its head turned; this could be an early example of a depiction using pose to indicate a hunted animal (Xanthoudides 1924: 62–3) (Figure 5.7). If so, it was an animal thing which brought the animal practice of hunting into the tomb, perhaps as a marker of identity; not everyone was buried with an object like this. Prepalatial seals depicting agrimia can similarly be seen as a means for particular hunters to identify themselves through an association with this animal practice, although hunting is not explicitly shown. Before the palaces and peak sanctuaries it is unlikely that the mountains were closely integrated with local collectives, but rather they were places inhabited by birds and wild goats. The hunting of these animals distinguished anyone who was prepared to travel out of the lowlands and into the mountains.

The way in which hunting was absorbed into the palatial collective can be traced on seals. Agrimia and boar appear on Protopalatial three-sided prisms; it

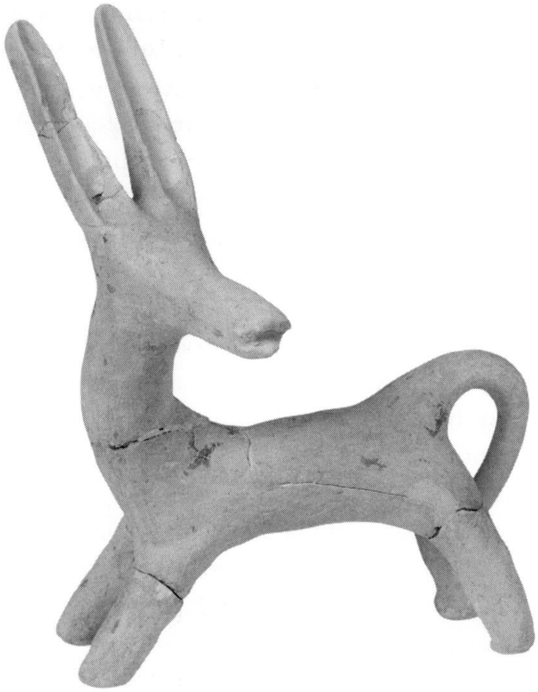

5.7. Clay figure in the shape of an agrimi from Porti, EMIII–MMI. Length: 11 cm (AMH Π5114). © Heraklion Archaeological Museum – Hellenic Ministry of Culture and Sports – 'Hellenic Organization of Cultural Resources Development (HOCRED)'

was suggested in Chapter 3 that these seals were part of an inclusive palatial collective based at Mallia and other centres in northern Crete, which incorporated the different activities of its inhabitants. Hunting agrimia would have been one such activity, and it is likely that these animals were contributions to commensal feasting. The deposition of model agrimi horns alongside figurines of domestic animals at the peak sanctuary of Petsophas also appear to demonstrate that hunting was still a relatively inclusive animal practice in this period because all of these clay offerings were made in the same simple way. Different types of hunting depictions on hard-stone seals start to appear at the end of the Protopalatial period. The Phaistos sealings and Knossos Hieroglyphic Deposit contain detailed depictions of hunting goats with dogs (see Figure 5.1c–d). These depictions suggest that using dogs as hunting companions did help to define a distinct palatial animal practice of hunting agrimia. Much hunting, particularly by individual hunters, would still have been carried on at the level of the domestic collective, only loosely connected with the palatial collective. The exception, discussed in Chapter 3, is hunting cattle, with speared cattle depicted from the end of the Protopalatial. This animal practice helped define the palatial collective of Knossos and was tied up with territorial expansion.

As more detailed hunting depictions, both of cattle and goats, appeared on seals there is evidence that the uplands of Crete were increasingly being absorbed into the palatial collective. One sign is the appearance of so-called guard houses, often in high places commanding views of the local area. At Choiromandres near Zakros one of these megalithic structures was built at the end of the Protopalatial period at the same time as a sanctuary site went out of use (Tzedakis et al. 1990; Alušík 2007: 27–9). The guard house too went out of use in the Neopalatial period and was succeeded by a number of similar structures, some of which were probably used seasonally (Vokotopoulos 2015). These guard houses could have been used as both garrisons for soldiers and hunting lodges. In many societies hunting has been an animal practice participated in by soldiers as a means both to procure food and practise their skills with weaponry. Large-scale hunting expeditions, perhaps driving animals into nets, would have acted as military training exercises as well as demonstrations of control over territory. Hunting redefined the upland landscape in the Neopalatial period by establishing the hills and mountains primarily as a hunting ground which was part of the palatial collective rather than the gathering place of local collectives for cult.

Hunting is not shown explicitly on Neopalatial Cretan frescoes or stone relief vases, with the possible exception of two fragments of relief stone vessels: one from Palaikastro showing a charging boar and another from Knossos showing a human wrestling an agrimi (Evans 1921b: 676; Warren 1969: 177). Nevertheless it could provide the context for one of the most famous relief vessels from Crete, the 'Harvester vase'. Found at Ayia Triada, it shows a procession of young men carrying long forked implements, apparently marching in time to a sistrum carried by one of the participants (Figure 5.8a). Despite there being no other scene related to cereal harvesting from Bronze Age Crete, even after 100 years of excavations, it is widely accepted as a depiction of some kind of agricultural festival, with the implements usually interpreted as threshing forks (Logue 2004: 165–6; Blakolmer 2007).[14] The hafted blade on the shaft of each implement has caused trouble for this interpretation, with others suggesting that they are hoes or tools for pruning olive trees (Forsdyke 1954; Koehl 2006: 90–1). Also difficult to explain from this perspective is that the participants seem to be wearing padded hats or helmets. As the first publication of the vase suggested, this could equally be a military marching scene with the sistrum player marking time (Savignoni 1903: 83–8). Unlike harvesting, there is an iconography of warfare and hunting in Bronze Age Crete (Molloy 2012: 98–112), as well as processions involving spears and other weapons, including a contemporary miniature fresco fragment showing men waving spears from Knossos (Evans 1900: 48) (Figure 5.8b).[15] Given the prevalence of spears rather than threshing forks on contemporary seals and frescoes it can be suggested that the fork-like implements on the

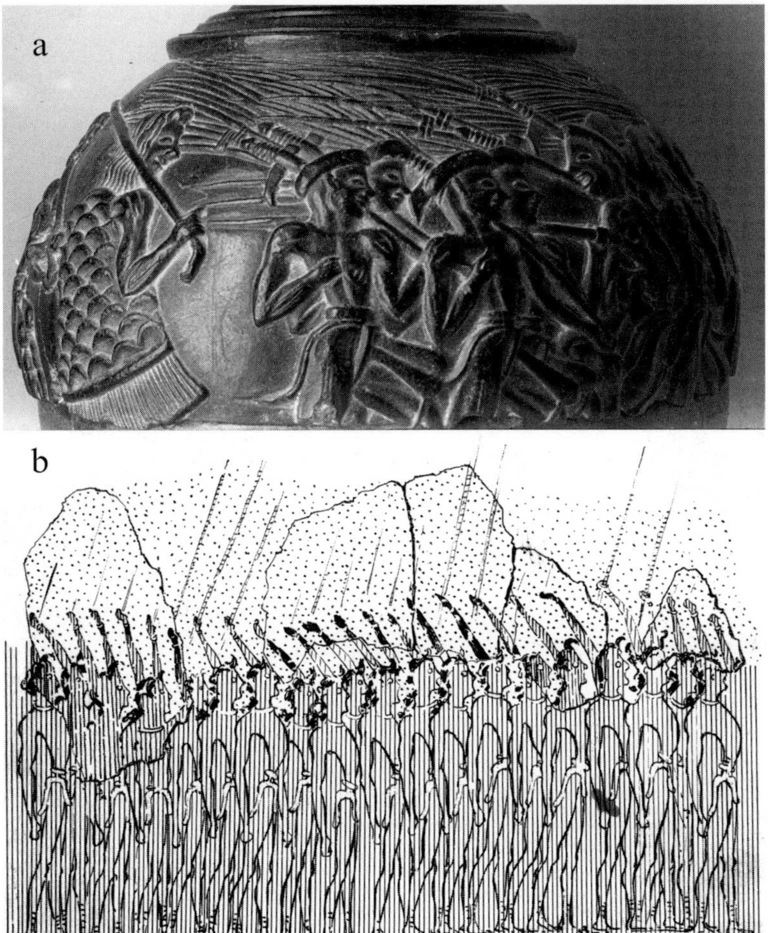

5.8. The Harvester vase and miniature fresco from Knossos showing men carrying spears. (a) The Harvester vase from Ayia Triada, LMI. Diameter: 11 cm (Sir Arthur Evans Archive); (b) the miniature fresco MMIII–LMI (Evans Fresco Drawing K/1). © Ashmolean Museum, University of Oxford

Harvester vase are bundles of throwing spears carried on a long-handled axe (poleaxe) for warfare or hunting. For hunting, spears could have been used to slow the animal down, before the incapacitating blow was delivered using the poleaxe. The hafted tools carried by the 'harvesters', with one long blade and one blunt end, are very similar to poleaxes found in other cultures used for stunning animals, particularly cattle.[16] They also resemble the weapons for stunning cattle discussed in Chapter 3. This would connect the Harvester vase more closely with other stone relief vessels from the same site, the 'Boxer rhyton' and 'Chieftain cup', which also show martial activities and animal practices related to cattle. Another related object is the ivory pyxis from a Final Palace period tomb at Katsamba, near Knossos, which shows a man aiming a spear at a bull while another jumps over the horns (Alexiou 1968: 34). This makes the relationship between hunting and bull games explicit.

Several animal things connected hunting and warfare in Late Bronze Age Crete and Mycenaean Greece. One is the boar's-tusk helmet which was made from the tusks of a number of the animals stitched on to a leather cap. As Christine Morris (1990: 155) suggests: 'Possession of such a helmet implies the bravery and hunting skills needed to acquire the tusks.' There is a variety of depictions from Crete, dating back to the Neopalatial period (Evely 1996) (Figure 5.9a–b). A pair of Final Palace polychrome vessels from Isopata Tomb 5 also depict both figure-of-eight shields and boar's-tusk helmets (Evans 1914: 25–6) (Figure 5.10).[17] Actual boar's-tusk helmets have been found in Neopalatial and Final Palace graves at Poros and Zapher Papoura near Knossos (Evans 1906b: 66, 1935: 867–71; Dimopoulou 1999b). One of the Zakro hybrid seals (Figure 5.9e) appears to show two dogs' heads, an agrimi horn and a boar's-tusk helmet, which encapsulates in one image different aspects of hunting: the ownership of hunting dogs and the collection of trophies. Boar are also carved on weapons; a Late Bronze Age dagger from Lasithi shows a hunter pursuing a boar with a spear (Long 1978), and a boar's head appears on a spear, possibly for hunting, from Anemospilia (Sakellarakis

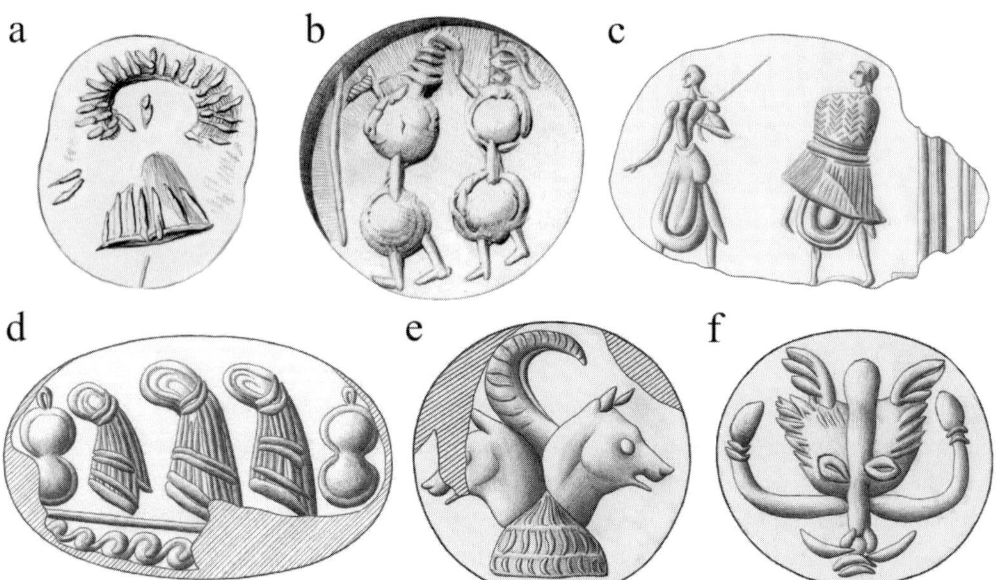

5.9. Drawings of seals showing animal things relating to hunting and warfare, LMI. (a) Boar's-tusk helmet (CMS VI 470); (b) warriors wearing boar's-tusk helmets and figure-of-eight shields, Knossos (CMS II 3 32); (c) men wearing hide skirts, Ayia Triada sealing (CMS II 6 11); (d) figure-of-eight shields and 'sacral knots', Knossos sealing (CMS II 8 127). Zakros sealings: (e) boar's-tusk-helmet-dog's-head-agrimi-horn hybrid (CMS II 7 168); (f) boar's-face-'snake-frame' hybrid (CMS II 7 204). © CMS Heidelberg

5.10. Clay jar decorated with boar's-tusk helmets and figure-of-eight shields, with replica, LMII. Height: 28.9 cm (AN1911.609 [original], AE.2393 [replica]). © Ashmolean Museum, University of Oxford

and Sapouna-Sakellaraki 1997: 597–8). Hunting boar for sport and as martial training had a long history in Greece, which can be traced back to the Bronze Age; this animal practice was encapsulated in the boar's-tusk helmet.[18]

Some of the things which are interpreted as religious symbols or paraphernalia can instead be drawn into this set of relations between hunting and warfare. Figure-of-eight shields are also associated with another object termed a 'sacral knot' by Evans which in at least some depictions was worn as an item of clothing like a cape, perhaps made of leather (Figure 5.9c–d).[19] Another garment, which appears on the same seal impressions, is termed a 'hide skirt', and its bulbous shape does seem to suggest leather rather than cloth (Forsdyke 1952: 14). All of these garments were shown being worn in processions, often involving double axes (Blakolmer 2007, 2010: 104–5; Haysom 2010: 42–8). The leader of the procession depicted on the Harvester vase wears a cape, with a scaled pattern which could be either cloth or leather (see Figure 5.8a). The short curved sceptre he carries could be another animal thing, a stick made from a dried bull's penis, an object attested in ancient Egypt and other cattle cultures (Gordon and Schwabe 2004: 138–45). The figure-of-eight shield is the most recognisable animal thing connected with hunting and warfare, but tracing connections between depictions suggests that there were other related animal things which would have had a similar significance in the Bronze Age Cretan Lifeworld.

Another of the Zakros hybrid seals shows a boar's head with an object conventionally known as a 'snake-frame' bisecting it (Figure 5.9f). Although it does appear on apparently cult scenes in the Final Palace period, it is likely that this is another animal thing derived from hunting and used in warfare: the composite bow (Evans 1935: 835). This was made from a combination of wood and agrimi horns, and was widely used in Eastern Mediterranean warfare. Agrimi horns were one of Crete's export products to Egypt in the Late Bronze Age (Murock Hussein 2015), and composite bows with similar finials as the Zakros sealing were found in Tutankhamun's tomb (McLeod 1970). The Linear B tablets preserved in the final destruction of Knossos show that agrimi horns, most likely used for bows, were being received as tax contributions (Ventris and Chadwick 1956: 301–3; Killen 2008a) (see Figure 1.2). It is not clear from the Linear B tablets whether the palatial administration maintained an interest in the territory where wild goats were hunted or whether it simply required a certain number of horns to fulfil a target. Like the woollen cloth industry which is detailed on these tablets, agrimi horns could have been part of the trade network which connected the palatial collective of Knossos to the Eastern Mediterranean.

Hunting iconography remained common on seals of the Final Palace period, appearing on some of the seal impressions preserved in the final destruction of the palace at Knossos. In this period, symmetrical design and additional symbols become common, with landscape indicators becoming more stylised. Soft-stone seals, not generally used for administration, also show hunting, indicating that it continued to be an important animal practice among a wide group of people. The cemetery at Armenoi provides a snapshot of the seals buried with people towards the end of the Final Palace period and after the destruction of Knossos (Figure 5.11a–b). One of these shows a woman in front of a goat which perhaps has a spear over its back. This could relate to the earlier depictions of women feeding goats, but made more explicit, since now the same scene is juxtaposed with what appears to be a hunted animal. In the Neopalatial period, as suggested above, female figures and hunted animals were generally depicted separately, although a connection between them can be traced.

In addition to the seals, which appear to be personal possessions, some of the larnakes (burial chests) from Armenoi are also decorated with hunting iconography. One larnax from Armenoi shows a goat and two cows with their young being hunted, again seemingly combining the kind of depiction seen in the Temple Repositories with the animal practice of hunting itself (Tzedakis 1971; Tzedakis et al. 2018: 10) (Figure 5.11c). As well as spears in their backs, nets are apparently being used by the hunters to round up these animals. One human figure is carrying an implement which looks like an axe, similar to those on the Harvester vase, and the other is apparently using a similar object to stun the speared cow. Although

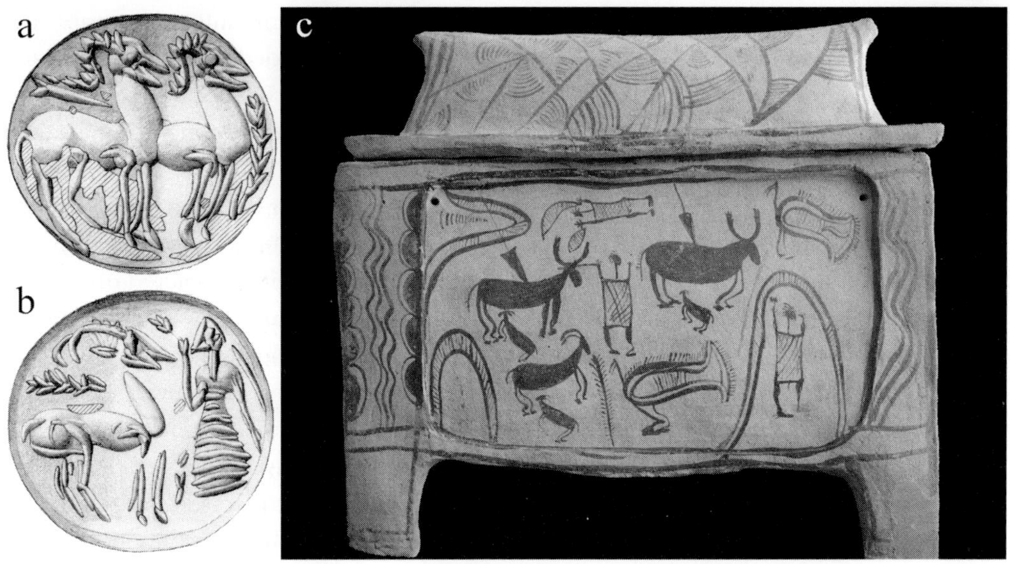

5.11. Seals and larnax from Armenoi, LMIII. Drawings of seals: (a) two goats with spear (CMS VS 1B 229); (b) woman and goat with spear (CMS VS 1B 261). © CMS Heidelberg. (c) Larnax with hunting scene (RM Π 1707). Courtesy of Rethymnon Ephorate of Antiquities

these larnax designs have often been seen in terms of beliefs about the afterlife (Watrous 1991), they are also evidence for the way in which palatial iconography was borrowed and adapted even after the end of the palaces. The set of associations established in the Neopalatial period between fertility and hunting seemingly continued in a different form in the Postpalatial local collectives.

With the collapse of the Mycenaean palatial collective on Crete, cattle, goats and deer would have been available for anyone to hunt. The remains from Chania restudied by Kerry Harris (2014a, 2014b), including agrimia and deer, indicate that the consumption of wild animals was of importance in the Postpalatial period. The display of agrimi skulls at Vronda, on the other side of Crete from Chania, is another indication that the animal practice continued to be of importance to local collectives which were established after the fall of the palaces (Preston Day et al. 2009). Although the iconography of hunting became rarer in the absence of seals, whose production declined once they were no longer needed for palatial administration, the display of trophies clearly continued. The skulls of agrimia at these Postpalatial sites could indicate that the provision of animals and display of their heads at consumption events remained important after the disintegration of the palatial collective.

CONCLUSION

This chapter has sought to reinterpret a range of archaeological remains and animal depictions from palatial Crete in terms of hunting. Although Arthur

Evans (1935: 523–90) discussed the connection between hunting and sealstones in *The Palace of Minos at Knossos*, it has never been regarded as a defining activity in Minoan Crete in the same way as in Mycenaean Greece. There are two reasons for this. The first is the contrast between peace-loving Minoans and warlike Mycenaeans which is dismissed as a cliché but continues to shape interpretation. The idea that sealstones are primarily formal exercises in depicting animal bodies rather than the animal practice of hunting is one example of this tendency. The second reason is that hunting is mainly depicted on seals in Bronze Age Crete, many of which do not show this animal practice explicitly for modern viewers used to reading naturalistic images as the products of a naturalist ontology. As has been argued, details of pose and direction would have been clearer indicators of hunting within a Bronze Age Lifeworld.

The history of the discovery of seals continues to have an influence on their interpretation. With so many unstratified examples, collected from Crete for over a century, it has become necessary to develop stylistic typologies in order to date them. This has resulted in an art historical approach which has sometimes resulted in the designs being considered as arbitrary, or reflecting the florescence and decay of artistic achievement. Only with the discovery of sealing deposits was it fully appreciated that these objects functioned in the Bronze Age as personal or official identifiers. One of the affordances of sealstones is to leave an impression in wet clay which preserves the presence of the user after the transaction. It is apparent, however, that Talismanic seals were rarely used for sealing. This has favoured interpretations which emphasise ideas about the efficacy of the design or stone, or a primarily typological approach which emphasises their repetitive designs. None of these interpretations, however, explains why animals, specifically hunted animals, were the most frequent subject of depiction during the Palatial period. Seals with animal depictions were not a functional necessity to the practice of bureaucracy, but were a means of participation in the animal practices which were integrated into the palatial collective.

Hunting was an animal practice in Crete from the time of the original settlement of Knossos, although its significance shifted as it came to define members of palatial rather than local collectives. It is possible that goats were brought to Crete by the first human settlers in order to be hunted; by analogy with other Greek islands agrimia could have been descended from the goats deliberately released to populate the uplands for hunting when needed. At the time of the formation of the palaces, hunting appears to have been one among many of the activities which people used to define themselves. This can be seen in the range of iconography on Protopalatial seals. Hunting, in most cases, can be inferred from the pose of the animal. This was also the time when mainly domestic animal figurines were deposited in peak sanctuaries. The reorganisation of the Cretan landscape, traced in Chapter 3, was hastened by the events which resulted in the destruction of the palaces at Phaistos and Mallia, and the temporary abandonment of other sites at the end of the Protopalatial period. Symptomatic of this breakdown was the hunting of domestic cattle, after which Knossos emerged at the centre of a collective which extended across Crete. The hunting of cattle resulted in the figure-of-eight shield; the boar's-tusk helmet and composite bow were other animal things which connected hunting and warfare. The structures termed guard houses, which were in use during this period, would have helped demonstrate the extent of the Knossos palatial collective, providing a base for hunting and territorial control.

This territorial control, although it involved the animal practice of hunting, was manifested at the centre through a set of analogies between elaborately dressed women, goats and the plants and rocks of the Cretan landscape. Crocuses, part of the iconography of these scenes, were of particular economic importance and, along with agrimi horns, were perhaps the main reason why control over the uplands was important. These analogies were revealed in restricted material culture at the palatial centres: frescoes, faience objects and carved stone rhyta. It is no accident that one of the clearest depictions of these analogies, including a crocus gathering scene, was at the villa at Ayia Triada, thought to be part of the Knossos palatial collective following the suppression of the palace at Phaistos. Protopalatial Phaistos, however, with influence from Near Eastern iconography, was the place where some of the iconographic conventions of Neopalatial seals appear to have emerged following the dissociation of script from seals. As well as the convention of direction, the addition of plants and rocks to depictions of animals helped to locate them in the space of the Cretan landscape. At Knossos the existing technology of cutting hard-stone seals was combined with these new conventions, resulting in both fine naturalistic depictions and more rapidly produced seals now known as 'Talismanic'. These could be distributed to a group of people who were attached to the palace at Knossos, but seemingly less directly involved in administration (although sealing deposits were only preserved by destructions and so there are few Neopalatial sealings from Knossos). The iconography of these Talismanic seals will be explored further in the next chapter because they frequently depicted marine subjects which were also linked with the palace at Knossos.

The depictions of deer and lions will be discussed in Chapter 7; deer were introduced to Crete in the Bronze Age and lion-hunting would have necessitated a trip to the mainland or a claim to have done so. As Knossos became part of a wider Mycenaean palatial collective, hunting continued to be depicted on seals. By this time it was no longer bound up with territorial control, which appears to have been unchallenged by this period, when large parts of Crete were home to the flocks of the palatial collective of Knossos. It must have continued to be a domestic animal practice, however, witnessed by its depiction on larnakes and seals, as well as its continuation after the fall of the palaces. Hunting was always part of the interaction between the people living on Crete and the animals which both inhabited and defined its marginal areas. It was perhaps never entirely controlled by the palaces but its depiction on seals brought this animal practice, and by extension the areas where it took place, into their orbit.

SIX

FISHING

INTRODUCTION

> The sea-weeds and the nautilus are drawn with such apparent realism that I had no thought but that they would be instantly recognized and identified by naturalists. This hope, however, proved delusive. While certain features are admittedly rendered with great force and truthfulness, yet, on the whole, the nautilus on these vases is not nearly an accurate drawing of the nautilus known to modern naturalists.
>
> (Murray 1890: 438)

In 1890 the British Museum acquired a vase from Armant in Egypt whose closest parallels were not with Egyptian ceramics, but with the stylised sea creatures already observed on Mycenaean pottery (Figure 6.1).[1] By coincidence the curator responsible for its purchase, Alexander Murray, had just returned from New York, where he had seen an almost identical design on a jug of the 'Mykenai type', also found in Egypt, which was 'far more naturalistic than any vase of the kind that had been hitherto observed and recorded' (Murray 1890: 437). Although the British Museum alabastron was indeed made on the Greek mainland, it is now regarded as an example of Marine-style pottery, a term which was used by Arthur Evans to describe marine depictions from Knossos. This style has become emblematic of Minoan pottery and is widely known through the widespread reproduction of particular vases in both art books and tourist souvenirs. Comprehensive corpora of Marine-style pottery have been assembled and production centres on both Crete and the mainland have been identified through a combination of both stylistic and scientific analysis (Mountjoy et al. 1978; Mountjoy 1984;

6.1. Marine-style alabastron found in Egypt, Late Helladic IIA. Diameter: 20 cm (BM GR1890,0922.1). © Trustees of the British Museum

Müller 1997). There have been attempts at even closer stylistic analysis, identifying the outputs of particular workshops and the hands of painters (Betancourt 1973; Müller 1997). It has also been seen as part of a broader set of designs originating from Knossian workshops termed the 'Special Palatial Tradition' (Betancourt 1985: 140–8). In this way Marine style has been absorbed into an art historical approach with its stylistic groups, workshops and even individual artists and their masterpieces. Marine style has become more about the artists than the animals and plants depicted. The current definition of Marine style marks a narrowing of horizons, a purification of over a century of archaeological discoveries into a set of interrelated pottery motifs.

The British Museum alabastron was accommodated within a much wider debate about the origins of art. Ultimately Murray concluded that the inaccuracies in depiction were a result of observation of these animals being adapted to fit pre-existing abstract forms; for them to have been drawn from life would be 'contrary to the regular process of invention in art' (Murray 1890: 442, 1892: 28). This appears to be a reaction against his colleague, Charles Newton, who regarded the naturalism of the gold octopus cutouts from the Shaft Graves at Mycenae and related designs in pottery as being directly inspired by observation (Newton 1880: 284).[2] Arthur Milchoefer (1883: 31) too suggested that these 'naturalistic' images gradually became more schematised as they were copied on pottery. The same debate continued into the twentieth century; Arne Furumark, who established the classification for Mycenaean pottery, distinguished between 'pictorial' designs which were copied from reality and 'pictorialised' designs whose origins lay in abstract forms (Furumark 1941: 133). He suggested that on pottery, unlike seals, pictorialised designs came first. Gisela Walberg (1983, 1987) followed this approach in her classification of

Middle Minoan pottery, defining the few depictions of marine animals in Protopalatial Kamares pottery as 'pictorialised' motifs. These motifs 'were deliberately chosen for the representation from the beginning', but were not primarily based on direct observation of an animal form (Walberg 1983: 58). Her definition of motif leaves little room for real-world observation: 'an artistic theme, an idea, or trend realized through a specific type of design, for example, a whirling movement, a radiation or a continuous surface *pattern*' (Walberg 1987: 47; italics in original). It relies instead on Friedrich Matz's (1928, 1951) argument that pottery design was essentially a geometrical solution to the problem of decorating a given vessel form. As Furumark (1941: 115) suggested, an octopus can readily be rendered as a radiating design, its arms embracing the sides of a rounded vessel. From this point of view, however, the octopus is primarily an appropriate abstract design and only secondarily an octopus. The idea that vessel shape determines the animals depicted has continued to be popular (Mountjoy 1972: 128; Berg 2011: 122–3).

Arthur Evans is credited with inventing the term 'Marine style', which fitted into his pottery scheme at Late Minoan IB (Müller 1997: 17). But unlike many subsequent authors he regarded this style, with its underwater scenes of marine life, as the culmination of centuries of marine imagery in other media. The now widespread use of upper-case for 'Marine style' marks a shift away from Evans's lower-case 'marine style', which denotes an interest in marine life rather than a typological term. For Evans (1928: 502), the 'finest known example of this marine style' was a stone relief rhyton from Knossos depicting an octopus, which he regarded as earlier than the similar depictions on painted pottery. His wide discussion of marine imagery also includes relief mouldings of shells in earlier ceramics, frescoes, sealstones and metal vessels. The origins of 'marine style' he located in the use of the bodies of marine animals: sponges and shells. This is one of the key characteristics of the 'naturalistic spirit', characterised by a close connection with animal and plant forms. Evans traced the development of these forms through their 'florescence' at the end of the Middle Minoan period into a period of 'decadence' in the Final Palace period. 'The tentacles of the octopus are finally resolved into a series of monotonous undulations. So, too, the tentacles of the delicate and beautiful argonaut shell [...] tend to become a mere pattern' (Evans 1928: 512). Although the idea of florescence and decay has become outmoded, Evans's account highlights the indexicality of marine designs in Bronze Age culture before they gradually lost their connection with the sea. This still provides an important alternative to accounts which simply regard marine designs in stylistic terms (Shapland 2014, forthcoming).

For recent authors, Marine style is iconic but not indexical; there is resemblance between designs and marine animals but not connection. This situates Marine style not in observations of marine animals, as Evans suggested, but in a pottery style which had long ceased to be produced.

> Small irregularities, along with naturalistic details (e.g. crossing tentacles and air bubbles) make the compositions appear full of life and as if taken directly from nature, but still they are altogether planned according to ornamental principles, and these principles are essentially the same as those which dominate the Kamares decoration. (Furumark 1941: 146)

Here Marine style becomes Kamares ware with tentacles; Wolfgang Schiering (1998: 194) locates an 'Uroktopus' in Protopalatial Phaistos, whose characteristics are taken over by the octopuses of this second phase of the design. This epitomises a view of artistic production in which designs or 'motifs' are invented once and then endlessly copied; realism becomes an artistic trick rather than a sign of interaction with the world. In addition, these Marine-style vases tend to be regarded as ritual vessels, further distancing them from interactions with animals (Mountjoy 1985).

Ever since Murray's publication, Marine-style depictions have been measured against the standards of natural history and have failed, leading to these suspicions that they are abstract pottery motifs in disguise. Comparing depictions of animals such as the argonaut (paper nautilus) against natural historical knowledge only emphasises their inaccuracy by modern standards (Gill 1985: 75–7; Müller 1997: 212–16). Yet the first accounts of Marine style are worth returning to because, unencumbered by a century of typologising, they record scholars' immediate reactions to these new discoveries: an impression of lifelike marine animals. To quote Murray again, the rocks and plants either side of the argonaut are 'a contrivance to give us a sort of bird's-eye-view down into the sea' (Murray 1890: 438). What persuaded him that these depictions are not lifelike is reference to a modernist standard: accurate natural history drawings, used in the delineation of individual species. Although it sometimes is possible to identify particular species with Cretan Bronze Age depictions, this is an anachronistic way to assess the intentions behind them.

The same nineteenth-century scholars were suggesting too that one of the unexplored parts of the Mediterranean where the origins of this art could be found was Crete (Newton 1880: 279–80; Milchoefer 1883: 134). When he excavated at Knossos, the emphasis was on Arthur Evans to discover these origins; he lived up to these expectations, showing a determination in his publications that Mycenaean art should find its prototypes in Minoan art. To some extent his location of the origins of Marine-style pottery decoration in 'nature-moulding' and 'nature-printing', using the bodies of real marine animals, can be seen in this desire to show that Minoan art was always in advance of the mainland. But rather than locating the origins of marine decoration in abstract designs he suggested that its inspiration was the animals themselves, whether through direct printing or observation. The importance of this suggestion is that it offers a different starting point for the discussion of Marine style. Rather than a closed corpus it can be seen as a point of connection with the underwater world. Marine-style pottery, in Evans's account, marks the end of a sequence which starts with direct contact between object and animal during manufacture, continues as imitation in relief and ends up as flat depictions on vessels and frescoes. Marine style becomes one of the ways in which material culture was used to establish a relationship with the sea. Its 'florescence' came at a time when the palatial collective of Knossos was expanding into the sea; its 'decay' came at a time when these connections had long been secured. This chapter will consider Bronze Age Cretan relationships with the sea along a continuum of real animal bodies to their reproductions and depictions in a variety of material culture.

MARINE LIFE

The marine animals of the Mediterranean Sea have not changed significantly since humans settled on Crete, although the opening of the Suez Canal in 1869 allowed Indian Ocean species to enter via the Red Sea for the first time. The underwater world of the Mediterranean is, however, increasingly visible as a result of developments such as scuba diving and modern aquaria. The Cretaquarium in Gournes, a major tourist attraction, is one of the key sites on Crete in which to encounter marine life today. It is run by the Hellenic Centre for Marine Research, which describes it as 'a site of marine science and aquaculture, a meeting point where science, discovery and recreation are equally accessible'.[3] The large tanks offer visitors an underwater view of a variety of Mediterranean marine animals. These include a range of commonly eaten fish and molluscs such as sea bream and octopus, as well as threatened marine animals including the loggerhead sea turtle and sand tiger shark which are the subject of conservation efforts. In one of the smaller tanks are a number of large sea snails of the *Charonia* genus known as tritons; unlike sharks or turtles, this is an animal which is depicted in Bronze Age material culture. Triton shells are found frequently in Minoan sites and were even imitated in other materials. Since these snails generally live at lower depths, it is unlikely that many people in the Bronze Age had a similar experience of seeing them underwater in this way. The aquarium offers the chance to see a variety of marine animals with which humans have interacted since at least the Bronze Age, but at the same time it offers a uniquely modern experience of watching fish and turtles swim by, separated by a thick pane of glass. The industrial technology which allowed the development of aquaria and fish tanks in the nineteenth century has transformed our relationship with marine animals (Brunner 2005). It has made them visible to an extent that they were not in the Bronze Age.

Nevertheless, Bronze Age depictions often present marine animals as if underwater, even if details would only have been known from observation on land. This suggests that interactions with these animals at sea or underwater were important.

> Shells and other shore debris were easily come by, some marine animals could be watched on and from the shore, and the fisherman's catch might be examined when brought to land dead or dying, but creatures of the open sea and the underwater habits of those nearer the shore were more difficult to observe; then the artist would have had to depend on chance sightings or on sailors' and travellers' tales. (Gill 1985: 63)

Even though some were more familiar, the distinctive bodies of marine animals were useful markers of a different environment (Riegner 1993). The underwater view that Marine style offers represents a different way of seeing marine animals compared with their appearance when caught or gathered. This emphasis on where marine animals came from indicates that they were important not simply as things to eat, but depictions of their bodies established a connection with the coast and the sea. The affordances of fish, molluscs and crustaceans as things for humans to eat have not changed since the Bronze Age but their accessibility has; modern technology makes these animals easier to see and to catch. Before looking at their depiction in the Bronze Age, it is necessary to look at how the animal practices of fishing and gathering have changed over time.

Fishing

Fishing is a human-animal relationship which extends on Crete from the Neolithic period to the present. In 2012 Greece as a whole had the largest fishing fleet in the European Union, although the catch is comparatively low (1.3 per cent of the total) compared to the states whose fleets are mainly Atlantic trawlers (EU 2016). The main species caught by Greek fishing boats are anchovies and pilchards, both small, oily fish. A wide variety of fish, molluscs and crustaceans are caught for human consumption, and can be sampled particularly in the fish restaurants, psarotavernas, found next to the sea. Nevertheless, there are important differences between Bronze Age and contemporary fishing as a result of modernisation. The industrialisation of fishing, even if Greek fishing is small-scale by European standards, blurs an important distinction which zooarchaeologists draw in discussions of fish bone assemblages between coastal and open-water fisheries. Many fish can be caught without venturing far from the shore, or even getting in a boat. Pelagic fish, by contrast, including anchovies, pilchards and larger fish such as barracuda, are usually caught out at sea. Another contrast is between hook-and-line fishing, which is now largely recreational, and the use of nets by the fishing fleet. These different fishing methods represent different types of interaction with the sea and marine animals using different types of material culture.

There is a range of material culture evidence for fishing from the Bronze Age Aegean (Powell 1996), but the evidence from fish bones has been lacking until recently (Rose 1994). The absence of fish remains in earlier faunal reports is unlikely to be evidence of absence, despite, for example, Peter Warren's (1972: 265) suggestion for Early Minoan Myrtos that the 'the sea was apparently little exploited, on the evidence of the comparatively few shells [. . .] and the absence of fishbones'. It is only recently that site reports have started to include fish bone studies as a result of systematic water-sieving.[4] As a result detailed evidence for Bronze Age fishing comes from a growing number of recently excavated sites (Theodoropoulou 2014b: 371–2). For Kommos, Mark Rose (1995: 204) suggests that the fish remains indicate 'a coastal fishery' which would have involved the use of nets. Fish bones can usually be identified to the family taxonomic level, and here sea breams and picarels constitute the largest group. These are small coastal species which could have been caught in shallow water. Hooks were also recovered from Kommos, and these, with the presence of a small number of deep-water fish such as scad and tuna, were used as evidence for fishing from boats. Dimitra Mylona (2004, 2020: 193–5) suggests a very similar picture for the Neopalatial buildings on the shore opposite Mochlos and at nearby Chalinomouri, with an assemblage of mostly small coastal fish, with barracuda and a few larger specimens possibly indicating the use of boats. There are a number of remains at various sites from cooking pot contexts; it has been suggested that these could have contained fish sauce or broth (Rose 1995: 223; Mylona 2004: 123), and the characteristic state of preservation of the remains at Palaikastro has been used as further evidence for this (Mylona 2007: 220). Some of the remains even show signs of digestion (Mylona 2004: 124, 2007: 217). The evidence from microscopic fish otoliths (ear bones) has the potential to provide a more complete picture of Bronze Age fishing, confirming, for instance, that picarels and other coastal fish formed the majority of the catch at Pseira (Rose 2009: 145–6).

Octopus and squid remains are only occasionally recognised in archaeozoological assemblages because, unlike other molluscs, they do not have a shell and so only tiny parts of the mouth survive (Reese 2016a: 407). Cuttlefish do have an internal shell (also called a cuttlebone) and these have been reported from various sites.[5] Both octopuses and cuttlefish are frequently depicted on vases and sealstones, and so it is likely that they were caught and eaten during the Bronze Age, as they are today. Although they can be caught with nets, a traditional method of fishing for squid, cuttlefish and octopuses is with a baited line dragged behind a boat (Powell 1996: 132–3). This type of fishing often takes place at night, with a light on the boat to attract these animals.

Octopuses can be caught by a variety of methods; traps are also widely used in the Mediterranean, and spearing is also practised (Rathjen 1991; Lloret and Font 2013). Argonauts are a type of octopus but they are found in deep water and are unlikely to have been caught for food.

The term fishing covers a number of different animal practices, and a wide variety of different types of animals. Although coastal fishing with nets was a significant source of food, depictions of humans with fish (and an octopus on a seal from Knossos) tend to show them held on a line (Powell 1996: 136–7) (Figure 6.2a–b). By extension, depictions of fish, octopuses and cuttlefish on Protopalatial and Neopalatial seals could indicate the animal practice of line fishing (Powell 1992: 311), although, as will be suggested below, many were shown underwater rather than in the context of fishing (Figure 6.2c–e). In the well-known fresco from Akrotiri on the island of Thera, contemporary with Neopalatial Crete, the fish held on a line by the 'Young fisherman' have been recognised as dolphinfish (Economidis 2000). Like dolphins, these fish accompany boats out at sea and so the care taken to depict them suggests that it was significant that this particular type of fish had been caught. A traditional method of catching them is with a metallic lure imitating a flying fish dangled from a boat (Jackson 2012: 19). The marine animals most frequently depicted tend to be

a b c d e f

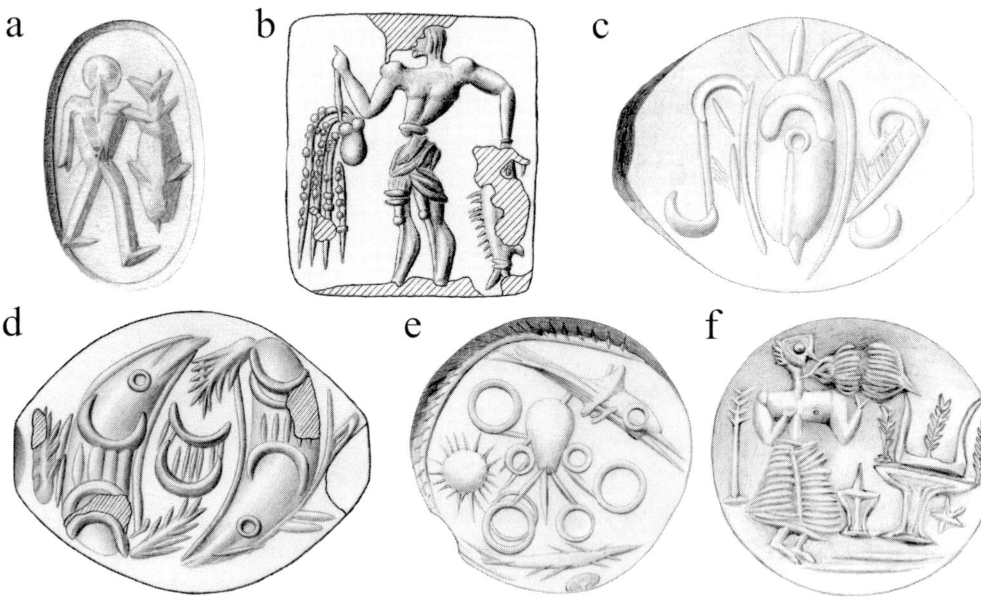

6.2. Drawings of seals showing marine animals and humans. (a) Human with fish from three-sided prism, Mallia, MMII (CMS II 2 174a); (b) man with octopus and fish, Knossos, MMIII-LMI (CMS VI 183); (c) Talismanic cuttlefish, Sphoungaras, MMIII-LMI (CMS II 3 244); (d) Talismanic 'fish protomes', Elounda, MMIII-LMI (CMS II 3 295); (e) Talismanic octopus and fish, Mochlos, MMIII-LMI (CMS II 3 251); (f) woman blowing shell, Idaean Cave, LMIIIA (CMS II 3 7). © CMS Heidelberg

those which would have been encountered by boat, from which some were caught by line. Whereas coastal fishing would have been a common activity in the Bronze Age, sailing in open water to catch fish would have distinguished its participants because of its potential danger and the skilled knowledge required (Mylona 2020: 202). As with land animals, the marine animal practices depicted or implied, particularly from the Late Bronze Age, are rarely mundane and everyday but suggest a more specialist or out-of-the-ordinary activity.

Gathering

The term shellfish covers a variety of marine animals which can be gathered close to the shore, usually to be eaten. Although overexploitation and coastal development can affect shellfish, many of the types eaten in the Bronze Age are still gathered on Crete. Shells from molluscs survive well archaeologically, and large numbers have been published from Bronze Age sites; other types of shellfish such as crustaceans and sea urchins are included in more recent archaeological reports as a result of more systematic recovery, particularly water-sieving. At the time of its publication Kommos had the largest Bronze Age marine invertebrate assemblage in the Mediterranean or Near East, with over 35,000 specimens (Reese 1995c: 240). Shells can be divided into bivalves and gastropods, which tend to inhabit sandy shores and rocky shores respectively, and are often reported to genus or species in archaeological publications. In many cases the common English or Greek name corresponds to a genus or family (e.g., *Patella/Patellidae* = limpet; Greek: petalida), as is often the case for folk taxonomies, although terms such as 'clam' or 'crab' can refer to a number of different families. Figure 6.3, which gives the composition of some of the larger published assemblages, uses the common name at the family level.[6] There are still regular reclassifications of marine invertebrates, however, meaning that the same type of shell is sometimes referred to using a different genus according to the date of the publication. This highlights the contingency of modern classification systems which are used in the practice of identifying shells in archaeological publications.

Another way to group shellfish is by their affordances. Throughout human history shellfish have been gathered to be eaten (Claassen 1998: 16). At Kommos, for example, the majority of shells were limpets (over 31,000 specimens; 84.8 per cent), which strongly suggests that they were gathered for food (Reese 1995c: 252). Top shells were another commonly found type, which were almost certainly eaten, although there are a handful of examples of top shells being used to make rings in the Late Bronze Age (Reese 1984). Limpets and top shells also make up the majority of the shell assemblage at other sites such as Final Neolithic Petras, Bronze Age Pseira and the Neopalatial buildings on the shore opposite Mochlos (Reese 2004, 2009: 132; Theodoropoulou

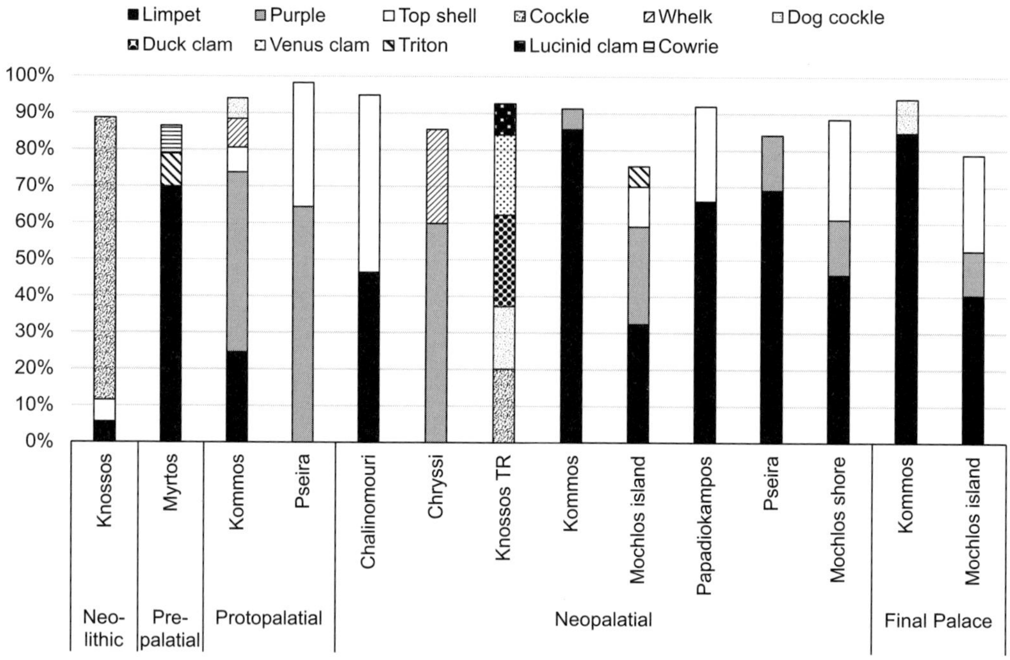

6.3. Shellfish assemblages by site and period. Shows shellfish family as proportion of assemblage where it comprises 5 per cent or more of total

2012: 92). At a recently excavated house at Papadiokampos in eastern Crete, abandoned suddenly at the end of the Neopalatial period, 50 kg of shells were recovered, of which 98 per cent were limpets and top shells (Mylona 2020: 198). Some were found in cooking pots where they had been stewing. Although purple shells (also known as murex shells) are usually associated with dye production, they are edible and are often found in the same context as limpets and top shells. In some cases, as in a Neopalatial house at Petras, the comparatively small quantities of purple shells, and the way they are broken, can be interpreted as consumption (Theodoropoulou 2016: 231). Nearly all types of shellfish found on Crete are edible, although in some cases would have been collected primarily for their shell or crushed for dye.

Shells were frequently used for adornment or craft production in the Bronze Age Aegean; these could have been collected on the beach with or without the animal inside (Karali 1999: 26–42). Cockles were found in the Neolithic levels at Knossos which had been pierced at the protruding umbo, and were presumably worn on a cord (Evans 1928: 10; Shackleton 1968). Dog cockles, which belong to a separate family from cockles, were often treated in a similar way in the Neolithic and Prepalatial period, and were often waterworn, meaning that they had been collected dead (Reese 1987).[7] The less frequent, or smaller, genera in any assemblage are often explained as aesthetic, such as tusk shells, cowries and whelks (Shackleton 1972: 325; Reese 1987; Ruscillo 2006: 801).

The appearance of exotic Red Sea genera too, such as *Nerita*, found at the Protopalatial burial cave of Hagios Charalambos, and *Tridacna* (giant clam), found at Neopalatial Knossos, suggests that shells were traded and used as ornaments (Reese 1991: 170; Chlouveraki et al. 2008: 576).

Not all shells found in Bronze Age contexts can be explained in such functional terms. Some shells were deliberately collected from the shore and were used as indexes of the sea. Deborah Ruscillo (2006: 805) has argued that a large number of waterworn dog cockle shells were collected and deposited at Final Palace Kommos as mementos of the cataclysmic event, such as a storm, that brought them to shore. The deposition of shells at the peak sanctuary of Juktas, including a variety of whelks, could also be a form of commemoration rather than eating (Kyriakidis 2005: 130). Sea pebbles were frequently deposited at peak sanctuaries, suggesting that other things could be used to index the sea too (Nowicki 1994; Peatfield 1992: 76). Another deposit, the early Neopalatial Temple Repositories at Knossos, contained thousands of seashells, mostly cockles, but thirty-seven different genera in total (Evans 1921b: 518–19; Panagiotaki 1999). These were found alongside clay sealings showing shellfish and a number of faience things which indexed the sea such as flying fish and argonauts. As will be suggested below, this assemblage brought the sea into the heart of the Knossos palatial collective, and shells were important as traces of the sea.

Two types of shell, purple shells and tritons, became particularly associated with the palatial collectives, and were gathered by various means, including baited traps and diving. Like open-water fishing, diving is a potentially hazardous activity which would have distinguished its participants in the Bronze Age (Apostolakou et al. 2014: 328). Triton shells and purple shells are often found together since they occur at similar depths around the coast of Crete. Purple shells were formerly referred to by the name of the genus *Murex*, but the two main purple species, *Hexaplex trunculus* and *Bolinus brandaris*, are now regarded as belonging to different genera. Large quantities of crushed purple shells have been found at a number of sites on Crete which were the waste from dye production (Stieglitz 1994; Ruscillo 2005, 2006; Brogan et al. 2012; Reese 2016a: 408–12). A range of evidence for dyeing has been found; at Protopalatial Alatzomouri-Pefka in eastern Crete there are dyeing vats, and purple (and other) dye residue in the associated pottery (Betancourt et al. 2012; Apostolakou et al. 2016, 2020; Koh et al. 2016). On the island of Chryssi, off the south coast of Crete, a specialist purple processing installation has been excavated, largely dating to the Neopalatial period (Apostolakou et al. 2012, 2014; Brogan et al. 2019). The purple shells were crushed and then boiled in cooking pots to produce the dye, which would then have been sent to Crete for use in dyeing woollen cloth. Given the size of Chryssi, it was probably occupied seasonally and reliant on a settlement on the opposite shore, most likely a palatial collective based at Ierapetra, although no palace has been found (Chalikias 2013).

Triton shells found on Crete have long been regarded as an indication of religious ritual (Åström and Reese 1990: 8–10). Arthur Evans provided an influential interpretation, describing a ledge with a triton shell on it as a 'domestic shrine', with the triton 'a usual concomitant of these sanctuaries'. He continued: 'Such conch-shells were in fact used as trumpets in Minoan cult for calling down the divinity to altars of offering' (Evans 1921b: 580–1). This argument was based on a seal from the Idaean Cave which appears to show a human blowing a large shell (Evans 1901a: 142) (Figure 6.2f); 'trumpet shells' have been identified at ancient Eastern Mediterranean sites (Reese 1985: 353–64). At Chryssi and Papadiokampos, where triton shells were processed, there were no trumpet shells but several had been pierced to serve as decanters. Others were found next to larger vessels, suggesting they were used as ladles or scoops, without any ritual associations. This contextual analysis of two recently excavated sites helps to undermine the circular argument that triton shells were found in religious areas, since here they were used as containers for various purposes (Apostolakou et al. 2014). Nevertheless, triton shells were processed at these sites and were used for ceremonial purposes in the palatial periods. Several triton shells from Protopalatial Phaistos were deliberately modified, since their surfaces were blackened and carved with decoration (Sanavia 2014; Weingarten and Sanavia 2016). Triton shells, and another large gastropod, the tun shell, were even reproduced in pottery and stone. To understand the significance of fish and shellfish in the palatial collectives it is necessary to turn to their depiction in other media.

NATURE-MOULDING AND NATURE-PRINTING

> It thus appears that from the very beginning of the Middle Minoan Age the development of ceramic Art in the 'marine style' was facilitated by a system of 'Nature-moulding' in the same way that at a slightly later date – M.M.I.I.a – we have seen the same style in wall-painting promoted by a process of 'Nature-printing' by means of small sponges dipped in paint.
>
> (Evans 1935: 109)

Modern publications of Cretan Bronze Age sites are often split into chapters by different specialists, with shells, pottery sherds and fresco fragments discussed separately. Although this reflects the growth of the discipline of Aegean prehistory since the *Palace of Minos at Knossos* was written, Arthur Evans's observations of the interaction between different materials, including ceramics, frescoes and stone vessels, and animal bodies such as sponges, shells and octopuses, have been neglected as a result.[8] As noted above, even the Marine-style depictions of octopuses, while acknowledged to be representational, are traced back to abstract whirling forms in earlier pottery. From this perspective such octopus depictions bear only an iconic, and not an indexical, relationship with real octopuses because they instead index earlier pottery styles. Although the late Neopalatial Marine style has received considerable attention, it comes at the end of a long period of interaction between marine bodies and forms and a variety of types of material culture. Fish were occasionally depicted on both Prepalatial pottery and seals,[9] but from the start of the Protopalatial period pottery was impressed with, moulded from or sculpted into a variety of marine forms as the sea was absorbed into the material world of the palatial collective.

Birth

Arthur Evans traced the origins of pottery designs in a variety of sources including textiles, metal vessels, plants, flowers and stone (Evans 1935: 100). One distinctive type of pottery decoration termed barbotine starts to appear on a small number of vessels including drinking cups at the end of the Prepalatial period (Foster 1982: 143; Momigliano 2007: 84–91). Evans saw the origins of this irregular relief effect in the imitation of certain marine forms such as barnacles and the thorny oyster (*Spondylus*) (Evans 1935: 102–5). This explanation of barbotine has fallen out of favour. Instead, it tends to be seen as a technique which refers ultimately to the manufacture of ceramics, exploring the potentials of this medium.

> [T]hree-dimensional slip design endows a vessel with an additional spatiality, increasing the potential of its surface. At the same time, this creates a tension between the vessel's two- and three-dimensional embellishments, resulting in products that are often visually disconcerting, but intellectually challenging. (Foster 1982: 144)

Walberg (1987: 41) too suggests that barbotine has 'a decidedly ceramic character'. One of the reasons for this non-referential interpretation is that barbotine is the term used for a wide variety of additions to the surface of pottery, from discrete nodules and lines to the completely irregular effect Evans also termed 'barnacle ware'. Regarding barbotine as a pottery technique results in the downplaying of its skeuomorphic potential. The alternative is to regard pottery as an inherently plastic medium which was used to imitate and reference a variety of forms and patterns, whether through shape or decoration. From this point of view some barbotine could be representational, such as in an unparalleled stand from Phaistos, with plastic dolphins circling it (Foster 1982: 92–3). The dolphins themselves show that ceramics can be shaped into a variety of forms. Other barbotine effects can be seen as imitating metal, particularly the added clay 'rivets'; the black paint of Protopalatial Kamares pottery is widely agreed to imitate blackened silver, and the impressed designs of 'eggshell ware' or the sharp carinations of some pots also seem to relate to metal vessels (Evans 1921b: 242; MacGillivray 1998: 57). It should also be emphasised that ceramic relief is a rare phenomenon considering that the vast majority of vessels are not decorated in this way. But accepting that barbotine can be a means of indexing marine forms opens up a different way of viewing Minoan pottery, as indexical rather than simply iconic (Knappett 2005b: 144–5).

A particular pottery form sometimes decorated with barbotine is a bowl with an interior handle usually interpreted as a 'spinning bowl'. Other spinning bowls from Quartier Mu at Mallia have depictions of fish and octopuses on the interior (Poursat and Knappett 2005: 224). The affordances of these bowls for spinning wetted thread, most likely flax, were realised by filling them with water (Barber 1991: 73–7). This phenomenon can be traced in later periods. In her description of a fish depicted on the interior of a Final Palace period bathtub larnax, Christine Morris points out that when filled the fish would be seen underwater. More generally she argues that depictions of fish are related to vessel forms associated with liquid, regarding them as 'iconographic reinforcement' which 'reflect and emphasize the function of the object they adorn' (Morris 1995: 193). From this point of view the association between ceramic containers and liquid explains the frequent occurrence of marine forms as ceramic depiction. Rather than reinforcement, however, the presence of animal depictions is transformative; it establishes an analogy between the water in the vessel and water from the sea.

Protopalatial pottery sometimes incorporated shells and other marine animals such as crabs in relief. Evans termed this 'nature-moulding', epitomised by the remains of a vessel found in a pottery dump from the end of the Protopalatial period at Knossos (Figure 6.4a). The crabs appear to be modelled in clay but some ceramic shell reliefs were moulded from actual shells (Evans 1921b: 521). Knossos was not the only site where this occurred; a small number of vessels found at Quartier Mu in Mallia were decorated with appliqués of crabs, fish and several types of shell (Detournay et al. 1980: 122–5; Poursat and Knappett 2005: 115). As will be discussed in the next chapter, these vessels from Mallia also bear reliefs of cats, suggesting that these vessels established a wider set of connections between exotic bodies and the sea. Sponges were also used to decorate vessels, as well as one Protopalatial wall painting fresco at Knossos, which is regarded as the earliest figural fresco (Evans 1930: 361–3). This was the technique Evans called 'nature-printing' (Figure 6.4b). A related technique was the use of stamps in the form of shells to create an impressed design on pottery (Figure 6.4c). Printing and moulding are two means to produce an index of the original form; in locating the origins of what he termed 'marine style' in these techniques, Evans highlighted the indexicality of marine depictions in the Bronze Age.

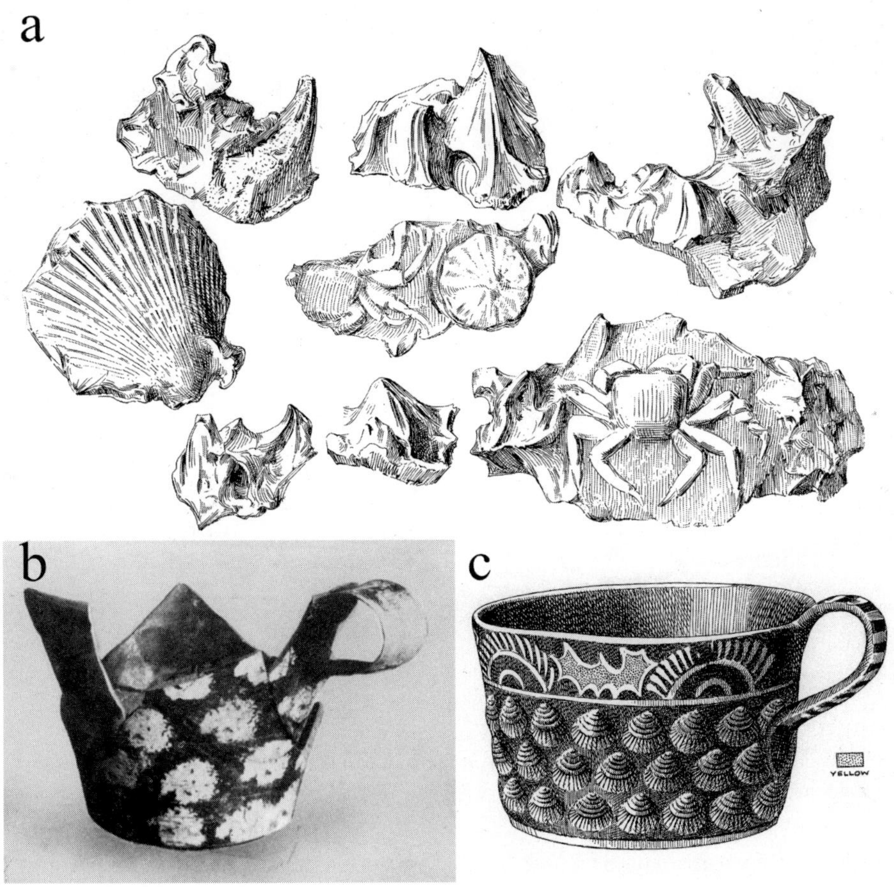

6.4. Examples of 'nature-moulding' and 'nature-printing' on pottery. (a) Relief decoration of marine animals from the Knossos kouloures, MMIIB-IIIA (after Evans 1921b: fig. 380); (b) cup from Knossos with sponge print, MMIB-IIA (after Evans 1930: fig. 240); (c) cup from Phaistos impressed with shell decoration, MMIIB (after Evans 1935: fig. 83)

Florescence

There was a shift at the end of the Protopalatial period to depictions of marine animals underwater. Sealings from Phaistos preserve the impressions of three different seals depicting tritons, either with plants or with a wavy line which appears to indicate an underwater setting. One of the depictions possibly shows the operculum, the barrier which the snail uses when inside the shell, indicating a live animal. In the same deposit are a number of sealings showing an octopus looking back at the viewer (Figure 6.5a–b). Another group of sealings produced at a similar time is the Hieroglyphic Deposit from Knossos, which includes one impression of a fish and squid in a rocky landscape (Figure 6.5c). This was one of the epitomes of naturalism for Evans, who

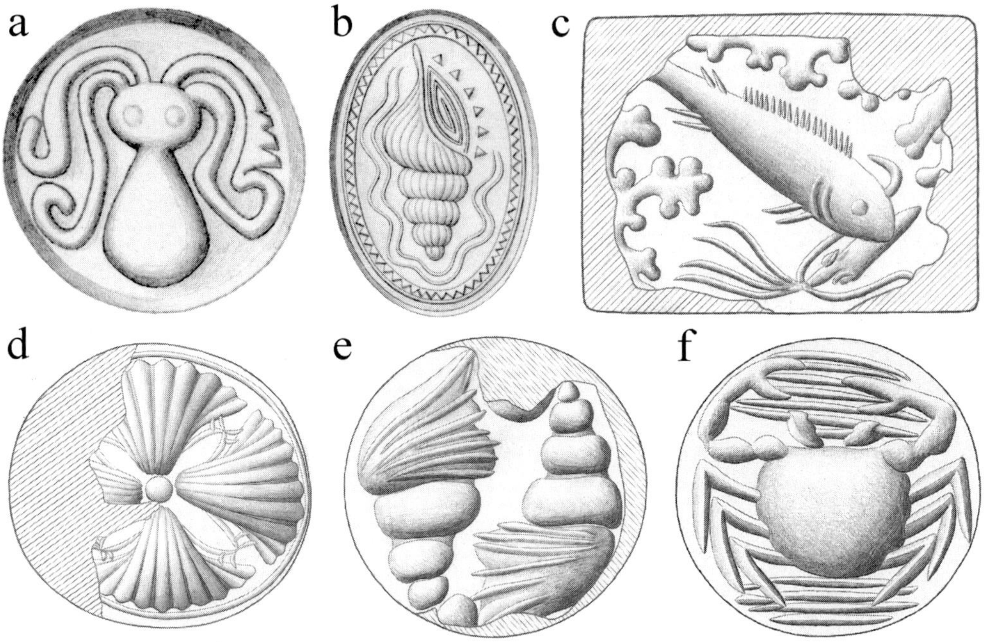

6.5. Drawings of seals showing marine animals. Phaistos sealings, MMII: (a) octopus (CMS II 5 304); (b) triton shell (CMS II 5 301); (c) fish and squid, Knossos Hieroglyphic Deposit sealing, MMIII-LMI (CMS II 8 157). Sealings from Knossos Temple Repositories, LMI: (d) four shells, probably cockles (CMS II 8 150); (e) two triton shells (CMS II 8 151); (f) crab (CMS II 8 154). © CMS Heidelberg

suggested that it displayed 'truest sympathy with natural forms and surroundings' (Evans 1935: 491). The depiction of fish and octopuses underwater was transferred from seals to other media in the Neopalatial period, resulting ultimately in the Marine style on pottery.

A variety of marine forms continued to be depicted on seals in the Neopalatial period. Sealings found in the Temple Repositories at Knossos, a closed deposit which can be dated to the early Neopalatial period, bore impressions of tritons, crabs and shells (Panagiotaki 1999: 104–6) (Figure 6.5d–f). These appear to have been made using metal rings, associated with the administration of the palaces. Marine animals were also found on contemporary 'Talismanic' seals, which, as discussed in the previous chapter, were rapidly produced colourful hard-stone seals and very rarely used for sealing. The technique of manufacture leads to some uncertainties of identification, although they have been convincingly grouped (Morgan 1985: 11–13; Onassoglou 1985, 1995). The basic 'cuttlefish' form, for instance, consists of an oval body, S-shaped tentacles and fins (Figure 6.2c). Since the Talismanic forms are based on a small number of drill cuts it is not possible to distinguish cuttlefish from squid based on body shape; the feeding tentacles, common to both animals, are the main feature of these depictions. Another Talismanic form conventionally termed 'fish protome' does appear to show fish because of the presence of the eye, but it is possible that these are triton shells because these are shown antithetically on more naturalistic seals (Figures 6.2d, 6.5e). Fish are also shown on Talismanic seals, although there is considerable variability within this group. It is difficult to identify the fish genus in most seals, although fins are marked on even the simplest depictions. Judith Powell (1996: 67–8) doubts that any identifications can be made confidently, pointing out the inconsistencies in various authors' attempts.

Like other animal depictions of this period, the waving tentacles of octopuses and cuttlefish, and the indications of plants next to fish, suggest that these animals on Talismanic seals are shown underwater. In some cases the colour of the seal seems to act as a material reinforcement of the underwater scene.[10] This is apparent on individual examples such as a clear rock-crystal seal showing an octopus and fish (Figure 6.2e) but can also be demonstrated statistically for Talismanic seals. Clear/blue/green seals are significantly more likely to be used for marine animals than land animals, for which red/yellow/brown stones are far more common (Shapland 2010b: 118–19). Clear rock crystal and green jasper in particular are predominantly used for marine animals, although a red/brown stone, carnelian, is also commonly used for seals showing fish and cuttlefish. Other factors besides colour, such as availability of material, date of manufacture or perhaps translucency could have affected the choice of design, but there is a clear association between marine animals and particular types of stone whose colour relates to the sea.

Marine animals continued to be associated with vessels in the Neopalatial period, such as the stone relief 'Ambushed octopus rhyton' found in the fill above the Throne Room at Knossos (Figure 6.6a). With the coral overlapping the octopus, it is regarded as another highpoint of naturalism: 'Strength, naturalism, humour and freshness are delightfully conveyed' (Warren 1969: 176). As Bernd Kaiser (1976: 186) suggests, the marine environment negates many of the difficulties of depicting the spatial relations of landscape. These relief depictions highlight an important component of 'naturalism', which is an attempt to make animals spatially present, albeit at a smaller scale. Their tactility would have enhanced the perception of presence. Another stone vessel from Knossos shows an alternative approach to depicting an octopus, with the plastic tentacles separate from the surface of the vessel (Figure 6.6b). This is more similar in approach to ceramic relief, although relief tentacles are not known from contemporary ceramics. In both cases the octopus exists beyond the flat surface of the vessel.

Stone was also used to reproduce triton shells in the Neopalatial period. Real triton shells were used as vessels throughout the Bronze Age, but the evidence from Chryssi and Papadiokampos suggests that they were gathered on a large scale and modified for use as pouring vessels in this period (Apostolakou et al.

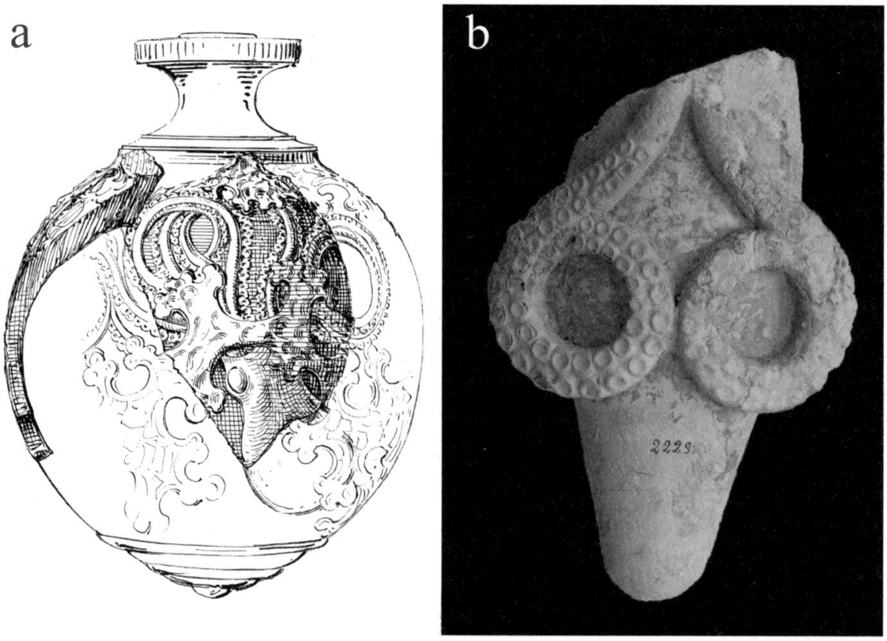

6.6. Stone vessels decorated with relief octopuses, LMI. (a) Drawing of Ambushed octopus rhyton from Knossos (Sir Arthur Evans Archive). © Ashmolean Museum, University of Oxford. (b) Octopus vessel from Mavro Spelio, height: 18.4 cm (AMH Λ2229). © Heraklion Archaeological Museum – Hellenic Ministry of Culture and Sports – 'Hellenic Organization of Cultural Resources Development (HOCRED)'

2014: 329). Some gathered shells were holed at the base so that they could be used as rhyta, although only one of the stone tritons was also a rhyton (Koehl 2006: 125). Stone tritons were found in Neopalatial and Final Palatial contexts (Figure 6.7a), although fragments of ceramic triton shells have been found at several Protopalatial sites, showing that the earliest triton skeuomorphs were made of clay (Binnberg 2013: 18; Sanavia 2014: 544). Two shell-shaped vessels have been found on Crete made of obsidian, a hard volcanic glass, which would have been exceptionally difficult to carve, meaning that these were the products of specialist workshops (Bevan 2007: 60). One, from Palaikastro, was in the shape of a triton, and the other, from Ayia Triada, was in the shape of another large shell found in the waters off Crete, a tun shell (Warren 1969: 91) (Figure 6.7b). The stone shells show a high degree of realism, with the fine striations and scalloped edge. They were metaphorical animal things from the bottom of the sea, made and used on dry land. The stone and clay skeuomorphs had the same form as gathered shells but were also the restricted products of a palatial workshop, reinforcing the links between tritons and the palatial collective.

Towards the end of the Neopalatial period there was a shift from depictions on stone relief vessels to painted pottery, although explanations for this differ.

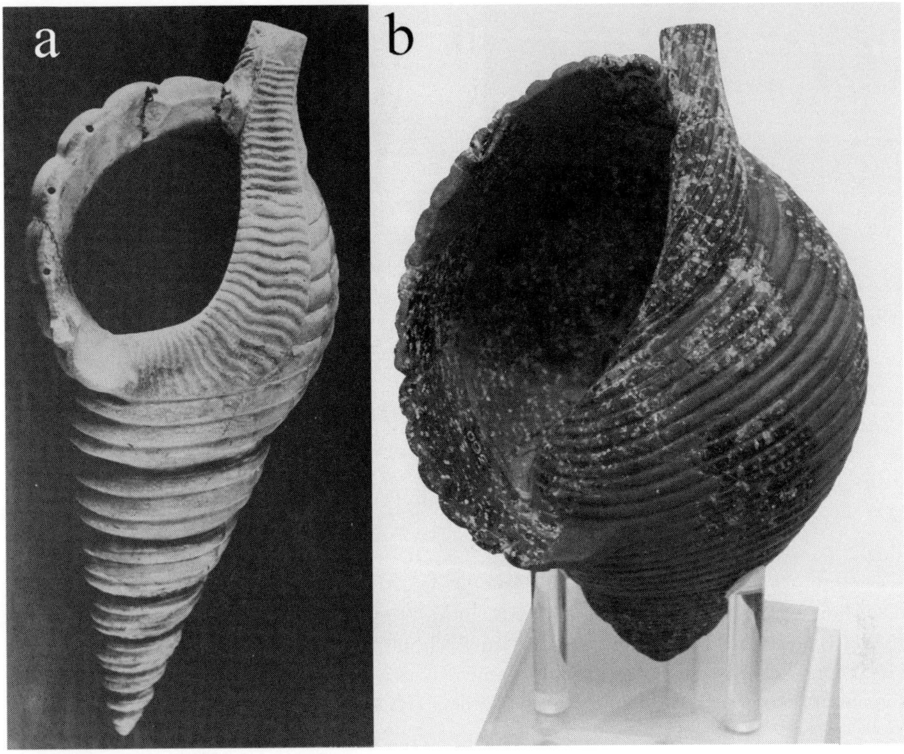

6.7. Stone zoomorphic shell vessels, MMIII-LMI. (a) Limestone triton from Knossos, height: 31.7 cm (AMH Λ45) (Sir Arthur Evans Archive). © Ashmolean Museum, University of Oxford. (b) Obsidian tun shell from Ayia Triada, height: 28.5 cm (AMH Λ360). © Heraklion Archaeological Museum – Hellenic Ministry of Culture and Sports – 'Hellenic Organization of Cultural Resources Development (HOCRED)'

Several authors have suggested that the disruptions to the Minoan psyche and marine environment caused by the Thera eruption explain the timing of the appearance of Marine-style pottery (Müller 1997: 322; Schiering 1998: 199; Bicknell 2000). A more nuanced suggestion is that the eruption caused disruptions to the production of palatial goods in other materials, either as a result of social breakdown or an interruption in the supply of materials, and resulted in the shift of focus to ceramics to fill the gap (Driessen and Macdonald 1997: 62). This helps to explain why naturalistic animal depictions on pottery only take off at this time, when a tradition of naturalistic depictions in luxury materials was combined with the existing tradition of pottery decoration. This raises the question of why marine forms were adopted on pottery rather than other scenes on relief vessels. Whether or not there is a direct connection, it is clear that the affordances of both stone and pottery vessels for containing and pouring liquids had already been explicitly related to the sea.

One of the most elaborate examples of Marine-style pottery is a jug found at Poros near Knossos with its floating argonauts and what are probably jellyfish

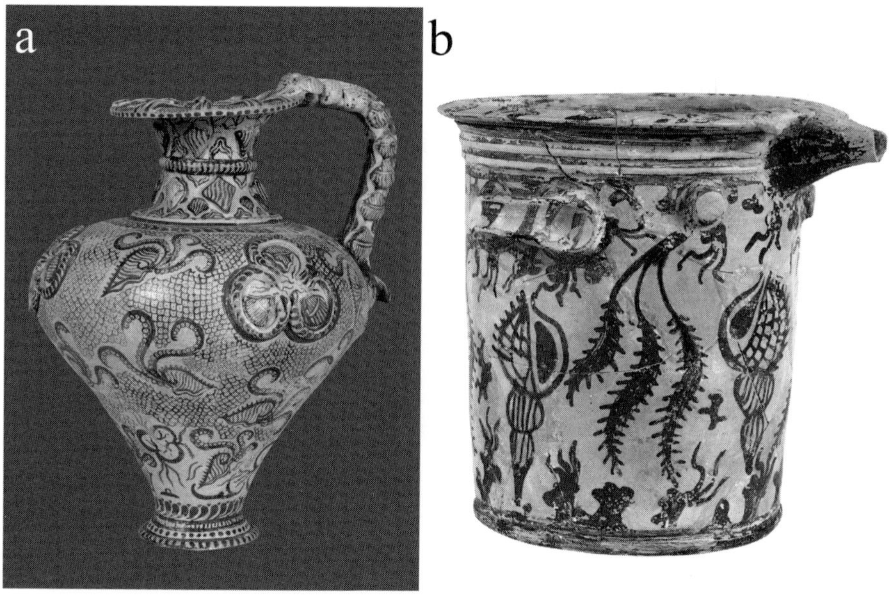

6.8. Marine-style vessels from Crete, LMIB. (a) Jug from Poros near Knossos, height: 26.2 cm (AMH Π26661). (b) Spouted bucket vase from Nirou Chani, height: 18 cm (AMH Π7572). © Heraklion Archaeological Museum – Hellenic Ministry of Culture and Sports – 'Hellenic Organization of Cultural Resources Development (HOCRED)'

(Legendart forthcoming), and modelled cockle shells on the handle (Dimopoulou 1999a) (Figure 6.8a). This is one of the latest examples of 'nature moulding' of shells, first seen on Protopalatial ceramics. The argonaut is rare in coastal waters but is found out at sea with its 'shell' floating on the water. It has been pointed out that the argonauts are shown swimming upside-down, and don't have enough tentacles, but the Marine style nevertheless reanimates animals which were rarely seen, or which were normally encountered out of water when caught or washed up on the shore (Müller 1997: 214–16; Berg 2011). The Marine style, following on from the naturalistic relief stone vessels, shows argonauts and shells swimming underwater, as well as octopuses with open eyes and curling tentacles. Whereas modern observers have all too often looked at these and seen Kamares syntax or zoological errors, in the Bronze Age these depictions indexed an underwater world.

The other way in which Marine style indexes the underwater world is by surrounding marine animals with rocks and plants. The same device is used on the Ambushed octopus rhyton, helping to locate the octopus underwater, behind coralline rocks. These rocks and plants are common to nearly every Marine-style depiction and can be compared with the landscape indicators which appear on seals in the Neopalatial period, and also the rocky landscape which surrounds the goats on the 'Sanctuary rhyton'. By modern standards these Marine-style depictions are inaccurate; shells, which could be either purple shell or triton, float between rocks and plants rather than crawling along the bottom (Figure 6.8b). If they are tritons, it was perhaps their affordances as vessels that resulted in them being depicted upright. Other objects including forms which are often identified as starfish, and also double axes, are shown in the same way to show that they are underwater and associated with the sea. Marine style establishes a set of associations between marine forms and provides them with an underwater location.

Decay

The ceramic stages succeeding Late Minoan IB, the last pottery phase of the Neopalatial period, are seen in terms of the dissolution of naturalism, with the standardisation and eventual abstraction of natural forms (Betancourt 1985: 159). Furumark (1941: 167) discerned a new interest in the ornate and grandiose: 'Without regard to nature designs are made symmetrical, elements belonging to one motive are added to another, and pictorial motifs are made ornamental.' Palace-style jars from Knossos, from the initial stages of the Final Palace period, show increasingly symmetrical octopuses with only six tentacles (Figure 6.9). From this period onwards, the copying of designs and

6.9. Palace-style jar from Knossos decorated with octopus, LMII. Height: 74.5 cm (AN1911.608). © Ashmolean Museum, University of Oxford

their dissolution over time can be seen; the referent is palatial designs rather than animal forms. Naturalism is a phenomenon which peaked in the Neopalatial period; after this, depicting human-animal relationships decreased in importance. Instead, copying Neopalatial designs became significant, resulting in a decrease in realism (Shapland 2014). The referents of Final Palace depictions of octopuses on pottery were Marine-style octopuses, and the compelling depictions of marine animals underwater were replaced by lifeless, symmetrical forms.

Marine animals became less common on seals after the Neopalatial period but they continued to be present at the centre of the palatial collective in the Final Palace period. The well-known 'Dolphin fresco' from Knossos was almost certainly a floor painting (Koehl 1986a). One piece of evidence for its position on the floor was part of a stone frame, similar in form to the rocky elements found in Marine-style pottery. It is likely that the Dolphin fresco was created in the Neopalatial period, and was broken up during the Final Palace period (Hood 2005: 54). Another marine floor painting was found *in situ* in the Final Palatial villa at Ayia Triada. This showed an octopus and fish (Militello 1998: 148–53). Margaret Gill (1985: 67) makes the interesting point that the red colour used for the octopus on the Ayia Triada floor suggests 'an awareness of the habits of the living animal not merely a pale corpse seen in the hands of a fisherman'. These frescoes brought the animals of the sea into the centre of the palatial collective. But unlike stone and pottery vessels depicting marine forms, floor paintings offered a different type of interaction with marine animals. They enhance the realism of what David Summers (2003: 43) terms the 'virtual space' of the picture by establishing the surface of the fresco as a metaphorical water surface and allowing the viewer to walk on water.

Marine floor paintings are also found in mainland Greece after the destruction of Knossos and Ayia Triada; an octopus appears on the floor of the Throne Room and elsewhere in the Pylos palace and there are other examples of marine subjects on floors at Tiryns (Hirsch 1977). But at Pylos octopuses were enclosed within one of the squares in a chequerboard design, with dolphins or fish in another (Blegen and Rawson 1966: 84, 214). What originated in an attempt to relocate the viewer had been replaced with an orderly, symmetrical design in which marine animals became filling elements. This indicates that marine floor paintings were important devices at Knossos and elsewhere since they appear to have been borrowed by the mainland Mycenaean palatial collectives. Another echo of the importance of marine depictions occurs on Postpalatial Cretan larnakes. At the site of Armenoi in western Crete these often combine marine imagery with depictions of hunting; both can be seen as relating to animal practices formerly under the control of the Knossos palatial collective. As with pottery it seems that marine depictions lost their indexical connection with the sea after the Palatial period and instead indexed the palatial collective itself. By extension, marine imagery came to symbolise the palace as a result of its prevalence in earlier palatial material culture.

The depiction of octopuses on these larnakes is typical of Postpalatial pottery; they became symmetrical or were reduced to tentacles without bodies (Watrous 1991: 294; Tsipopoulou and Vagnetti 1997). Octopuses and other marine forms were widely used on Mycenaean pottery, becoming part of a wider Late Bronze Age collective centred on the mainland, and continued to be used to decorate vessels after the end of this palatial collective in around 1200 BCE. The Octopus style of Late Minoan IIIC shows highly stylised octopuses but its origins can be traced back to the Marine style of Crete and earlier relief vessels (Paschalidis 2009; Alberti 2013). The endurance of marine imagery after the end of the Neopalatial period can be attributed to the effective way in which the palatial collective deployed the bodies and shells of marine animals as a demonstration of its associations with the sea. These forms became referents of the palatial collective, and were repeated over and over again until, as the link with the sea was lost, they became abstract forms. In order to understand how marine imagery became so prevalent it is necessary to return to the palace at Knossos at the start of the Neopalatial period.

BRINGING THE SEA INTO THE PALACE

The deposit known as the Temple Repositories was discovered in two cists hidden under the floor of a room to the south of the Knossos Throne Room, and dates to the early Neopalatial period. It consists of a range of ceramics, sealings, things made of faience and thousands of seashells, seemingly deliberately buried together (Panagiotaki 1999; Hatzaki 2009) (Figure 6.10). As Evans records:

> The shells from the Repository [. . .] bushels of which were taken out, were the ordinary seashells of the neighbouring coast, many varieties being included, though cockles were the most abundant. But they had been streaked and banded with brilliant artificial tints – crimson, Venetian-red, orange, brown, green and black – tastefully applied in unison with the natural lines and hues. (Evans 1921b: 518–19)

The Temple Repositories assemblage is different in composition from other Neopalatial shellfish assemblages because it is different in intention (see Figure 6.3). Rather than being eaten and discarded, the shells here were gathered together, and some were painted. The majority of shells come from sandy beaches, probably nearby, instead of the rocky shores where limpets and top shells were gathered. There are a few top shells in the deposit, all of which

6.10. Objects from the Knossos Temple Repositories (Sir Arthur Evans Archive). © Ashmolean Museum, University of Oxford

are painted. It is the most diverse (in terms of different genera) Cretan Bronze Age shellfish assemblage, with thirty-seven genera represented. This too differentiates it from collections of shells gathered for food, since only a few types were normally collected. Cockles were turned into animal things at Knossos from the Neolithic period onwards through piercing, which enabled them to be worn. At the Temple Repositories they were further elaborated, either through painting, or through imitation in faience and depiction on sealings. This assemblage contains animal things which are both traces, the gathered shells, and metaphors, the faience imitations. The formation of the assemblage helps to reveal a set of connections at the heart of the Knossos palatial collective.

Faience, a glazed material which was produced across the Bronze Age Eastern Mediterranean, was used in the Temple Repositories to create a variety of forms including twelve cockles, six argonaut 'shells' and two flying fish. Neither argonaut nor flying fish remains have been identified in Bronze Age contexts on Crete. Evans interpreted these faience objects as part of a panel, along with faience rocks, but at least some are likely to have been free-standing. Another faience vessel had an embossed cockle-shell pattern. Of this collection, Panagiotaki (1999: 81) writes: 'It is significant that such a variety of marine creatures is portrayed. Their existence in the TR may point to the veneration of the sea, perhaps in the form of a sea goddess' (Panagiotaki 1999: 81). The Temple Repositories have long been associated with goddesses because of Evans's identification of a faience figure found here as a 'Snake goddess' with her votaries (Evans 1903: 74–87, 1921b: 500–6). Whatever their identity, what distinguishes them is their elaborate clothing, whose designs indicate that they were made of dyed textiles. There are also two faience dresses with a crocus design, showing the source of one of the dyes used in textile manufacture (Evans 1921b: 506). This is not the first use of faience in Crete but its sophisticated use suggests that these things were made in the palatial workshops (Foster 1979: 60). The diverse forms created from the same material acted to link together these various different parts of the palatial collective, particularly elaborately dressed female figures and marine forms.

Evocations of religious significance do not in themselves explain why these animal remains and objects were deposited. Vesa-Pekka Herva (2005: 217–18) prefers to see such deposits in terms of an animistic conception of human-environment relations, in which buildings are to be seen as dynamic entities, engaged in a process of interchange with their inhabitants. He acknowledges that such deposits could involve social relationships between people too, and the manipulation of power relations. But his prime concern is with the way in which deposits such as the Temple Repositories should not be seen as containing religious objects in storage, but in themselves are an important part of the fabric of buildings. As argued in Chapter 2, however, it is not clear that this animistic worldview is appropriate to Minoan Crete. Eleni Hatzaki (2009) argues that the Temple Repositories can be seen as an example of structured deposition; buried deposits need not be seen as waste but can gather together objects for a purpose, in this case to take them out of circulation. The shells in the Temple Repositories confound the usual reason for collecting shellfish, that of consumption. Instead, sea shells were brought to the centre and turned into artifacts, both through the addition of paint and through manufacture in a different material. The painted shells also indicate the use of various colouring agents, analogous to (or perhaps identical to) the dyes used for cloth. Marine animals which were more difficult to collect, and which would lose their distinctive forms when on dry land – the fragile argonaut and the flying fish – have been materialised in faience instead. Although these objects could have had a variety of meanings, one can suggest that both the shells and faience objects share an indexical link with the sea.

The Temple Repositories demonstrate how the Knossos palatial collective was gathering together things from the sea and connecting them with the production of dyed cloth. The Talismanic seals, which appeared at around the same time, show the opposite process: the dissemination of colourful things associated with the sea and the land, by the same palatial collective. Talismanic seals appear to be made in palatial workshops because although they were quick to produce, they were made using specialised technology and imported hard stones. Matthew Haysom (2011) has suggested that Knossos was the initial centre for designs which are focused on pouring liquids (various pouring vessels are also Talismanic designs) and sea creatures, given that a similar iconography, although not style, is seen in the Temple Repository sealings and other objects. As argued above, this watery connection was reinforced by the colour of the stone. On the basis that these seals were not apparently used within the palace because they were rarely used for sealing documents, and the large number surviving, it appears that these seals were not for palatial officials, but for others who interacted with the palace (Murock Hussein 2018). Such seals could have been rapidly produced and distributed widely at the start of the Neopalatial period to establish an association with the Knossos palatial collective.

Talismanic seals can be seen as the successor to the soft-stone Protopalatial three-sided prisms.[11] Both were produced in large numbers, but the Talismanic seals adopted the material and drill carving technique of the more exclusive hard-stone Cretan Hieroglyphic seals. As Anna Margherita Jasink (2016) has argued, many of the Talismanic designs were borrowed Cretan Hieroglyphic signs, including the pouring vessels and bucranium, although there are not clear antecedents for the marine animals.[12] Cretan Hieroglyphic went out of use at around the time Talismanic seals started to be produced, marking a separation between seal designs and the script (Linear A) used in administration. Sealstones continued to be the point of contact, both literally and metaphorically, between the seal user and the palace, even if Talismanic seals were rarely used in this way. Instead, they acted as index of the palatial collective, through designs which were associated with earlier palatial administration and animal practices such as fishing and hunting. The Temple Repositories show an association between the palatial collective and dyeing, and the use of carnelian for a large number of Talismanic seals could have been linked with the orange-red colour of saffron. In contrast to Talismanic seals showing fish and cuttlefish, there is a strong association between goats and red/brown seals (Shapland 2010b: 119). As discussed in the previous chapter, goats and crocuses appear together on the Sanctuary rhyton and Ayia Triada fresco, which, like the Temple Repositories, were found at the heart of palatial buildings.

Just as sheep (or crocuses) were rarely shown on seals, purple shells also appear to be absent despite their close connection with dyeing. The few shells depicted appear to be triton shells, which were gathered in the same places as purple shells, either by diving or baited traps. These were different practices from the coastal gathering and fishing for food that are apparent in the fish and shellfish assemblages from Bronze Age Cretan sites. The marine animals frequently depicted in Bronze Age Crete were encountered away from the shore, either by diving underwater or going out in a boat. These two places were of particular economic importance to the palatial collective because purple shells and tritons were gathered underwater and sailing ships crossed the sea as part of a trading network which spanned the Eastern Mediterranean. In both cases the bodies of marine animals acted as analogies for these parts of the world: unfamiliar argonauts and flying fish for sea travel; octopuses and triton shells for the underwater world. The animal things imitating their forms in stone or faience helped to demonstrate the expansion of the palatial collective. The unfamiliar bodies of exotic animals also came from over the sea and these will be discussed in the following chapter.

The analogy between purple shells and saffron crocuses as dye sources is significant because the territories in which these were found were defined by particular animal practices. For crocuses this was the hunting of wild goats; for purple shells this appears to have been diving for octopus and triton shells. Both were risky pursuits offering status to their participants. Octopuses are more difficult to catch than triton snails because they are more intelligent; they can move fast, hide and fight back. The Ambushed octopus rhyton need not be seen as a generic underwater depiction of an octopus, but the moment of encounter with a diver. By controlling the means of depicting this practice, the palatial collectives came to be associated with the animal practice, and by extension the place it occurred. It is no accident that Neopalatial pottery showed crocuses and the underwater world because these were associated with the palatial economy. The site of Alatzomouri-Pefka provides concrete evidence for the large-scale dyeing of cloth with a variety of colours in the Protopalatial period, while Neopalatial Chryssi shows the specialised production of purple dye (Apostolakou et al. 2016, 2020). Both seem to have been linked to nearby palatial collectives.

Both octopuses and triton shells appear among the Protopalatial sealings at Phaistos; triton shells were another important means of demonstrating connections with the underwater world. Triton shell vessels, found at the same site, were enhanced animal things, a combination of the animal form and skilled crafting (Sanavia 2014). They could be manipulated in palatial ritual, bringing these members of the palatial collective into the hands of those at the centre. The same can be said of Marine-style pottery in general, which brought the underwater world onto land. Both were distinguished by naturalistic depiction, reproducing the forms of marine animals. As Evans suggested, this naturalistic depiction can be traced back to the Protopalatial period, with the moulding of cockle shells and crabs as appliqués to pottery. It was in this period that control over the sea became increasingly significant as the trade in woollen cloth began to increase in importance, entailing control over not just sheep and human cloth workers, but the sources of the coloured dyes that distinguished Cretan textiles from others produced in the Eastern Mediterranean and made them desirable. In this context, it seems likely that the abstract patterns on seal impressions from Phaistos were particular textile patterns, the act of sealing transferring them from the domestic collective where they originated to the palatial collective.

The expansion of the Knossos palatial collective, and the material culture forms developed at the start of the Neopalatial period, can partly be explained by a desire to take over this growing textile industry. Talismanic seals, many showing marine animals carved on stones the colour of the sea, were distributed to members of the collective. The Temple Repositories gathered together shells and the forms of other marine animals in faience, as well as depictions of elaborately clothed women, wearing the products of the textile industry; as if to emphasise this, two dresses are adorned with crocuses. Cycladic pottery, in the form of jugs painted with birds, points to overseas contacts. Among the sealings are abstract designs which closely resemble weaving patterns.[13] The Temple Repositories deposit demonstrates the connection between the marine world and the palatial industries of Knossos. At Ayia Triada, closely associated with Knossos, there were crocus-gatherers on the walls in the Neopalatial period, and an octopus staring up from the floor in the succeeding period. These objects, often associated with religious ritual, demonstrated the extension of the palatial collectives over the parts of the environment that produced dyes, among other things.[14] By the end of the Final Palace period this palatial collective extended to the south coast of Crete and included at least 100,000 sheep whose wool could have been dyed purple or saffron-yellow and woven into cloth by the workers also numbered in the Linear B tablets, before being sent across the sea.

CONCLUSION

The two main approaches to Marine style outlined above are iconic and indexical. Some authors have regarded the relationship between Marine-style pottery and the bodies of marine animals as one of resemblance only, tentacles adapted to the whirling forms of earlier linear designs. In tracing back the origins of Marine-style pottery to ceramic relief (nature-moulding) and the imprint of sponges (nature-printing), Evans provided an alternative, indexical origin. From this point of view the naturalism of Late Minoan IB pottery depictions was one of several ways of making marine animals present to the viewer on the surface of pottery. Marine floors took this one stage further by rendering the floor as the water's surface and allowing the viewer to look down to the marine creatures below.

The indexical approach is favoured here because it explains the timing of naturalistic depictions of marine animals. The products of the sea, particularly purple shells, became important to the palatial collective at the same time as octopuses began to appear on seals and shell reliefs on pottery. Marine forms came to be used as analogies for the exploitation of the sea by the palatial collectives. The appearance of triton shells and octopuses on sealings at Phaistos appears to relate closely to the underwater world rather than the coastal fishing and gathering which was a normal part of subsistence. The palatial collective was involved only with particular underwater products, particularly purple for dyeing woollen cloth. This came from the domain of the octopus, whose hunting was perhaps regarded as an animal practice on a par with bull games since both were associated with muscular male figures.[15] This animal practice helped to define the underwater world as connected with the palatial collective.

The sea was not just important to the palatial collective for its products; trading ships provided the palatial collectives with exotic bodies and materials in return for Cretan trade goods. The next chapter will consider how these bodies were accommodated within the palatial collective. It is no accident that a faience cat was found among faience marine animals in the Temple Repositories, just as a relief cat was placed alongside relief seashells on a vessel from Mallia. The sea became a vital resource for the Cretan palatial collectives and came to be connected with the bodies that travelled across it as well as those found under it. The importance of the sea to the palatial collective of Knossos can be seen in its echoes after the destruction of the Knossos palace; octopuses continued to adorn larnakes, the floor of the Throne Room at Pylos and numerous Mycenaean vessels. But over time they became lifeless forms, no longer members of the palatial collective whose reach extended across the sea and under the waves.

SEVEN

COLLECTING

INTRODUCTION

> The whole tone of the painting differs from that of the mature Mycenaean style of the Knossos frescoes, and the tint of the boy's body, here a pale blue, differs from the regular Mycenaean convention, in which male figures are painted a reddish-brown, while the women are white.
>
> (Evans 1900: 45)

There are two versions of the 'Saffron-gatherer fresco' on display at Heraklion Archaeological Museum. One shows a 'Blue boy' picking a crocus from a flower pot, and the other shows a monkey performing the same act (Figure 7.1). The Blue boy was illustrated in the first volume of the *Palace of Minos at Knossos* but when a more complete fresco showing monkeys stealing birds' eggs was found in the House of the Frescoes at Knossos in 1923, it became apparent that the Blue boy was a monkey too (see Figure 2.4). The colour blue was a convention in some frescoes showing humans, but for hair rather than skin. A fresco that Arthur Evans, and his restorer Émile Gilliéron *père*, assumed transgressed norms of depiction instead defied their expectations of the sort of bodies that they would encounter at Knossos. The detailed and naturalistic depiction of monkeys in the House of the Frescoes, and also at Thera, makes it likely that monkeys were brought to the Aegean during the Bronze Age (Greenlaw 2006, 2011). The frescoes do not just show monkey bodies but demonstrate knowledge of the ways monkeys move. No Aegean Bronze Age monkey bones have been found to date, however, making these frescoes the best evidence for the presence of monkeys at Knossos.[1] But does

7.1. The Saffron-gatherer fresco, MMII–III (Evans Fresco Drawing L/3 a). Watercolour drawing by Émile Gilliéron, *père*, undated. © Ashmolean Museum, University of Oxford

the fresco also provide proof that these monkeys picked crocuses? The naturalistic style of the fresco helps persuade the viewer that they did, but picking crocuses was not an innocent act in Bronze Age Crete. The recognition that the body of the Saffron-gatherer belonged to a non-human animal solved one problem of interpretation, the blue colour, but raises more fundamental questions about the role of exotic animals in the palatial collective.

Another animal painted on the walls of the palace at Knossos was the griffin. Found, like the Blue boy, in the first season of excavation, the Throne Room griffins were readily recognisable because some were complete, and preserved *in situ* (Figure 7.2).[2] Evans (1900: 40) used the term 'monster' to describe the griffins, noting their possible Egyptian influence. As he noted, unusually for griffins, these ones lack wings. Griffins are found widely in palatial depictions, sometimes pulling chariots like horses or attacking animals in the same way as lions. Yet they are often not described in the same terms as lions or horses; they are 'supernatural', 'mythological' or 'fantastic' (Crowley 2013). Evans's term 'monster' usefully echoes Classical and Mediaeval ideas about the monstrous forms which inhabited the edges of the world; lions were perhaps equally monstrous in Early Bronze Age Crete since they would only have been known on Crete from depictions and travellers' tales. The boundary between existent animals like monkeys and lions and non-existent animals like griffins is difficult to sustain for Bronze Age Crete and so griffins will be discussed alongside these other exotic animals. The detailed depiction of the griffins' bodies, like those of the monkeys, shows that the painters were familiar with

7.2. Reconstruction of the Throne Room at Knossos. Watercolour drawing by E. J. Lambert (after Evans 1935: pl. 33)

their form. Like the monkeys, these wall paintings made their bodies present at Knossos, even if the living body was not present.

A third fresco fragment, found in the early years of the twentieth century not at Knossos but at Ayia Triada, shows a more familiar animal: a cat (now regarded as part of the Park fresco; see Figure 5.5). Unlike monkeys or griffins, cats are still found across Crete, although it is not clear how today's stray cats relate to those of the Bronze Age. It is widely accepted that cats were imported to Crete in the Bronze Age but it has proved difficult to identify cat remains from Bronze Age levels. There has been a tendency to describe scenes involving cats as 'Nilotic', and so an iconographic transfer rather than observations of cat behaviour. This is despite, as Jacke Philips (1995) points out, depictions of cats hunting birds being uncommon in Egypt at the time they appear in Crete. As she also notes, a depiction of a cat hunting a bird could easily result from familiarity with cats and their hunting behaviour. In some ways cats follow a similar trajectory in Bronze Age Cretan depictions to monkeys: stiff depictions of the animal in the Protopalatial period on seals, followed by naturalistic images of their behaviour and interaction with the landscape in the Neopalatial period. The appearance of a faience cat in the Temple Repositories, whether or not the reconstruction of its position on one of the female figures' headdresses is correct, suggests that cats were still important animals at the end of the Middle Minoan period. Cats too were exotic animals on Crete in the Bronze Age, despite being a common sight on Crete now.

All of the land mammals found on Crete in the Bronze Age, with the exception of two species, a spiny mouse and a shrew, were brought to the island by humans. Cattle, goats and sheep, as well as animals such as badgers and martens, were imports to Crete in the Neolithic period, some of which had formed populations outside human control by the time of the Bronze Age. This chapter focuses instead on the animals imported to Crete during the Bronze Age, including cats, deer, monkeys and horses, and, most likely, some breeds of dog. The term 'exotic' is used by ecologists and environmental historians to describe animals and plants imported from outside of a particular biome or ecosystem. J. Donald Hughes (2003) has traced the consumption of 'exotic biodiversity' in the Mediterranean from ancient Greek temple menageries to the animals imported to the Roman arena for slaughter. In the absence of contemporary written sources it is more difficult to establish which exotic animals were imported to Crete. In some cases the animal appears to have become known from depictions before its physical presence; for animals such as lions and deer, it is likely that skins and teeth and other body parts were also in circulation. These animal things, both body parts and depictions, were incorporated into the collectives of Bronze Age Crete from the Prepalatial period onwards. These exotic bodies frequently provided a tangible connection with the world beyond Crete, and over the course of the Bronze Age the significance of these different bodies shifted, particularly when the animals themselves arrived from over the horizon.

The bodies of these animals were depicted alongside other animals such as sphinxes, 'Minoan geniuses' and griffins. These forms are often lumped together as fantastic creatures but each exotic body had a different biography. Lions and monkeys were monstrous forms in the Early Bronze Age but their depiction changed as related animal things like pelts, and perhaps the animals themselves, arrived on Crete. Monsters such as the dragon, sphinx and griffin were absorbed into the palatial collective as animals; they are depicted in the same way as other animals, sometimes led by humans or drawing chariots. Egyptian deities including Tawaret, the hippo goddess, and Bes were adopted and given human-like roles. On Crete, where Evans renamed her the 'Minoan genius', Tawaret is often shown holding a jug or hunting an animal (Evans 1921b: 200). This exemplifies the way in which analogist collectives deal with the outside world, characterised by Philippe Descola (2006: 400) as 'muddled collections of unfamiliar existing beings whose sporadic growth has to be contained and whose chaotic nature must be reduced by absorbing them into the sociocosmic order in which a place has already been prepared for them'. The drawing of analogies allows the unfamiliar to be assigned its place within the collective, whether unknown bodies are brought into the centre or always located somewhere over the horizon. This chapter will examine how these exotic bodies and animal things from overseas were incorporated into the expanding collectives of Crete.

EXOTIC ARRIVALS

It is difficult to pinpoint the arrival of exotic species using zooarchaeology alone. Even in cases where the remains of animals have been found and identified at Bronze Age sites, iconographic evidence tends to suggest earlier knowledge of the animal. Cats are widely depicted from the Protopalatial period onwards but very few cat bones have been found at Bronze Age Cretan sites. It was only with the reanalysis of the Knossos faunal assemblage that the first Bronze Age cat bone was identified at this site (Isaakidou 2004: 133).[3] Cat remains were found at Postpalatial Chania and also Kastro, although these could be post-Bronze Age (Hallager and Hallager 2003: 137; Snyder and Klippel 2000: 68). The horse provides another example of an animal which definitely arrived in the Bronze Age; a combination of horse bones, iconographic evidence and Linear B documents points to the presence of horses on Crete in the Final Palace period. The earliest securely identified horse remains on Crete also date to the Final Palace period, although there are less well-documented reports of horses and donkeys from early excavation reports going back to the Prepalatial period.[4] The majority of horse remains come from the Minoan Unexplored Mansion near the palace at Knossos (Bedwin 1984). A complete but disarticulated horse burial was found at the entrance to a side-chamber in Tholos Tomb A at Archanes-Phourni (Sakellarakis 1970). But there are also depictions of horses from the preceding Neopalatial period which help to support the identification of horse bones at Neopalatial Tylissos (Hazzidakis 1921: 76). Uniquely the transport of a horse by ship is depicted on a Cretan seal from Knossos, although it can only be given a broad stylistic Late Bronze Age date (Evans 1928: 243–4).[5] For both cats and horses it is likely that the depictions provide a better indication of the time of arrival than the first bones because the actual remains of the presumably small numbers of imported animals would be difficult to find archaeologically.

There is some scope for genetic testing of present-day animal populations in order to trace them back to the Bronze Age or earlier, as has been attempted for human populations on Crete (Martinez et al. 2007). There have been many waves of cats and horses brought by various human invaders to Crete, contributing to the ancestry of present-day animals. It is possible that the wild cats found in the mountains of Crete, which have been identified as closely related to the ancestor of domestic cats, *Felis sylvestris libyca*, rather than the European wildcats, are a remnant of a Bronze Age population (Masseti 2012: 128–32). Genetic testing of cats has suggested that the dispersal of cats in the Eastern Mediterranean from the Neolithic period onwards was most likely linked to trade routes (Ottoni et al. 2017). Some of the first cats brought to Crete could have rapidly formed a feral population outside human control, as has been demonstrated for agrimia, and provisionally dated, using DNA sequencing (Horwitz and Bar-Gal 2006). The curly tailed hunting hound still bred on Crete today could be an example of an animal whose roots can be traced back to the Bronze Age, although at the moment this is based on iconographic evidence (Wall-Crowther 2007). As noted in Chapter 2, nearly every land mammal living on Crete today is the result of human introductions at some point in the past, although some of these introductions are still difficult to date (Vigne 1999: fig. 2).

Detailed depictions of animals, particularly on large-scale frescoes, raise questions of whether they were observed directly in the Aegean or drawn from observations or depictions made overseas. In the case of monkeys, there is no consensus about which species is depicted, and hence whether they were from North Africa, as is usually assumed, or elsewhere (Evans 1928: 448; Cameron 1968: 5; Greenlaw 2011; Pareja et al. 2020). Although there are morphological differences between monkey species, the debate hinges on the problematic assumption that Aegean Bronze Age fresco painters were trying to depict an animal body accurately by modern standards. Inaccuracies either show that an animal was not drawn from life, or that what are now seen as species markers, such as coat markings or tail position, were ignored. Another animal which is depicted throughout the Bronze Age, although it was not found wild on Crete, is the lion. Late Bronze Age depictions on seals demonstrate detailed knowledge of lion morphology and behaviour but, like monkeys, no lion bones have yet been found on Crete. Unlike monkeys, lions were to be found wild in mainland Greece and Anatolia in the Bronze Age, providing a source of knowledge of lions independent of imported depictions (Thomas 2004, 2014). It is possible that lions were brought to Crete in the Late Bronze Age, as John Younger (1988: xi) puts it, as a 'palace showpiece', or even kept in hunting parks (Blakolmer 2020b). The same can be said of leopards, since a fresco fragment from Knossos seems unmistakeably to show a leopard (Evans 1921b: 540; Evely 1999: 219). Animals like these, if they were brought to Crete, would have been kept in captivity and so would never have formed a feral population. The bones of exotic animals kept captive in menageries have been found elsewhere (O'Regan et al. 2006), but such remains are likely to be few in number and easy to overlook.

Conversely, the presence of particular body parts does not indicate the presence of the live animal. Remains of three species of deer, roe, fallow and red, have been found at Bronze Age Cretan sites, but antlers and bones from the extremities could have arrived as trophies. The small number of remains of fallow and roe deer at Protopalatial Kommos can be explained in this way, although the presence of live animals cannot be excluded (Reese 1995b: 190–1;

Ruscillo 2006: 795–6). The same can be said of the sporadic deer remains reported at various sites, including Final Neolithic Knossos, Palatial Ayia Triada and Neopalatial Chalinomouri near Mochlos (Jarman 1996; Wilkens 1996a, 1996b, 2003; Vigne 1999: 301; Reese 2004: 120). There were three roe deer antlers in the Temple Repositories at Knossos, at least one of which could have been collected after shedding (Panagiotaki 1999: 118). By contrast with these non-meaty body parts, the range of fallow deer bones at Neopalatial and Final Palatial Knossos is clear evidence for their hunting and consumption on Crete (Isaakidou 2007b: 16). Valasia Isaakidou suggests that these animals could have been kept in a deer park, since deer remains are not found in large quantities on Crete. There is also evidence for consumption of deer at Postpalatial Chania; the high proportion of deer limb bones in pits indicates large-scale events defined by the consumption of both fallow and red deer (Harris 2014b). These were short-term introductions since deer are no longer present on Crete, but it is possible that a feral population had emerged.

Other animal things were traded primarily as exotic materials. Ostrich eggs, elephant tusks and hippo teeth were all found in the Late Bronze Age Uluburun shipwreck, found off the coast of Turkey, which appears to have been heading for the Aegean (Pulak 1998: 203–4). Whole elephant tusks, probably from the extinct Syrian elephant, and ostrich eggs were found in the Neopalatial destruction levels at Zakros (Platon 1971: 136, 245). Both elephant and hippo ivory, and to a lesser extent ostrich eggs, were used as raw materials in Bronze Age Crete but it is unclear how much knowledge of the animal travelled with them. Hippo tusks were frequently used for Prepalatial Cretan seals, whose distinctly Cretan iconography show that they were carved on the island (Krzyszkowska 1988). The same can be said of later sculpture and containers made from elephant ivory, including finely crafted objects such as the Palaikastro Kouros and Mochlos pyxis (MacGillivray et al. 2000; Soles 2016). Ostrich egg fragments have also been found on Crete at Prepalatial Palaikastro and a late Protopalatial tomb at Knossos (Dawkins and Currelly 1904: 202; Hood et al. 1959: 260; Reese 1985: 372–4). A Kamares Ware skeuomorph of an ostrich-egg rhyton was found at Knossos, dating to around the same time as the tomb find, suggesting that ostrich eggshells were being turned into rhyta by this time (Evans 1921b: 594–5; Sakellarakis 1990; Warren 1995: 6). One such ostrich-egg rhyton from Shaft Grave V at Mycenae had a faience neck and added faience dolphin appliqués, whose closest parallels come from the Knossos Temple Repositories (Karo 1930: 146). Here the addition of dolphins appears to reinforce the overseas connections of the egg.

A useful analogy for the role of depictions of monsters in the Bronze Age Aegean comes from Mediaeval maps, which showed monstrous forms at the edge of the known world (Mittman 2006). As John Friedman (2000: 35) notes, 'Extremes of form and of place were closely linked in antiquity and the Middle

Ages'. The same argument has been made for all agricultural societies by Mary Helms (1993, 1999), who suggests that things from a long distance away are invested with power which can be harnessed by those who bring them to the centre. In the case of exotic materials, their unfamiliar properties act as markers of distance but the question arises of how the forms and affordances of unseen monsters come about. The exotic bodies existing on the edges of Mediaeval maps were often recognisable and clearly defined entities such as unicorns or Blemmyes, headless humans with eyes on their chests. Mediaeval map-makers relied on the accounts of previous authors, particularly Pliny, providing textual authority for the existence of monstrous forms, even though these had never been seen in person. Images were also copied, resulting in the forms of monsters becoming stable (Mittman 2006: 79–80). Images of exotic creatures such as unicorns, which are now regarded as non-existent, were circulated and printed in natural history books into the early modern period (Mezzalira 2001: 33).

The same process has been seen at work in early state societies by David Wengrow (2011, 2013), who suggests that craft specialisation resulted in the first era of mechanical reproduction, resulting in forms such as sphinxes and griffins becoming stabilised. Following the work of cognitive scientists he suggests that these hybrids were popular because they were counter-intuitive, combining parts from different animal bodies. The island of Crete was at one remove from this process of invention; the arrival of these hybrid forms on objects such as sealstones and amulets allowed local groups to 'engage with and encompass the foreign' (Wengrow 2011: 144). But of the various monstrous forms circulating on Mediterranean trade routes, only a few were copied on Crete and fewer still were given a prominent iconographic role (Zouzoula 2006; Wolf 2019). The following sections examine how exotic and monstrous bodies were incorporated into local and palatial collectives.

LOCAL COLLECTIVES

Crete became increasingly linked with a wider Eastern Mediterranean network at the end of the Early Bronze Age, resulting in an influx of exotic forms. The contact between Crete, the Near East and Egypt from this period onwards has been extensively discussed, and the surviving imports catalogued (Pendlebury 1930; Buchholz 1999; Aruz 2008; Philips 2008). The development of sailing technology facilitated the arrival of new materials such as hippo tusks and new forms such as scarabs, lions and monkeys. Many of these imports were deposited in communal tombs, particularly in the south of Crete (Xanthoudides 1924; Legarra Herrero 2014). Here they were mingled with the bones of the human members of the collective, whereas in life they could have provided the basis for differentiation by demonstrating overseas contacts and knowledge of unfamiliar materials and bodies. It is also likely that at least some of them arrived with people from overseas since some of the most common imports are closely associated with care of the female body (Wengrow 2010). But there is scarce evidence for the import of exotic animals in this period, only depictions and perhaps stories preserved in oral poetry (Anderson 2016: 129, 2020).

Hippo ivory was used to make seals in the Prepalatial period but there are no depictions of hippos associated with this material or any other, making it unlikely that knowledge of the animals travelled (Krzyszkowska 1983, 1988, 1989). Ivory was used for the most elaborate and finely carved seals, including cylinders whose ends were frequently carved with circular arrangements of animals including lions (Yule 1980: 208–209) (Figure 7.3a). These lions, shown in profile, are easily recognisable for the modern observer but would not have been seen in Crete. It is likely that knowledge of the lion form, like the material of these seals, was imported, but the seals themselves are a distinctively Cretan product (Anderson 2016: 20). Lions are shown with other, more familiar Cretan bodies including scorpions and spiders. These animals could have been apotropaic because of their poisonous bite, and their association with lions suggests an analogy for the lion as another apotropaic animal. Such harmful animals could have reinforced the affordance of the seal to protect goods; a small number of impressed sealings are known from the Prepalatial period (Schoep 1999: 269; Vlasaki and Hallager 1995). In general the

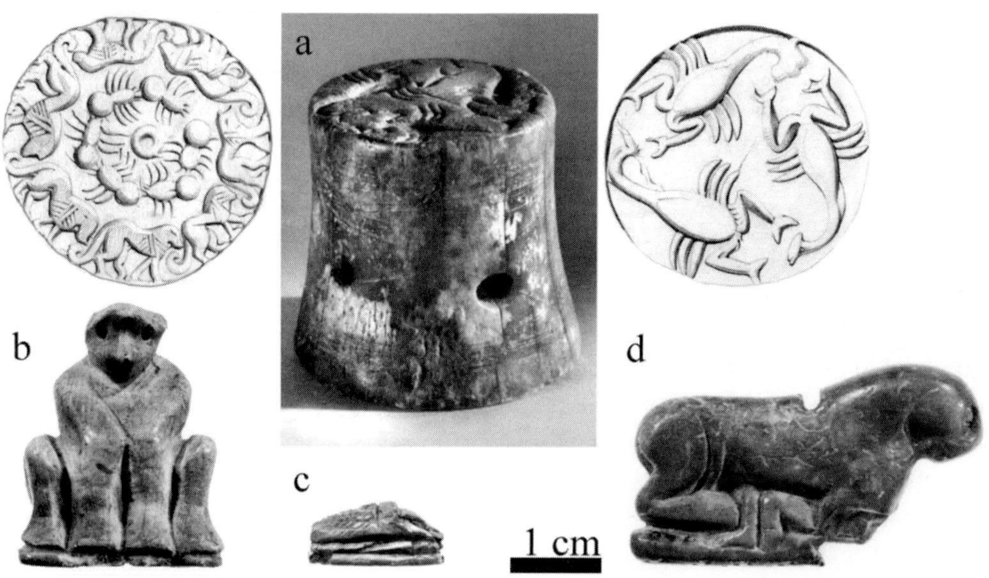

7.3. Prepalatial cylinder and zoomorphic seals, EMIII-MMIA. (a) Bifacial cylinder seal, Platanos, with drawings of two faces (drawings not to scale) (CMS II 1 248); (b) seated monkey seal, Messara (CMS III 2); (c) scarab seal, Aspri Petra (CMS II 1 1); (d) lion and human, Kalathiana (CMS II 1 130). © CMS Heidelberg

symmetrical or rotational designs create a sense of perpetual motion rather than any feeling of a specific time or place, unlike the seal designs of later periods (Wiencke 1981: 254). These designs, like the abstract knots and spirals of Prepalatial seals, can be described in Alfred Gell's (1998) term 'technologies of enchantment', aimed at trapping the viewer rather than indexing something else. A few seals instead juxtapose human and lion in a symmetrical design which has been interpreted as a lion-hunt scene, but these lack the violence of later hunts, making the relation between human and lion ambiguous (Krzyszkowska 2005: 67; Anderson 2016: 210–35, 2020: 353–4).

Another type of seal, sometimes also made using imported ivory, was zoomorphic (Sbonias 1995: 44–6). One of the more common forms is the 'seated ape' figure which resembles a baboon and has close analogies with Egyptian and Syrian forms (Aruz 2008; Phillips 2008: 180–1) (Figure 7.3b). The scarab was another imported zoomorphic form, sometimes made using an imported technology similar to faience, the so-called 'white pieces' (Pini 1990, 2000b) (Figure 7.3c). For both the baboons and the scarabs the designs on the seal face were distinctly Cretan, again emphasising that these could be used for sealing in a local context as well as personal adornment. These zoomorphic seals do not indicate familiarity with the depicted animal, although one seal from Kalathiana in the shape of a lion is crouching over a human, indicating an affordance of lions as dangerous animals (Figure 7.3d). This, however, is an exception; the seated apes, based on Egyptian depictions of baboons, continued to shape the way monkeys were shown until the Neopalatial period, indicating sparse knowledge of this animal. Similarly, the scarab form continued to be used for seals into the Protopalatial period, but the way scarabs were carved shows little knowledge of their connection with beetles, despite there being similar dung beetles on Crete (Phillips 2004: 166, 2008: 132). Unfamiliar bodies and materials were absorbed into domestic collectives in the Prepalatial period but these objects appear to index the world beyond Crete rather than having any specific relationship with exotic animals themselves in this period.

One possible exception which depicts a newly imported animal in the Prepalatial period is a unique fourteen-sided bone seal from the cemetery of Archanes-Phourni (Krzyszkowska 2005: 70–2). Five of the seal faces depict quadrupeds, and two of these have been identified as horses because of their long necks (Sakellarakis 1970: 209–10). Horse teeth were reported from another Prepalatial tomb, Tholos A at Ayia Triada, but these are more likely to be either imported or from a donkey (Banti 1933: 216; Preston et al. 2013: 63–7).[6] There are no securely dated donkey remains from this period either but there is circumstantial evidence for donkeys in the Aegean from the third millennium (Whitelaw et al. 1997; Brodie 2008). Firm evidence for donkeys on Crete belongs to the Postpalatial period in the form of a few remains and, possibly, a vessel in the shape of an equid carrying two pots from Postpalatial Phaistos (Koehl 2006: 79).[7] Whether this is a donkey or a mule, its rarity highlights that equids were absorbed into domestic and palatial collectives as beasts of burden largely without trace.

PALATIAL COLLECTIVES

New Arrivals

New materials, technologies and bodies continued to come to Crete as the increasingly differentiated collectives centred on the palaces became involved in Eastern Mediterranean trade. Two of the most significant technologies to arrive at the start of the Palatial period involved rotational energy. The first was the fast wheel used for making pottery and the second was the bow drill used particularly for seal-cutting (Knappett 1999b; Krzyszkowska 2005: 83). Each technology established a new set of relations between craftsperson and material. The affordances of different materials were extended; the plasticity of clay meant that it could be spun into new, more symmetrical shapes while previously unworkable hard stones could be carved with intricate seal designs. Seals, made of imported, colourful hard stones, often showed the bodies of new arrivals such as cats or griffins. Over time the diversity of animals on seals grew, peaking in the Neopalatial period (see Figure 5.3). The following section seeks to explore the appearance of some of these new bodies and their incorporation into the palatial collective.

Cats, Monsters and Monkeys

Cats, either whole or just the face, first appear on colourful Protopalatial hard-stone seals and vessels. Cretan Hieroglyphic seals showing cats brought together the imported technologies of writing and gem-cutting, imported raw materials and, in the case of cats, imported animals. The cat face is not accepted by all scholars as a sign in the Cretan Hieroglyphic script because it has not been found inscribed on clay tablets but it does appear together with accepted signs on seals (Olivier 1981: 109; Younger 1999: 387; Jasink 2009: 140) (Figure 7.4a). Complicating the debate further are two seals which show a whole cat seated in profile, but with a frontal face, alongside Cretan Hieroglyphic signs (Figure 7.4b). This raises the question of whether cat faces are abbreviated cats, and whether either is acting as a syllabic sign or modifying the meaning of the inscription in some other way (Krzyszkowska 2015; Ferrara et al. 2016: 88–90). One way to understand the significance of cats on these seals is to trace the development of the cat face through the Linear A and Linear B scripts, where it became the sign for the syllable 'ma' and, with the addition of

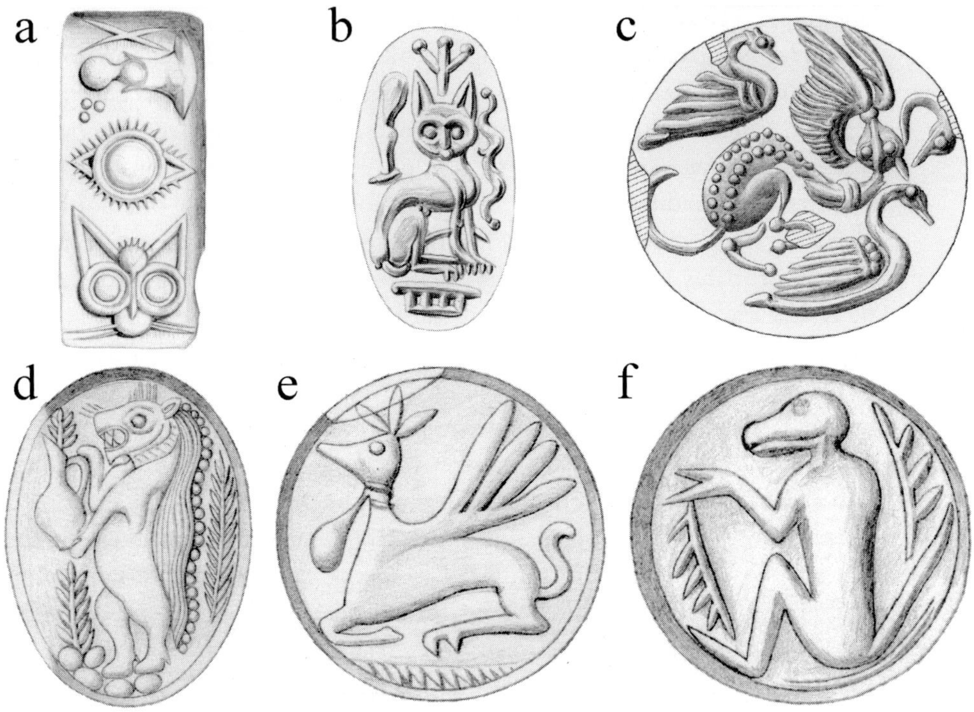

7.4. Drawings of seals showing cats and other exotic animals. (a) Cretan Hieroglyphic signs including cat face from four-sided prism, MMII (CMS II 2 316d); (b) cat with Cretan Hieroglyphic signs from three-sided prism, MMII (CMS VI 93a); (c) cat and waterbirds, Archanes, LMI–II (CMS VI 367). Sealings from Phaistos, MMII: (d) Minoan genius (CMS II 5 322); (e) griffin (CMS II 5 317); (f) seated monkey (CMS II 5 297). © CMS Heidelberg

another sign between the ears, the ideogram for wool (Killen 1962; Younger 2005).[8] This suggests an analogical association between cats and wool, both related to overseas trade. At the same time as the sign was in use, relief appliqués in the form of seated cats were attached to vessels from Quartier Mu in Mallia. These two cups, and a bridge-spouted jar, were also decorated with appliqués of plants and marine forms (Detournay et al. 1980: 122–5; Poursat and Knappett 2005: 115). These vessels can be seen as establishing an analogy between the bodies of animals from under the sea and another which arrived by sea on a sailing ship.

These cat depictions have often been seen as borrowings from Egyptian iconography rather than evidence for the presence of cats on Crete (Immerwahr 1985: 41; Karetsou and Andreadaki-Vlazaki 2001: 56–7). As Jacke Phillips (1995, 2008: 202–6) argues, the lack of plausible Egyptian prototypes makes it more likely that cat depictions are based on the animal itself. Another argument for the importation of cats in this period is that cats are shown both crouching and sitting on Protopalatial seals, in addition to the cat faces. This suggests that Cretan seal-cutters had to develop their own iconography of cats based on their characteristic postures. There are Egyptian parallels for Neopalatial seals showing cats hunting waterbirds, but this does not preclude the importation of cats to Crete as this could be based on observation of cat behaviour (Morgan 1988: 147; Militello 2000) (Figure 7.4c). Cats are not depicted catching rodents in Crete; it is possible that martens or other small predators were traditionally kept in houses for this purpose.[9] But household activities are generally not depicted in Bronze Age Crete and so even though cats were probably useful rodent-killers, they are shown hunting birds instead. The Ayia Triada 'Park fresco' shows a cat stalking a bird in the same landscape as wild goats. Since goats were related to the animal practice of hunting, the cat stalking the bird can be read as an analogy for another animal practice, fowling. As noted above, a faience cat was also found in the Knossos Temple Repositories. Here, as at Ayia Triada, it was depicted in the same medium as goats, crocuses and elaborately dressed women, helping to show the position of the cat in a broader assemblage of things with connections beyond the palace.

Other exotic bodies were imported from the Eastern Mediterranean and incorporated in similar ways to the cat. A ceramic appliqué of a sphinx, probably based on a Levantine form, was also found at Quartier Mu in Mallia (Detournay et al. 1980: 116–18; Winkler-Horaček 2011). The Egyptian god Bes/Beset was carved on hard-stone seals, most commonly the head, but is also shown frontally on a prism from Petras (Krzyszkowska 2012; Weingarten 2015). The Minoan genius was imported at the same time as cats and sphinxes and is based on the Egyptian hippopotamus goddess Ta-urt/Tawaret, (Evans 1935: 433; Gill 1964; Weingarten 1991; Rehak 1995a; Phillips 2008: 157–67). The process of incorporation was different to these other animal forms since the Minoan genius is often depicted holding a jug. This appears to establish an analogy between this body and a liquid of some kind. Minoan geniuses are shown pouring from jugs on a chlorite triton-shell rhyton from Mallia (Baurain and Darcque 1983). This object establishes a set of connections between imported technologies of stone-cutting, the imported body of Tawaret and the sea. The Minoan genius with jug becomes a stable form which continues to appear in the Late Bronze Age, notably on the Tiryns gold ring from mainland Greece. Exotic bodies imported by the first palatial collectives became representatives of the palace in later periods.

The Phaistos Room 25 sealing deposit, from the end of the Protopalatial period, includes two different depictions of a Minoan genius with jug, as well as winged griffins or sphinxes (Figure 7.4d–e). This sealing deposit has been discussed in previous chapters because it also includes marine animals and some of the first depictions of hunted animals in a landscape setting. Like the lions which also appear on sealings in this deposit, the sphinx, griffin and Minoan genius all appear to derive from the imported material culture of the Eastern Mediterranean (Rhyne 1970: 33–52; Crowley 1989b: 40–53;

Weingarten 1991). Protopalatial Cretan ceramics in Egypt and the Levant suggest that a two-way trade was developing which almost certainly also included less archaeologically visible goods such as textiles (Kemp and Merrillees 1980; Merrillees 2003). Neopalatial frescoes from Knossos appear to show textiles with designs of both a seated griffin and sphinx. Although these have been seen as imported textiles, it is more likely that they are indications of the skilled textile production of Crete by this period, and potential export goods (Frankfort 1937: 117; Hood 2005: 58–60). Before its apparent absorption by Knossos, the palatial collective of Phaistos, on the south coast of Crete, was well placed to send Cretan goods directly to Egypt. The exotic bodies shown in the sealing deposit allowed seal users to demonstrate knowledge of faraway places, extending the palatial collective across the sea.

Exotic bodies of griffins and, most likely, a sphinx were made present at Neopalatial Knossos using stucco relief, the same technique that was used to make bulls and bull-leapers present at the North Entrance. A collection of similar reliefs of human limbs, one holding a bull's horn, as well as griffins seemingly associated with columns were found in the East Wing of the palace. Evans (1930: 497–517) used these in the reconstruction of an 'East Hall' decorated with relief frescoes. This combination of subjects in one area of the palace helps to explain how a human relief torso came to be combined with a sphinx's headdress in order to create the figure of the 'priest-king' from fragments found in the south of the palace. The fresco has been deconstructed by a number of authors (Coulomb 1979; Niemeier 1987, 1988; Shaw 2004), but the existence of the crown shows that sphinxes were depicted at Knossos earlier than a Final Palatial ivory mirror handle from Zapher Papoura (Evans 1906b: fig. 69). Like the later Throne Room griffins, these frescoes brought exotic bodies into the centre of the palatial collective and showed the extent of its overseas connections.

Another exotic body appears among the Phaistos sealings, the monkey (Figure 7.4f).[10] Since it is shown seated, it appears to be associated with the imported 'seated ape' depictions. The shift from knowledge through material culture to imported bodies appears to occur early in the Neopalatial period at Knossos, the most likely date for the painting of the Saffron-gatherer fresco. The House of the Frescoes frieze which includes monkeys stealing eggs in a rocky landscape was also painted before the end of the Neopalatial period (Cameron 1968: 19–24). The eggs and blue birds, probably rock doves (pigeons), connect these monkeys with the mountains; in the Bronze Age these birds would have been found nesting on cliffs and other rocky places.[11] Another fresco from the House of the Frescoes shows crocuses in a rocky landscape (Chapin and Shaw 2006). Although the monkey depicted in the Saffron-gatherer fresco from the nearby palace is picking a crocus from a pot, this too establishes a connection between this commodity and these exotic animals (Shaw 1993: 668–75). These exotic animals were most likely kept in captivity but they were set free in these frescoes, their interactions with the rocky ground and birds which were hunted there acting as an analogy for the extension of the Knossos palatial collective. The frescoes, found at its centre, revealed that these exotic animals belonged at its mountainous edges. Control over exotic animals, which were perhaps only rarely seen, was translated into control over the Cretan landscape by deploying their bodies in palatial material culture.

The Thera frescoes are better preserved than those of Knossos because of the volcanic eruption and appear to be part of a shared iconography with Crete. A fresco from Xeste 3 at Akrotiri shows a monkey presenting crocuses to an elaborately dressed female figure flanked by a griffin. There are differences between Cycladic wall paintings, which often relate to the local landscape, and those of the surviving Cretan frescoes (Morgan 1990: 258). Nevertheless, as Lyvia Morgan (1990: 261) suggests, the connection of monkeys, high-status women and crocuses appears to be part of a wider iconographic cycle. On the Ayia Triada Park fresco, cats hunting birds are in an analogous position to the House of the Frescoes monkeys stealing birds' eggs; at both places crocuses were depicted nearby. Monkeys, griffins and cats were brought into association with women dressed in elaborately woven textiles, who gathered together exotic bodies as a way to demonstrate their connections with the world beyond Crete. This collection of bodies and things acted as an analogy for the trade in dyed textiles, which resulted in a variety of exotic imports flowing into the Aegean.

Hunters and Hunted

New bodies became part of the palatial animal practice of hunting. Some animals were brought to Crete, others were sought overseas, while the animal things related to hunting multiplied. As discussed in Chapter 5, animals on sealstones can often be linked to hunting, whether depicted individually or with dogs or human hunters. Hunted animals on sealstones were primarily cattle, wild goats and lions, with boar and deer less frequently shown. All could have been hunted on mainland Greece or in Anatolia, but fallow deer had joined cattle, feral pigs and wild goats on Crete by the Neopalatial period, with red deer arriving by the Postpalatial period. As the palatial collectives extended spatially, there was more potential for hunters to go in search of animals on the mainland, and for trophies and even the animals themselves to be brought to Crete. Depictions of animals also continued to circulate, with lions appearing on a wide variety of objects across the Eastern Mediterranean during the Late Bronze Age (Rhyne 1970; Feldman 2002: 18, 2006). The depiction of lions on these International-style objects as predators, along with griffins, presented them as analogous to human hunters. As depictions of lion-hunting became widespread on Crete, it was this international imagery of predation that the palace drew upon to establish a hierarchy of hunters. Here the focus is on things found on Crete but it becomes increasingly difficult to disentangle Cretan hunting imagery from that found overseas, to be discussed in the following section.

As well as new prey animals, new hunting companions arrived on Crete in the Palatial period. Particular types of hunting dogs with different capabilities were depicted in Egypt from the start of the Bronze Age (Osborn and Osbornová 1998: 57–68), and there are signs of their import by the Neopalatial period, if not earlier. Seal depictions from the Protopalatial period onwards show dogs with collars, sometimes keeping animals at bay. Apart from indicating ownership and control, the collar emphasises dogs' role in hunting, since they would have been unleashed in order to chase and corner prey (Figure 7.5a). Dogs with pricked ears and curly tails were depicted on Neopalatial seals, closely resembling Egyptian hunting dogs from this period, known as 'tjesm' (Rice 2006) (Figure 7.5b). A large number of dog skeletons were found in two Neopalatial wells at Palaikastro, associated with horn cores, probably from wild goats. Sheilagh Wall-Crowther (2007: 195–7) suggests that these could have been hunting dogs and compares them with modern curly tailed Cretan Hounds. A different type of dog, a heavy-set mastiff with floppy

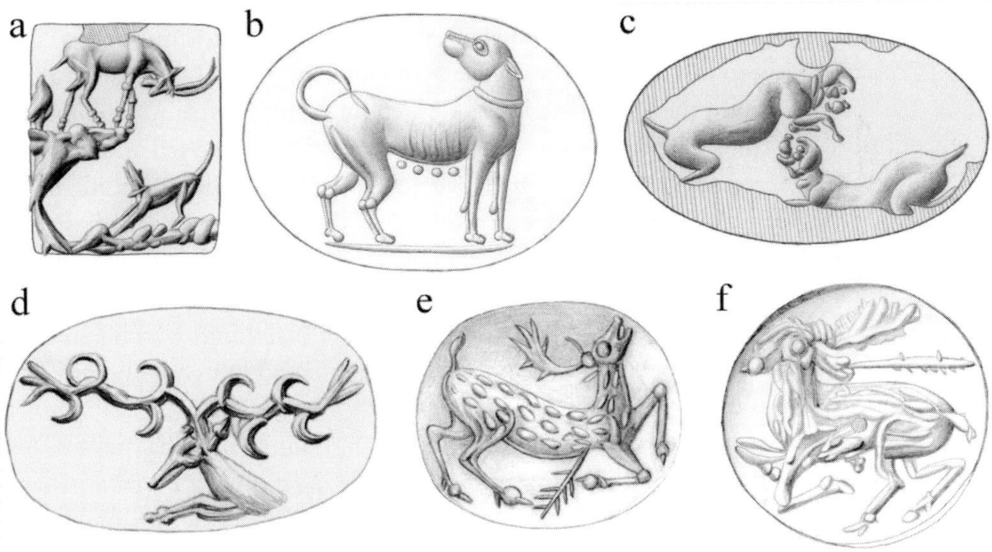

7.5. Drawings of seals showing dogs and deer. (a) Dog and goat, Archanes MMIII-LMI (CMS VI 180); (b) dog with curly tail, Knossos sealing LMI-II (CMS II 8 287); (c) two mastiff-like dogs, Ayia Triada sealing, LMI (CMS II 6 77); (d) deer head from three-sided prism, Rotassi, MMII (CMS VI 97a); (e) fallow deer with arrow, Knossos LMI (CMS II 3 74); (f) fallow deer with spear, Knossos LMI (CMS II 4 113). © CMS Heidelberg

ears, is depicted on seals, including a ring impression from Ayia Triada (Figure 7.5c) and a gold discoid from Poros near Knossos (Dimopoulou 2010). Figurines of mastiff puppies made from a variety of materials including gold and hippo ivory were found at Juktas (Karetsou and Koehl 2014).[12] These detailed depictions appear to be pointing to particular breeds of dog, arguably a way of distinguishing the dogs of the Knossos palatial collective from the more common type of dog in Bronze Age Crete.[13] The depiction of both mastiffs and curly-tailed dogs on metal seals emphasises their associations with the palatial collective and makes it likely that their ownership was restricted.

Deer remains consistent with trophies were found at Protopalatial sites but deer were not commonly depicted on seals until the Neopalatial period. The exceptions are two Protopalatial three-sided prisms, one of which shows a deer running, and the other a deer head, almost certainly red deer because of its elaborate antlers (Figure 7.5d). Given its similarities to earlier Anatolian figurines of red deer, it could be based on an imported depiction (Collins 2003: 73–4). On later seals the antlers tend to be less diagnostic of species, but one Neopalatial amethyst seal found at Knossos clearly shows a fallow deer with spotted coat and palmate antlers (Figure 7.5e). The deer's pose, and the arrow in its chest, indicates that it is being hunted. Since this is the period that fallow deer were eaten at Knossos, the seal appears to be connected with hunting in this area and would have associated the user with this animal practice. Other Neopalatial seals from Knossos depict deer with spears over their backs. Some of these seals were made from soft local stone (Figure 7.5f), a more widely available material than an imported hard stone like amethyst. This suggests that participation in this animal practice was not closely restricted. Deer continued to be depicted in the Postpalatial period, including on seals from Armenoi and a potsherd from Chania which depicts an animal with branched antlers characteristic of red deer (Hallager and Hallager 2000: 116). As at Knossos, it is likely that this depiction relates to deer-hunting in the vicinity. By contrast, the Protopalatial seal emphasising the deer's antlers needs only relate to knowledge of the animal through trophies or depictions, given the rarity of deer remains in this period on Crete. Both trophies and seals are animal things which extended this animal practice beyond the moment of the hunt and subsequent consumption of venison.

Lions continue to be depicted on seals throughout the Palatial period, but the form and context of the lion changes over time. Prepalatial and Protopalatial seals show clearly recognisable lions with stylised manes which were most likely derived from imported depictions of lions (Figure 7.6a). They are rarely given a definite context, although a small number of Prepalatial seals, as noted above, could show a lion hunt, while lions on the late Protopalatial Phaistos sealings are shown amid plants.[14] This changes in the Neopalatial period, when lions are often shown as hunted animals, with spears over their

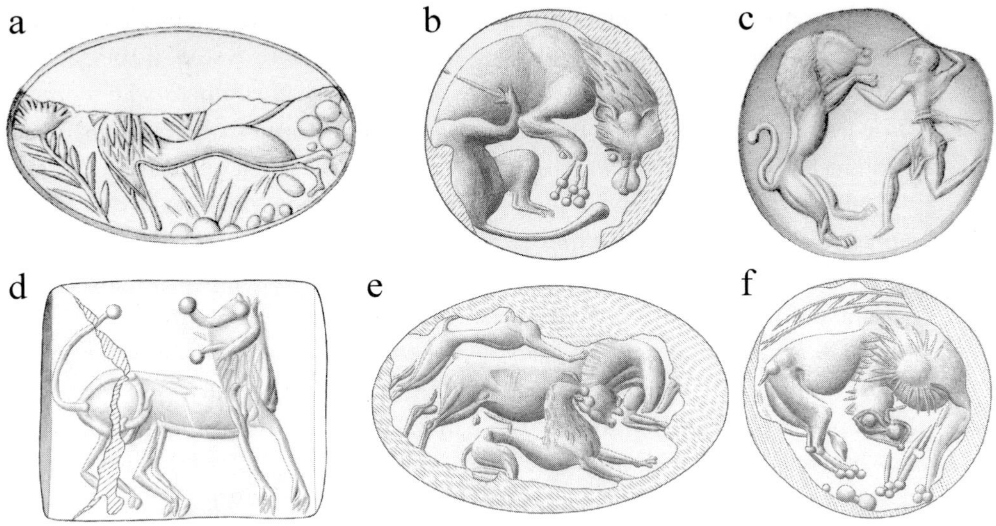

7.6. Drawings of seals showing lions. (a) Lion with plants and rocks, sealing from Phaistos, MMII (CMS II 5 270); (b) lion with arrow in side, Ayia Triada sealing, LMI (CMS II 6 91); (c) lion and man with spear, Chania sealing, LMI (CMS VS 1A 135); (d) lion with head turned, Knossos, LMI (CMS II 3 18); (e) two lions attacking bull, Tylissos sealing, LMI (CMS II 6 274); (f) lion with hair star, Knossos sealing, LMII–IIIA (CMS II 8 307). © CMS Heidelberg

backs or arrows in their sides, but rarely with human hunters (Figure 7.6b). An exception is a seal impression from Chania showing a man approaching a lion, spear in hand, which could relate to Mesopotamian seal iconography (Figure 7.6c). These hunted lions can be contrasted with the generic depictions of lions in earlier periods. From the Neopalatial onwards, lions are shown in a variety of poses and contexts, suggesting a broadening knowledge of them. This knowledge appears to come from two related sources: encounters with lions overseas, resulting in lion pelts coming to Crete, and an 'international' imagery of lions as predators which was increasingly shared across the Eastern Mediterranean.

Many Neopalatial seals showing lions belong to a widespread style known as 'Cretan Popular', usually made from locally available soft stone rather than imported hard stone (Pini 1995). Although the mane and tufted tail is indicated, the head is often shown with two protrusions (Figure 7.6d). Their peculiar appearance is often attributed to the competence of the engraver, 'certainly not one of the greatest artists of LMI glyptic' (Betts 1981: 7; see also Pini 1995: 196). This surprising decrease in realism compared with earlier periods illustrates that the engraver knew two things about a lion, that it had a mane and tufted tail, but was unclear about the shape of the head. Rather than the continuation of earlier imagery, these depictions appear to be based on lion pelts which retain the mane and tail, but distort the shape of the head when the skull is removed, leaving protruding flaps of skin from around the jaw (Shapland 2010a). Even finely crafted seals of lion-hunting such as a Neopalatial impression from Ayia Triada (Figure 7.6b) could relate to a pelt because of the way the feet are depicted, arguably more like toe bones. These seals of varying quality connect the user with lion-hunting but this would have required a trip to the mainland and so it is unlikely that everyone who had such a seal had killed a lion. Cretan Popular-style seals allowed people to connect themselves with the animal practice of lion-hunting, whether or not they had taken part.

A continued interest in lion anatomy and behaviour is apparent in Neopalatial and Final Palace seals. Near-identical metal-ring impressions from Myrtos Pyrgos and Tylissos depict two lions attacking a bull in a way which appears to show an understanding of lion behaviour (Pini 1985: 154; Ballintijn 1995: 23) (Figure 7.6e). A Final Palatial sealing from Knossos shows a 'hair star' on the shoulder of the lion (Rhyne (1970: 186) (Figure 7.6f). Although this could be an imported observation it points to a continuing interest in the details of lion morphology since this is an anatomical feature of male lions seen on various Eastern Mediterranean depictions (Kantor 1947b; Arkell 1948; Bate 1950). Other seals show antithetical arrangements involving humans and lions which can be related to Near Eastern imagery (Crowley 1989b: 182). These arrangements, termed 'Master' or 'Mistress of Animals', nevertheless show Cretan features, as in the architecture on the famous 'mother on the mountain sealing' from Knossos (Evans 1928: 809). They demonstrate the extensive exchange of imagery between the Aegean and Eastern Mediterranean in this period.

The three limestone lion's-head rhyta from the Knossos palace could be as early as the Neopalatial period, but are usually dated later (Koehl 2006: 122). The most complete is usually identified as a lioness, despite the mane, because of the proportions of the head (Figure 7.7). Notably this object shows another detail of lion anatomy: the superciliary tufts above the eyes. These eyes would have been inlaid with rock crystal, resulting in a strikingly realistic effect, as with other animal-head rhyta discussed above. This was an object which could act as a lion substitute, a metaphorical animal thing which could have been manipulated during drinking ceremonies, standing in for a lion's-head trophy. Compared with Cretan Popular-style seals these things show a greater degree of knowledge of lions, perhaps reinforcing closer connections with these animals.

The depiction of griffins on seals closely parallels that of lions, and they are sometimes shown together (Morgan 2010a: 318–21). Whereas the Phaistos sealings show lions and griffins as stationary animals, from the Neopalatial period onwards lions and griffins are shown attacking other animals, and sometimes one another. On one Final Palace seal, a hierarchy can be seen, or perhaps a pecking order, with a griffin attacking a lion attacking a deer (Figure 7.8a). Unlike lions, griffins are not shown being hunted by humans,

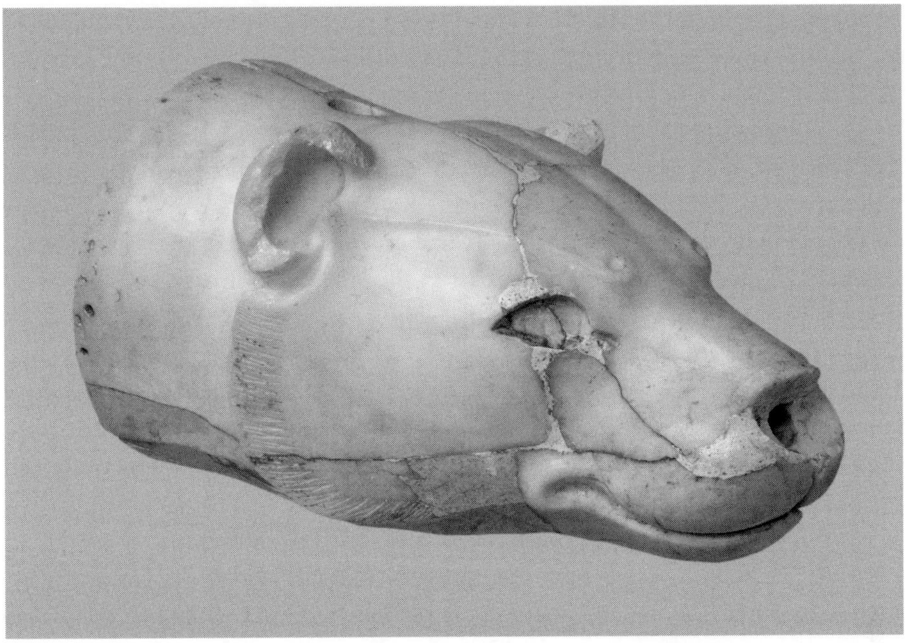

7.7. Stone lion's-head rhyton from Knossos, LM. Length: 29.5 cm (AMH Λ44). © Heraklion Archaeological Museum – Hellenic Ministry of Culture and Sports – 'Hellenic Organization of Cultural Resources Development (HOCRED)'

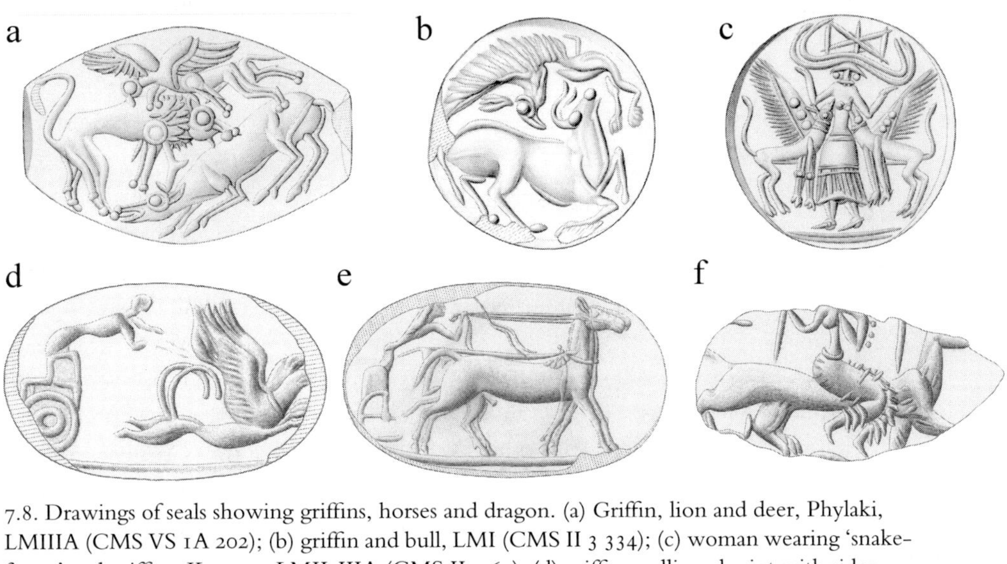

7.8. Drawings of seals showing griffins, horses and dragon. (a) Griffin, lion and deer, Phylaki, LMIIIA (CMS VS 1A 202); (b) griffin and bull, LMI (CMS II 3 334); (c) woman wearing 'snake-frame' and griffins, Knossos, LMII-IIIA (CMS II 3 63); (d) griffins pulling chariot with rider, Knossos, LMI (CMS II 8 193). Sealings from Ayia Triada, LMI: (e) horses pulling chariot with rider (CMS II 6 19); (f) woman riding dragon (CMS II 6 33). © CMS Heidelberg

but do appear in antithetic schemes such as the Mistress of the Animals, where they appear either side of a human figure (Figure 7.8c). The griffins either side of the throne at Knossos would have provided a physical manifestation of this arrangement when the throne was occupied, and illustrate that being flanked by griffins was a significant statement of authority (Reusch 1958; Galanakis et al. 2017: 87–91). Griffins were not simply Near Eastern imports by this period, but had developed particularly Cretan features such as the body spirals (Morgan 2010a). They had also become the top predator in the palatial collective and, by analogy, were identified with its human rulers.

The Minoan genius also takes part in hunting on a small number of seals, where it is shown spearing a bull or carrying dead animals, or, on a seal impression from Ayia Triada, fighting a lion. Its use of a spear connects it with human hunters, although they are rarely shown explicitly on Bronze Age Cretan seals. Fritz Blakolmer (2016) downplays the involvement of hybrid creatures such as the Minoan genius or griffin in animal attack scenes because they make up a small proportion of Aegean Bronze Age seal depictions and could be derived from Near Eastern iconography. Yet, as he points out, there is a clear hierarchy apparent, with lions and griffins occupying a dominant position compared with the usual prey animals depicted on Cretan seals, including cattle, goats and deer. As argued in Chapter 5, hunting became a restricted animal practice by the Neopalatial period, reinforced by depictions of elaborately dressed women feeding goats. The expansion of hunting imagery at the same time to include lions, griffins and Minoan geniuses as predators drew upon overseas imagery in order to emphasise the association between hunted animals and the palatial collective. This diverse group of bodies, through analogy and hierarchy, helped to define relations between the palatial collective and human hunters.

Drawing Chariots

Depictions of horses and chariots on Crete predate the earliest identified horse remains, from Final Palatial Knossos. The traditional idea that chariot-driving Mycenaeans conquered Crete is complicated by examples of the chariot in Neopalatial iconography. The impression of the same ring was found on sealings at two sites on Crete, and also Thera, showing a person driving a chariot drawn by at least two horses (Weingarten 2010: 396–9) (Figure 7.8e). Its appearance at Thera dates it to the earlier part of the Neopalatial period, and it is likely that it sealed a message from Knossos. These impressions were found on packet sealings, which enclosed a message on parchment. As discussed above, other packet sealings with such repeated impressions show bull-leaping, which was central to the identity of Knossos in the Neopalatial period; it indicates that the horse and chariot first arrived in Crete in this period and became closely identified with the palatial collective. A contemporary ring from the Mycenae Shaft Graves shows two men hunting a deer from a chariot which, along with remains from the Middle Bronze Age, suggests that horses had arrived in mainland Greece at this time too (Crouwel 1981: 32–3). There are no such scenes of hunting from chariots on Crete, and indeed depictions of horses and chariots are comparatively rare. One explanation for this is that the ownership of horses was carefully controlled and so few seal users were connected with them. There was also less need to make horses present at Knossos through depictions, since they were highly visible in the Final Palace period, as the Linear B tablets demonstrate.

The 'chariot tablets' from the Linear B archive at Knossos record up to 110 individual horses and various parts of chariots, including wheels and frames (Vandenabeele and Olivier 1979: 65). This group of tablets forms a discrete part of the archive, probably dating several generations earlier than the tablets preserved in the fire which destroyed the palace of Knossos (Driessen 2000). They are traces of the early Mycenaean administration of Knossos, during which Linear A was adapted to represent an early form of the Greek language. The ideograms for horses, though, appear to be newly invented; unlike the schematic ideograms used for the other domesticates, which carry on in use from Linear A, they are recognisable horses' heads, some with tufted manes.[15] The horses, together with chariots, were a new addition to the Knossos palatial collective at a time when it had become absorbed into a broader mainland collective. Later chariot kraters, large open vessels probably used for the mixing of wine, were commonly depicted with scenes of horses and chariots (Recht and Morris 2021). They are a common Mycenaean export but are scarce on Crete; a locally made example occurs at Chania (Hallager and Hallager 2000: 59, 148). A similar scene of horse and chariot appears on a fresco on the wall of the Final Palace at Knossos (Cameron 1967). These exotic animals were incorporated into the palatial collective as part of a ready-made assemblage of warrior-chariot-horse whose presence at Knossos is confirmed by the Linear B archive.

There is another Neopalatial ring impression from Knossos showing griffins drawing a chariot (Figure 7.8d). The same scene is repeated on one side of the Final Palatial 'Ayia Triada sarcophagus', mirrored by a scene of a horse and chariot on the opposite side. Both chariots are made using bull's hide, connecting them with the cattle offerings on the other two sides. There is no evidence that horses were ridden in the Bronze Age Aegean; the only animal which was ridden, by women, was the dragon. This appears to be an imported Babylonian dragon, usually referred to as the Minoan dragon when it is depicted on Cretan seals (Gill 1963). One seal impression from Ayia Triada shows a woman in an elaborate skirt riding one of these creatures (Figure 7.8f). Although Minoan dragons are rarer than griffins, their treatment is analogous since they are shown under the control of elaborately dressed women.

Like cats earlier, horses were animals whose physical presence on Crete helped to demonstrate the expansion of the palatial collective over the world beyond the sea. By this period griffins had come to share the affordances of lions as predators of animals like cattle and deer, and so were top predators associated with the leading human or divine members of the palatial collective. The interchangeability of griffins and horses as pullers of chariots indicates that they were analogous animals in the Bronze Age. The difference between griffins and horses was ultimately that horses were present bodily on Crete and griffins were made present through their depiction.

THE AEGEAN ANIMAL STYLE

The import to Crete of exotic animals, first as depictions on objects like sealstones and amulets, and then as animal things like pelts, or living bodies, increased in intensity over the course of the Bronze Age. The Cretan local and palatial collectives were not passive recipients but contributed to the flow of bodies and things with, to name some of the most visible, painted pottery, dyed textiles, perfumed oil and a new type of griffin. By the time the Eastern Mediterranean was at its most connected, the International Age of the fourteenth and thirteenth centuries BCE, Crete had been absorbed into a multicentred collective focused on the Mycenaean palaces of mainland Greece. The relations of this collective with the Eastern Mediterranean have been extensively discussed both at an empirical and a theoretical level. Empirical accounts have documented the imported Eastern Mediterranean objects in the Aegean, accompanied by discussions of trade and the influences of imports on Aegean artistic styles (Pendlebury 1930; Kantor 1947a; Crowley 1989b; Cline 1994). The most generally accepted theoretical perspective has been Andrew and Susan Sherratt's application of the World Systems model, tracing the shifting core-periphery relations of the Aegean and Eastern Mediterranean (Sherratt and Sherratt 1991; Sherratt 1993). This model has been particularly influential among Aegean prehistorians, who have applied it to the Aegean, particularly in discussions of Mycenaeanisation (Gorogianni et al. 2016). The role of imported objects in the Mycenaeanisation of the Aegean has also been the focus of attention (Burns 2010). More radical approaches have been taken to Minoanisation, recasting the similarities between sites across the Aegean as a multiscale and multidirectional network rather than a symptom of colonisation (Broodbank 2004; Knappett 2018). Looking out from the Levantine coast, Marian Feldman (2002, 2006) has examined the way in which the hybridised objects of the Late Bronze Age International style help to undermine traditional boundaries between cultures. Attack scenes involving lions and griffins are one component of the International style whose origins have traditionally been located in the Aegean. By tracing the intertwining paths of lions and griffins across the Bronze Age Aegean it becomes possible to focus on the relations within and between collectives, and recast observations from traditional stylistic approaches as descriptions of networks of people, animals and things.

One network, the Aegean Animal style, was initially defined by Helene Kantor (1947a) in opposition to the Egyptian Animal style in order to characterise contact between the Aegean and Eastern Mediterranean in the second millennium BCE. Whereas Egyptian animals tended to keep their feet on the ground, even when attacked by other animals, Aegean animals, particularly when attacking one another, adopted a variety of contorted poses and galloped after one another. The introduction of such scenes in Egypt became a sign of contact between the two regions. The similarities between a dagger showing a lion in flying gallop found in the early Eighteenth Dynasty tomb of Queen Ahhotep I and other inlaid daggers from the Mycenae Shaft Graves had already been noted by Arthur Evans (1921b: 714–21), but he regarded all as ultimately Minoan. Kantor preferred the term 'Aegean' because she regarded Mycenaean metalwork as a distinct phenomenon: 'It is true that the swift movement characteristic of the Aegean Animal style first developed in Crete, but the tradition had been adopted on the mainland in LH I. The traders who introduced the new style into Egypt at the beginning of the XVIIIth Dynasty must have set forth from harbors on the mainland of Greece' (Kantor 1947a: 66). The subsequent discovery of lions in flying gallop on the Thera West House miniature fresco complicated the picture further (Morgan 1988: 44–9). This, and a shift in scholarship away from purely art historical studies, led Nancy Thomas (2004: 162) to conclude that 'The old desire to sort lion art into pristine Minoan and Mycenaean camps has been largely abandoned as a flawed endeavor'. She also highlighted the identification of lion remains at a number of Mycenaean sites, demonstrating that observation of lions contributed to the Aegean Animal style (Thomas 2004, 2014). Lions also occurred in Egypt, and had long been depicted there, but at the start of the Late Bronze Age lion bodies began to move from the Aegean to Egypt on luxury items suitable for royal tombs.

Whereas lions could be observed in Greece in the Bronze Age, the distinctive form, poses and behaviour of the Aegean griffin have generated debates about the origin of this animal. As with the lion, the recognition of an Aegean-style griffin on another piece of Egyptian metalwork from an early Eighteenth Dynasty tomb showed the interconnectedness of the Late Bronze Age Eastern Mediterranean. The inlaid gold ceremonial axe, inscribed with the name of Pharaoh Ahmose I, juxtaposes typically Egyptian scenes with a seated Aegean griffin with crest and body spiral (Morgan 2010a: 308). Although Henri Frankfort accepted it as a Cretan griffin, he located its origins in the Mitanni kingdom of Syria (Frankfort 1937; Crowley 1989b: 46–51). There are also earlier griffins in Egyptian iconography, from the Predynastic period onwards, but the Aegean griffin had become a distinct form by the time it completed its circuit around the Eastern Mediterranean. The griffin occurs first in the Aegean in the Protopalatial Phaistos sealings but it first appears in flying gallop on the Thera miniature fresco and a number of objects in the Mycenae Shaft Graves (Morgan 2010a: 304).[16] The griffin also appears as a predator on the Thera miniature fresco, as on a number of Neopalatial and later seals from Crete. The griffin continued to be depicted as a predator in the International style (Crowley 1989b: 272–3).

The Aegean Animal style originated around the start of the Late Bronze Age and has been traced into the International style of the fourteenth and thirteenth centuries BCE in the Eastern Mediterranean, a term used for objects whose cultural origin is unclear (Crowley 1989b: 287–8; Feldman 2002: 6). As Feldman has argued, the search for artistic hands or cultural origins distracts from a recognition of the extensive interchange between palaces across the area in the fourteenth to thirteenth centuries BCE which resulted in the manufacture of luxury objects at multiple sites. The examples she gives are from Ugarit, on the Levantine coast, and include a gold bowl and ivory furniture panels.

> Object type, material, technique, iconography, and composition link these pieces together. On them, animal attack and hunt scenes amid lush, stylized vegetation are hybridized to such a degree that discerning individual cultural identities becomes extremely difficult. The combined repertoire of themes and motific elements derived from numerous traditions warrants using the term *international*, while the extensive sharing of these motifs suggests a visual koine. (Feldman 2002: 18)

International-style ivory panels, showing lions and griffins attacking one another, have long been recognised as having Aegean affinities, although the individual pieces could have been made anywhere (Kantor 1956; Poursat 1977; Crowley 1989b: 216; Feldman 2002: 22). For the past century, scholars have been invoking travelling craftspeople and imported objects in order to explain these similarities, while continuing to argue about where objects came from. By regarding things like the Mycenae 'Lion hunt dagger' or the ivory furniture panels from Ugarit as hybrids, they instead become objects embedded in a network of connections which spanned the Late Bronze Age Eastern Mediterranean. The International style, and the Aegean Animal style which relates to it, was composed of a variety of interacting entities including human makers and consumers, Syrian elephants whose tusks were carved into ivory panels, Mitanni cylinder seals depicting griffins and encounters between lions and hunters in mainland Greece.

By the time the International style had completely coalesced, the palace of Knossos and its associated collective had ceased to exist, but there is a clear relationship between the Aegean Animal style and some of the depictions of lions and griffins discussed above from Neopalatial Crete. These cannot be understood in isolation, and identifications of a common style help to trace the connections of Cretan lions and griffins with others from the Aegean and Eastern Mediterranean. As well as griffins in flying gallop, the Mycenae Shaft Grave assemblage includes a variety of lions including a gold lion's-head rhyton, the Lion hunt dagger and a gold cushion seal showing a man and lion in single combat (Thomas 2004; Anderson 2020: 371–4). None of these can be exactly paralleled on Crete, but there are similar depiction on seals and the stone lion's-head rhyta from Knossos. Similar but different again are two terracotta lion's-head rhyta found on Thera (Thomas 2004: 169–70). This has long been seen partly as an accident of preservation; the recycling of metals and the lack of rich burials on Crete have precluded finds like the Mycenae Shaft Graves (Kantor 1947a: 44). Until recently, no inlaid daggers had been found on Crete, but an example with an inlaid griffin found at Knossos shows what once might have been present in Bronze Age Crete (Kanta 2019: 249). Even so, the scene on the Lion hunt dagger, one side of which shows a group of men with shields attacking lions, remains unknown on Crete, and suggests that there were differences between lion depictions on Crete and the mainland. At the same time, the Mycenae Shaft Grave assemblage contains a range of things whose material and iconography connect them to contemporary Crete and Thera.

Another network is described by accounts of the 'Aegean style' or 'Minoan' frescoes found at various Cycladic and Eastern Mediterranean sites (von Rüden 2014). They are defined by the use of the fresco technique, in which paint was applied to wet plaster (Brysbaert 2008), and a shared iconography. A Protopalatial fresco of a swallow from Knossos (Roussaki 2014) shows the early occurrence of Aegean frescoes at Knossos, but many of the frescoes found outside Crete have no close parallels at Knossos or elsewhere on Crete. As with parallels for the Shaft Grave assemblage, one reason for this is the lack of a destruction horizon at Knossos at the end of the Neopalatial period, which would have preserved frescoes. Instead, various fresco dumps provide clues of the frescoes present at Knossos before its final destruction, although these fresco fragments can often only be dated on stylistic grounds (Hood 2000, 2005: 52–4). This gap in the evidence from Knossos need not be seen in terms of missing origins; instead, it was one of a number of interconnected sites across the Eastern Mediterranean with Aegean-style frescoes.

Although Cycladic frescoes from Thera and other sites have often been seen as part of the process of Minoanisation, a distinct island iconography has increasingly been recognised (Morgan 1990). The term 'Minoan' has also been used to describe wall paintings from Tell el Dab'a in the Nile Delta, and Levantine sites including Tel Kabri and Alalakh (Niemeier and Niemeier 1998, 2000; Marinatos 2010: 347). The scenes of bull-leaping at Tell el Dab'a relate them closely to Knossos (Bietak et al. 2007), but another scene known as the 'Hunt frieze' showing dogs, lions and griffins hunting cattle and other animals is not known from Cretan frescoes (Marinatos 2010; Morgan 2010a, 2010b). There is one possible stucco relief fragment of a lion from Knossos, from an unknown scene, but this was made using a different technique from the Tell el Dab'a frescoes, and the composition of the fresco is unknown (Evans 1928: 332–5; Marinatos 2010: 347). Instead, the closest parallels for the Hunt frieze are the lion and griffin from the miniature fresco from the West House at Thera and another fresco from Ayia Irini on Kea showing hunting dogs chasing fallow deer (Marinatos and Morgan 2005; Morgan 2020: 149–58). The terms 'Aegean style', used to describe frescoes outside the Aegean, and 'Minoanised', used for frescoes on the Cycladic islands, can instead be seen as part of a broad continuum of animal-hunting scenes which differ from place to place. This network intersects particularly with Cretan seals, which perhaps preserve scenes of frescoes at Neopalatial Knossos. It is a smaller part of the Aegean Animal style network, linking together a series of similar but different frescoes on walls across the palatial centres of the Eastern Mediterranean.

Unlike some other palatial sites destroyed in the Neopalatial period, no figural frescoes were found at Zakros (Platon 1971: 83). It did, however, have close connections with the Eastern Mediterranean, as finds from the palace demonstrate. Its location, at a sheltered harbour on the east coast of Crete, made it an ideal trading post, but the awkward location of the palace suggests that it was imposed on the town rather than developing from an earlier gathering place like some of the other palaces on Crete. This has been used to suggest that it was closely connected with Knossos, which used Zakros as its port in eastern Crete (Platon 2012). One piece of evidence for this is a packet sealing found in an adjoining house with an impression depicting two lions in flying gallop, typical of the Aegean Animal style (Figure 7.9a). The same impression occurs at Knossos and Ayia Triada, and it belongs to a group of metal rings which were used to stamp clay sealings containing tightly folded parchment whose distribution across the island suggests they came from administrators at Knossos. Another packet sealing shows a griffin in flying gallop attacking what is almost certainly a lion (Figure 7.9b). It is not attested elsewhere but it belongs to the Aegean Animal style too. These sealings, which draw on the imagery of lions and griffins circulating around the Aegean at this time, appear to connect the palace of Zakros with the palatial collective of Knossos.

In the same sealing deposit, among the many depictions of animals are a number of other lions. One seal impression from a stone seal shows two men carrying a trussed lion (Figure 7.9c). Rather than showing lions in a landscape setting, it appears to relate closely to the animal practice of lion-hunting since it shows a particular activity following the kill. Another impression comes from one of the Cretan Popular-style seals discussed above, showing an idiosyncratically Cretan lion with a spear over its back (Figure 7.9d). This too relates to hunting,

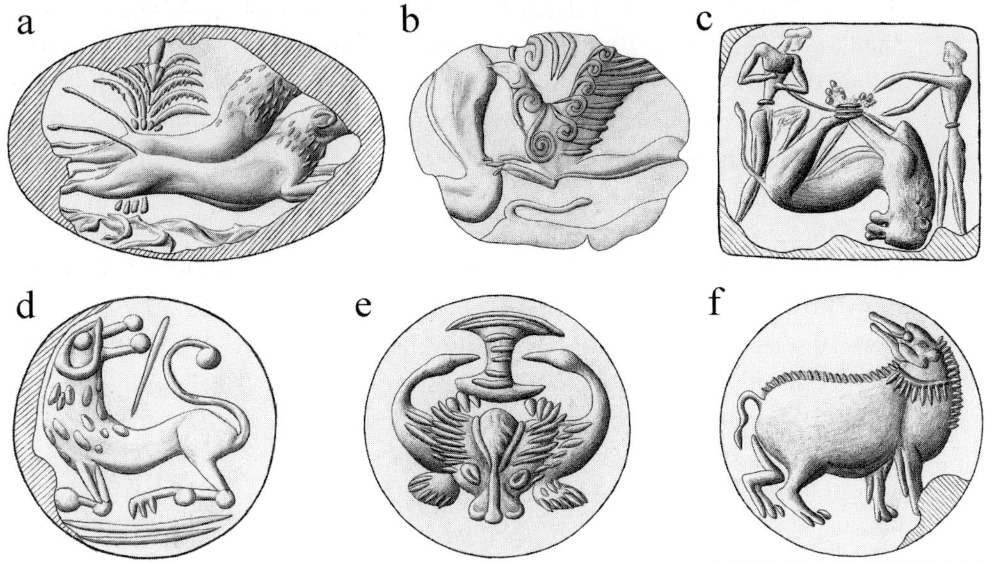

7.9. Drawings of sealings from Zakros, LMI. (a) Two lions in flying gallop (CMS II 7 71); (b) griffin attacking lion (CMS II 7 96); (c) men carrying trussed lion (CMS II 7 33); (d) lion with spear (CMS II 7 70); (e) lion's-head-waterbird hybrid with double axe (CMS II 7 196); (f) lion-boar hybrid (CMS II 7 115). © CMS Heidelberg

but is a more common type of depiction which could relate to the more extended animal practice of lion-hunting, distributed throughout Crete by means of seals and probably pelts and oral poetry (Shapland 2010a; Anderson 2020). Other sealings showing lion parts belong to a large group of seals attributed to a local seal-carver, 'the Zakro Master', whose characteristic seals show a combination of a lion's head with birds and a double axe, or a boar-like creature with leonine legs and a bird's head (Figure 7.9e–f). Other seals include parts of griffins, deer and other animals and show the systematic combination of various forms in order to create a thing which allowed participation in the local administrative system (see Figure 2.5a–d). The Zakro Master seals were stamped in combinations on so-called 'nodules' which marked the co-presence of two or three people in clay, probably to record an agreement (Weingarten 1992). These lion bodies have been absorbed into the workings of the Zakros palatial collective, while the packet sealings connect it to the distant palatial collective of Knossos. Other lions connect their users to the animal practice of lion-hunting, which could have been carried out on the mainland in this period, via the kind of boats depicted on the Theran frescoes that connected the islands.[17]

The assemblage of animal things found at Zakros, including the 'Sanctuary rhyton' and a fine bull's-head rhyton, have been discussed in previous chapters. They show links with other practices, such as goat-hunting and communal consumption. They are the traces of the collective which was once focused on this site. The Zakros palatial collective had links with the Eastern Mediterranean, preserved in the form of elephant tusks and ostrich eggs, animal things that could be refashioned into more local objects such as the ivory butterfly also found in the palace (Platon 1971: 148). Some of the sealings describe the activities of the Zakros palatial collective, and many are carved by a local manufacturer whose output combines the various different forms circulating in the collective into single objects. Other seals were probably not carved locally, since they show similarities with seals from across Crete, and show how lions were incorporated into people's lives on the island. Others materialise connections with the Knossos palatial collective, and these belong to a group of depictions of lions and griffins which was formed from interactions between Crete, the Cyclades, mainland Greece, the Levant and Egypt. What have been termed 'styles' describe interrelated depictions circulating in various media and the Aegean and beyond. The packet sealing showing lions in flying gallop was a trace of this wider network, but also part of the Minoanisation of Zakros through its connection with the palatial collective of Knossos. The Zakros palatial collective articulated with these different networks, at different scales, from the local to the international.

CONCLUSION

The affordances of exotic animals and animal things emerged as part of the process of importation to Bronze Age Crete. Materials such as ivory could be carved into seals. Cats could hunt the local wildlife, both household rodents and birds. But the process of importation also involved finding how these new bodies and things related to others in the existing collective. As the collective became more differentiated over the course of the Bronze Age, more specific niches were found for imports based on their perceived affordances. Whereas seated apes were exotic forms which acted as indexes of overseas contact in the Early Bronze Age, monkeys became associated with the saffron industry in the Palatial period. Cats too were given a specialised role in later depictions as bird-hunters. An analogy was found between the mountain edge of the collective, where birds could be hunted and crocuses gathered, and the exotic bodies of monkeys and cats arriving from the horizon's edge. The incorporation of the mountains enabled the palatial collective to gather resources to manufacture products for trade overseas and so monkeys and cats were recruited for this task. Their bodies were different to the local fauna but their affordances fitted into existing palatial concerns. Frescoes at Knossos and Ayia Triada enabled the palatial collective to shape the perception of these affordances.

Other imported animals' affordances favoured their incorporation into the animal practice of hunting. Deer were hunted on the mainland before they were brought to Crete during the Palatial period. Like Cretan wild goats, they afforded a chase and also distinctive trophies in the form of their antlers. The chase itself was affected by new types of hunting dogs whose bodies were given prominence on metal sealing rings used at the palatial centres. Hunters also travelled to chase more dangerous animals in the form of mainland lions. The combination of danger and travel ensured that lion-hunting was a significant animal practice and a popular subject for sealstones as a result. But the depictions of lions hunting other animals on contemporary metal rings show how those at the centre of the palatial collective were increasingly tied into a developing iconography of lions which emerged across the Aegean. The Aegean Animal style was the result of the interaction between the production of luxury goods such as metal daggers and frescoes and the hunting and observation of lions on the mainland. Whereas deer could be imported to Knossos, hunted and eaten, lion bodies were brought to Crete using depictions instead. Hunting certain animals remained an animal practice closely connected to the palatial collective.

Horses arrived on Crete with a specific affordance as pullers of chariots. The assemblage of horse, chariot and human driver provided a demonstration of the reach of the palatial collective which could be driven along the roads of Crete. Horses would have been expensive to maintain, and chariots were complex assemblages of different technologies and materials, and so they were probably limited to the palatial centres in the Final Palace period. The chariot tablets provide a snapshot of the relatively small number of these exclusive animals at this time. More exclusive, however, were animals which were never seen, but only revealed through viewing depictions of them. Pre-eminent among these appear to have been griffins, which borrowed affordances from different animals: predators like lions and chariot-pullers like horses. The griffin was an animal from beyond the edges of the palatial collective which came to occupy a central position, none more so than the Throne Room at Knossos.

EIGHT

THE NATURALISTIC SPIRIT

The naturalistic spirit of the Minoan animal forms is no new theme in this Work. This is strongest in the great Transitional Age M.M.III–L. M. I*a*, some intaglios of which are here included, such as the remarkable seal impression depicting a flying dove, the perspective rendering of the parrot wrasse, or the flying fish that might have been taken from the Phylakopi fresco. One of the most strange examples is the calf's head as dressed for the table.

(Evans 1935: xvii)

The phenomenon of naturalism, as described in this book, combines an increase in the quantity and diversity of animal depictions in Bronze Age Crete with an increasingly realistic style. The highpoints of naturalism defined by Arthur Evans in *The Palace of Minos at Knossos* included, along with the seals mentioned above, the relief bull paintings at Knossos, the 'Cat and bird fresco' from Ayia Triada and the 'Harvester vase' (Evans 1921b: 28) (see Figures 1.4, 5.5, 5.6b, 5.8, 8.1). Evans regarded all of these as roughly contemporary, belonging to the florescence of Minoan culture. Although such evolutionary thinking is no longer fashionable, it renders the production of highly realistic animal depictions part of a historical process. By contrast, the 'nature-loving Minoans' stereotype, which was an outgrowth of Evans's ideas, dehistoricises these animal depictions, making them a product of a love of nature or the worship of a nature goddess. As has been argued, this is the product of a modern ontology which separates culture and nature. This book has sought to locate these animal depictions within a premodern, analogist ontology in which the world is always expanding rather than one which is cut in two. But the question remains why animals came to be depicted in a realistic style at a particular period in

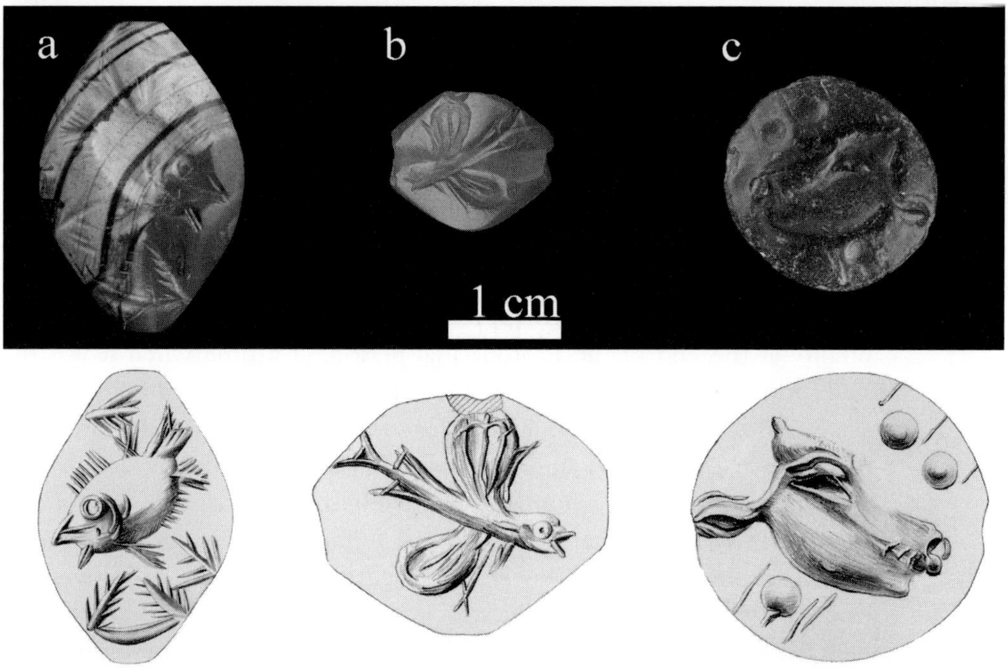

8.1. Seals showing the 'naturalistic spirit', with drawings. (a) Banded agate seal showing fish, Lappa MMIII-LMI (AN1938.973/CMS VI 257a); (b) carnelian seal showing flying fish, Kleitoria, Achaea, LMI (AN1938.972/CMS VI 462); (c) schist seal showing cattle head, Mirabello, LMI (AN1938.970/CMS VI 421). © Ashmolean Museum, University of Oxford/ CMS Heidelberg

the Bronze Age. The answer proposed in this book is that naturalism emerges from the animal practices that defined the palatial collectives.

These animal practices have been examined in previous chapters. The most significant were bull games, hunting/fowling, open-sea fishing and diving and the collection of exotic animals. These practices either involved animals which were largely outside the world of the domestic collectives or, in the case of bull-leaping and cattle-hunting, involved significantly different interactions with domesticated animals. The butchery and consumption of domestic animals was a practice shared by both domestic and palatial collectives but animal heads, whether displayed at consumption events or replicated in the form of rhyta, became more closely associated with the palatial collective. Only in the Final Palace period did formalised sacrifice become the animal practice which distinguished the Mycenaeanised palatial collective of Knossos. The other animal practice to appear around this time was riding in horse-drawn chariots. The same period saw a decrease in the realism of depictions, suggesting that animal things were less important in the organisation of the palatial collective by this time.

The naturalistic spirit, as defined by Evans, emerged around the end of the Protopalatial period, at a time when the palatial collective of Knossos was expanding across Crete, either incorporating or destroying the collectives which centred on the other palaces. The term 'Knossocentrism' has been used as a critique of the applicability of Evans's vision to the rest of Crete but it is an appropriate term for the role of Knossos in the large palatial collective which emerged in the Neopalatial period. This does not mean that every other palatial collective was ruled directly from Knossos, but rather that Knossos was central to the flows of humans, animals and things which crisscrossed the landscape of Crete and the sea routes beyond (Knappett and Nikolakopoulou 2005; Knappett 2018: 983–5). Knossos established its centrality at the end of the Protopalatial period, at a time when it was the largest settlement on Crete (Whitelaw 2012). There is limited evidence for open warfare, although the evidence for militarism and defensive architecture has been overlooked until recently (Alušík 2007; Molloy 2012). The animal practice of cattle-hunting, depicted on seals at the end of the Protopalatial period, appears to be part of raiding and if so provides one of the few clear instances of the depiction of aggression between collectives. An earthquake which devastated Phaistos at the end of the period could have shifted the balance in favour of the Knossos palatial collective, resulting in its expansion into southern Crete.

Among the earliest naturalistic seals are those from the Room 25 deposit at Phaistos. This deposit provides a snapshot of the seals in use at the time of the Protopalatial destruction event and includes a diverse collection of animals, often in a landscape setting. Roundels first appeared at Phaistos at the same time. From this time onwards, clay and seals were increasingly combined to produce traces of a complex set of transactions, which included the collection of wool and its processing into textiles, or the gathering of animals for consumption. Writing had emerged from seal iconography at the start of the Protopalatial period as a means of organising the growing complexity of the collective, as a result of the increase in scale. The script in use in eastern Crete and Knossos in the Protopalatial period, Cretan Hieroglyphic, was used on seals and the distinction between script and image was blurred. By contrast, at Phaistos, the more schematic Linear A was the script of clay documents alone, with seals freed to carry images with more narrative detail. The seal impressions of Room 25 showed wild goats being hunted, swimming octopuses and monsters such as griffins which belonged across the sea. Not all of the seals in use at Phaistos at that time were naturalistic; a variety of abstract designs appear to relate to textile patterns, perhaps the characteristic designs of particular domestic collectives whose production was gathered by the Phaistos palatial collective. In turn, the adoption of Linear A across Crete in the Neopalatial period, accompanied by an increase in the distribution and diversity of naturalistic depictions of animals on seals, can be seen as an indication of the absorption of southern Crete by the palatial collective of Knossos. Whereas the finest Cretan Hieroglyphic seals probably defined seal users by name, these naturalistic seals frequently defined their user with an animal practice.

Animal practices helped define activities of the palatial collectives against those of domestic collectives. By the time of their destruction at the end of the Protopalatial period, palaces had become central places for the surrounding domestic collectives, co-ordinating communal consumption events and organising overseas trade. Although some stages of textile production probably remained at the level of the domestic collective for much of the Palatial period, the palaces were likely centres of trade, and perhaps the production of dye. The close association between women and dyed textiles in the Neopalatial period suggests that women continued to organise this activity, at the level of both the domestic and palatial collective. The centrality of textile production to the palatial collectives meant that the sea became increasingly important, both as the underwater source of purple dye and also as the means of connection with overseas trade networks. At Phaistos the Room 25 seals showing marine animals and monsters demonstrate these underwater and overseas connections. At Mallia, in Quartier Mu, a different technology was used to make similar associations: clay appliqués of cats and underwater creatures combined on a single ceramic vessel. In the same period at Knossos the processes Evans termed 'nature-moulding' and 'nature-printing' similarly established an indexical link between ceramic vessels and the bodies of marine animals. The sealings of Phaistos and the clay appliqués of Quartier Mu were found in destruction layers, as were the nature mouldings of Knossos. Perhaps all were the result of earthquake destruction but as Knossos was rebuilt it incorporated these various ways of producing animal things and combined them in the phenomenon Evans termed the naturalistic spirit.

One of the seals Evans chose to typify the naturalistic spirit is the fish in the Talismanic style (Figure 8.1a). Like the latest Cretan Hieroglyphic seals, these were made of imported hard stones and cut using a drill. Although many are schematic, this example is detailed enough for Evans to suggest an identification with a particular type of fish, and also shows it underwater, surrounded by plants. What is notable about Talismanic seal designs is that they are a combination of the southern Cretan seals preserved among the Phaistos Room 25 sealings and northern Cretan hard-stone prism seals; some show animals in a landscape setting and others carry adapted Cretan Hieroglyphic signs (Jasink 2016). Talismanic seals could have been distributed as tokens of allegiance to the palatial collective of Knossos from the start of the Neopalatial period. Although quicker to produce than many of the more finely detailed Neopalatial seals which came to be used in sealing deposits, Talismanic seals were indexes of overseas connections, since both the cutting technology and colourful stones were imported to Crete. The colour of the stone also appears in many cases to have acted as an index of the sea when marine animals were depicted. Another class of depiction was the group of goats with spears over their backs, clearly scenes of hunted agrimia rather than domestic goats. The Knossos palatial collective expanded across land and sea, and the bodies of non-domestic animals became a means of demonstrating the incorporation of different areas. Fishing from boats and hunting in the mountains both took place beyond the familiar world of the domestic collectives and it was these animal practices which the palatial collective absorbed, partly by depicting them on the seals which connected seal users with the palatial administration at the centre.

A deposit at the centre of Knossos shows an analogous process, in which the imported technology of faience was used to create models of marine animals, including flying fish like the ones depicted on one of Evans's seals (see Figures 6.10 and 8.1b). The Temple Repositories also included faience plaques showing an agrimi suckling her kid and a cow suckling her calf. Associated with these were the famous Snake goddesses; similar elaborately dressed female figures appear widely in Neopalatial iconography. Here they are constituents of an assemblage of material culture which brought together both marine animals and the two land animals most frequently depicted in Late Bronze Age Crete: cattle and wild goats. Another component of the assemblage was the seashells, some of which were painted, which also acted as indexes of the sea. Crocuses too appear on clothing in the Temple Repositories, the source of another of the dyes used in cloth production (Day 2011a: 364–7). The production of such clothing involved a network of sheep, wool, weavers and dyestuffs which extended across the uplands of Crete and under the sea. The Temple Repositories brought animal things from these different parts of the landscape together, establishing the palace at Knossos at the centre of the collective.

The Ayia Triada 'Park fresco' shows a similar set of connections between elaborately dressed women, crocuses, wild goats and a cat stalking a bird. Although it is now regarded as dating to the end of the Neopalatial period, later than Evans suggested, it is part of the same phenomenon of naturalism. Ayia Triada appears to have been part of the Knossos palatial collective and acted as a subsidiary centre in southern Crete instead of Phaistos. The seal impressions of Ayia Triada include other depictions of similar female figures seemingly feeding agrimia and wild boar. These appear to show the taming of, and hence control over, these non-domestic animals by those at the centre. A ring showing male goats rutting, used in the administration at Ayia Triada, also appears to show knowledge of these animals' mating behaviour. As with the Temple Repositories, there is an interest in the fertility of these animals. The debate over whether these figures are goddesses or priestesses distracts from the possibility that women's control over domestic textile production gave certain women a leading role in the Knossos palatial collective as it expanded into the areas of Crete where crocuses were gathered and animals hunted. The cat hunting a bird depicted on the Park fresco, another highpoint of naturalism, can be seen as an analogy for bird-hunting, an animal practice absorbed by the palatial collective. The metal ring impression from Ayia Triada which Evans thought showed a flying dove actually shows a bird struggling in a net. The fine detail of the carving, epitomising naturalism, also allowed the accurate depiction of this animal practice; its carving on a metal ring allowed it to be brought into the centre of the collective.

Although the date of some the frescoes at Knossos is more difficult to establish, relief frescoes of cattle appear to be part of a wider Neopalatial phenomenon of relief depiction (Kaiser 1976). Relief depictions were a key part of Minoan naturalism, making animals physically present in the space of the viewer. The bulls at Knossos were life-size, but relief stone (and metal) vessels used the same technique on a smaller scale. The plaster cattle head found at the North Entrance was part of an animal thing which gave cattle a permanent presence in the palace. In Chapter 3 it was suggested that this coincided with a shift to ranching in Crete, with the Knossos palatial collective taking over cattle ownership from the domestic collectives and bringing them together in herds which were allowed to roam the lowlands of Crete. By analogy with the American rodeo, the round-up would have brought these cattle to Knossos, or perhaps Ayia Triada, and provided the opportunity for young men to show their prowess by jumping over bulls. These activities would have demonstrated the centrality of Knossos in the Neopalatial period. As well as frescoes in the palace, bull-leaping was depicted on Neopalatial metal rings whose impressions were found distributed across the island. These almost certainly emanated from Knossos, the animal practice of bull-leaping coming to signify this collective.

Finely crafted objects such as gold rings, bronze figures and stone relief vessels were produced in palatial workshops as a means of reproducing the animal practices of the palatial collective in durable form. Experimentation in different media resulted in a realistic style which reproduced these animal practices effectively. The bronze and ivory sculptures captured the moment of the bull-leap, just as the 'Ambushed octopus rhyton' froze the moment of encounter with an octopus underwater. Bull-leaping appears on one of the finely carved stone relief vessels found at Ayia Triada, the 'Boxer rhyton', connecting this animal practice with scenes of young men boxing. Hunting is less frequently shown, although the Harvester vase appears to show a hunting expedition and the 'Chieftain cup' appears to show cattle hides which are likely to be the product of hunting. There is an overlap on these two vessels with warfare; the cattle hide seems destined to be turned into shields, and the marching 'harvesters' could be training for war. Similarly the guardhouses of Crete could have functioned both as garrisons and hunting lodges, helping to establish control over territory. Individual hunting is very rarely shown on these most finely crafted objects, and not found among the surviving frescoes. The 'Sanctuary vase', like the Park fresco, instead helps to establish the palatial collective's inclusion of hunted animals by showing their association with peak sanctuaries, which had themselves come under palatial control in this period.

Communal gatherings involving the consumption of meat and wine predated the palaces, but both the courts of the palaces and also peak sanctuaries acted as new foci for these feasting events in the Palatial period (Borgna 2004: 256–60; Haggis 2007; Girella 2008; Driessen 2018). Animal heads appear to have been specially treated on these occasions, either deposited at the bottom of pits or retained and put on display as defleshed skulls (Shapland 2021). The significance of animal heads is reflected in their frequent use on Protopalatial seals, often as part of the Cretan Hieroglyphic script, but then on more naturalistic Neopalatial seals. Cattle-head rhyta also appeared in the Protopalatial period, becoming highly realistic depictions of cattle heads in stone in the following period. The finest Neopalatial cattle-head rhyta can be connected with the transformation of feasting from a communally organised event into one which was, quite literally, in the hands of a few. These objects condensed meat and drink into one durable animal thing, animated when it was filled with metaphorical blood. Whereas real animal heads would have temporarily marked the provision of an animal for a feast before they decomposed, stone bull's-head rhyta were durable things which transcended single events. Like the relief frescoes of bulls at Knossos they made transient events permanent and were associated with the palace, whether they were part of the fabric of the building or passed round at communal gatherings.

The organisation of such communal consumption events was clearly important in the emergence of the palaces. It can be suggested that one of the reasons there is not a clear ruler iconography in Crete was because the palatial collectives were initially corporate entities comprised of domestic collectives. Even as hierarchies emerged, and Knossos absorbed other palatial collectives, feasting was one means of maintaining a sense of collective identity. Animals were brought to palatial centres and their bodies shared in consumption. In the early Neopalatial period some of these consumption activities appear to have shifted to sanctuary sites, perhaps restricting them to a smaller group. There are indications that this group was composed of the male hunters and warriors whose activities helped to reorganise the Cretan landscape in this period. In the Final Palace period the sacrifice of cattle is depicted for the first time. This can be seen as a new animal practice which defined a new group at the centre of the palatial collective. This animal practice formalised the slaughter of animals and not just their butchery and consumption. It marked a shift from an emphasis on the act of provision signified by animal heads to the act of slaughter itself. Like the shift from Linear A to Linear B, this need not be seen as a radical break; rather, an existing practice, feasting, was associated with a different animal practice, sacrifice, as part of a closer association with, or incorporation into, a mainland palatial collective.

Another defining animal practice of this palatial collective was riding in horse-drawn chariots. This activity is depicted on a metal ring preserved on Neopalatial sealings and is shown occasionally on later seals. Linear B tablets from the Final Palace period at Knossos record the presence of a large number of horses and various parts of chariots. Given that they must have been high-status animals, it is striking that they were not depicted more frequently. Like cattle and sheep, horses were counted by the palace as well as depicted. The Linear B tablets suggest that horses were just one part of a suite of military equipment controlled by the palatial administration. The animal practice of driving a chariot would have been part of a broader warrior identity, defining an exclusive group comprising warriors, horses and hide-covered chariots, at the centre of the palace. It was depicted, like bull-leaping, in a fresco in the palace of Knossos but not as widely distributed in other media. Once Knossos had become the only palace on Crete it seems that depicting the animal practices of the palatial collective was less important in its maintenance.

Depictions of animals grew more stylised in the Final Palace period; seals became more symmetrical and additional symbols were sometimes added next to animal depictions. Seals increasingly focused on hunting scenes, with fewer types of animals depicted compared to the Neopalatial period. Octopuses and argonauts too lost their free-flowing tentacles in favour of more regular arrangements. Marine-style pottery at the end of the Neopalatial period saw some of the designs of finely crafted stone vessels being transferred to pottery. Later vessels continued to replicate these designs but the lifelike appearance of these marine animals was gradually lost. Burial chests, larnakes, from the Final Palace and Postpalatial period continued to show marine animals and hunting scenes, sometimes on the same larnax, but the sense of realism was lost. The remains of deer and wild goats at various Postpalatial sites show that hunting continued after the fall of the palaces; it is likely that there continued to be feral cattle on Crete once the palatial collective disintegrated and these too are shown hunted on larnakes. But the way they were depicted had changed; the hunted animals are still and the marine animals are constrained by symmetry. These lifeless bodies of marine animals no longer indexed animal practices but referred back to earlier depictions.

Whereas Evans suggested that naturalism peaked in Middle Minoan III, in order to fit his evolutionary scheme, it can be suggested that it was in fact the whole of the Neopalatial period that saw an emphasis on the realistic depiction of animals in a variety of media. This does not significantly affect Evans's argument that naturalism appeared and disappeared within a relatively short time period. This time period coincided with the expansion of the palatial collective of Knossos across the whole of Crete. The naturalistic depiction of certain animals was important because they inhabited the different areas of the land and sea into which the palatial collective of Knossos extended. Wild goats lived in the mountains, where crocuses also grew; octopuses and triton shells lived in the underwater world of the purple shell; cats, monkeys and griffins came from across the sea where other important commodities and luxuries could be obtained. Their bodies were markers of places and the realistic depiction of these bodies allowed them to be brought into the centre of the palatial collective and used as analogies for the palace's activities. Naturalism was partly a product of the palatial workshops, whose experiments with materials and style allowed these bodies to be brought into the palace ever more effectively. Carved stone triton-shell vessels epitomise the way in which skilled crafting allowed animal forms to be brought into the palace as animal things.

As has been argued in this book animal things were the product of animal practices. The affordances of particular animals resulted in their becoming important members of the palatial collective. The affordances of animal things helped demonstrate and materialise this relationship. Wild goats and octopuses were not simply distinctive bodies encountered in certain parts of the landscape but were also challenging to pursue and catch. Sheep, because of their role in the textile industry, were arguably more important to the palatial economy than cattle but it was a set of animal practices involving cattle that came to define the palatial collective of Knossos. Bulls ranged around the lowlands of Crete throughout the Bronze Age. They were hunted when collectives were competing over who could control them and the land they grazed. They were netted and leapt over when the Knossos palatial collective had established control over the lowlands. They were sacrificed when the animal practices of the mainland became established at Knossos. The relief fresco of a charging bull which rises above Arthur Evans's restorations of Knossos is a reminder that cattle have all but disappeared from Crete, and with them these various animal practices. Yet it continues to do what it was intended to: give cattle a permanent presence in the palace, even though today it is a member of a different, globalised, collective, formed of the hundreds of thousands of visitors who every year encounter this life-size bull among the ruins of Knossos.

NOTES

1 INTRODUCTION

1. Schinz in fact appears to have relied on Belon's observations, particularly that the horns of the Cretan wild goat were different from those of the wild goat of Cyprus.
2. The term 'seal' refers to an object used for sealing, and includes both sealstones and signet rings.
3. The name apparently comes from one particular agrimi, named Kri-Kri, who was given to President Truman and subsequently donated to Washington Zoo (Fielding 1953: 43).
4. The term 'minoische' had already emerged in German publications and was used as a counterpart to the 'mykenische' finds from the mainland (Momigliano and Karadimas 2004).
5. Minos Kalokairinos also called the building he excavated the 'Royal Palace or Megaron of King Minos I' (Kopaka 2015: 143). The term 'palace' had previously been applied to monumental Bronze Age buildings by Schliemann, deriving from the descriptions of palaces in the Homeric epics, and is still commonly used for any similar Bronze Age building, particularly with a central court, found on Crete or the mainland.
6. In his excavation diary for 16 May 1900 [Evans AE/NB 1900: 75], after finding remains of the relief bulls in the North Entrance and another bull fresco *in situ* at the West Porch, Evans asked, 'Was not some one or other of these creatures visible on the burned site to the early Dorian days which gave the actual tradition of the bull of Minos?'. This illustrates how Evans's immediate reactions to finds sometimes became the basis for arguments in subsequent publications.
7. For example: 'La nature a toujours été une des principales sources d'inspiration pour les artists minoens' ['Nature was always one of the principal sources of inspiration for Minoan artists'] (Vanschoonwinkel 1996: 351); 'The principal significance of such a popular motif is that Capra aegragus cretensis provides additional insight into one of the most basic features of Minoan culture: a profound interest in and enjoyment of the natural world' (Bloedow 2003: 2).
8. The relationship between the collection of skins, cartography and colonialism in the Victorian era is explored by Ann Colley (2014).
9. Rather than being defined in opposition to the knowing subject, for Latour (2013: 288), the term 'object' is interchangeable with Michael Serres's ([1980] 2007) term 'quasi object'. The term 'quasi object' describes the way in which entities or beings are embedded in semiotic networks rather than existing as isolated inanimate objects. Donna Haraway's term 'figure' is closely aligned with this way of conceiving the beings which emerge out of human-animal relations: 'Figures are not representations or didactic illustrations, but rather material-semiotic nodes or knots in which diverse bodies and meanings co-shape one another' (Haraway 2008: 4).
10. Descola (2013a: 361–4) uses the term 'collective' in a more general sense than Latour, to describe the extent of the use of a particular mode of relation and mode of identification. Here Latour's use to describe an interconnected group of humans and non-humans is preferred.

2 LOOKING

1. www.damienhirst.com/a-thousand-years.
2. In biological terms this is the same as a 'nondimensional species', that is, a population of interbreeding individuals encountered in a particular geographical area (Mayr 1949). It is worth noting that Mayr's

definition of a species was partly a response to studying the birds of New Guinea from the point of view of the local classification systems (Atran 1990: 260).

3. 'The frieze gives an anticipatory assurance of good cheer, and recalls, indeed, the walls of an old Dutch dining-room, hung with pictures of game or domestic fowls, or of still-life more immediately referring to the table' (Evans 1928: 114).
4. Similarly, the English/modern Greek term 'partridge'/'πέρδικα' refers to a variety of birds from the genus *Perdix* ('true' partridges) as well as *Alectoris* (rock partridges). As a result, contemporary folk biological terms do not coincide with the 'generic-species'.
5. The same can be said of depictions of people (German 2005).
6. Weingarten (1983: 50–4) describes how a number of scholars regarded the Zakro Master as mad.
7. For Gibson, ecology meant something akin to being-in-the-world; central was the study of interactions between animals (including humans) and environment (Cutting 1982, 1993: 235–8).
8. This is implicit in Gibson's work, although he is often criticised for suggesting that the affordances of cultural objects such as postboxes can be directly perceived (Noble 1981; Knappett 2005b: 46–7). Arguing against the idea that postboxes change in meaning according to whether one wants to post a letter, Gibson suggested that 'I prefer to say that the real postbox [. . .] affords letter-mailing to a letter-writing human in a community with a postal system' (Gibson 1979: 139). The implication, although this was never developed by Gibson, is that a large amount of cultural knowledge is required before this affordance of a postbox is revealed, but that this affordance is not reducible to the user's intentions.
9. Bryson (1983: 51) explicitly rejects the Peircean scheme because he denies the possibility of interpreting icons or indexes apart from social convention. Sonesson's (1989) account overcomes this problem by suggesting that signs are interpreted in a sociocultural Lifeworld without suggesting that the connections between them are arbitrary.
10. The 'Lifeworld', although a phenomenological concept, can be understood more broadly using the semiotics of Jacob von Uexküll (1982), who argued that each organism had its own 'Umwelt' within which perception and semiosis occurs. Importantly, concepts such as affordances and indexicality are relationships between organisms and their environment and are not reducible to human consciousness.
11. This act of understanding which establishes the intended link between sign and object formed the third part of the sign relation for Peirce, what he termed the 'interpretant'.
12. In Summers's (2003: 289) terms, such life-like sculptures resembling Christ or a saint are 'images with the value of effigies', meaning that they gain their efficacy from their supposed connection with the figure they depict. In Pre-Reformation Europe such lifelike images came to be regarded as similarly potent to relics themselves (Bynum 2011).
13. The contemporary art market is predicated on paintings and other artworks preserving the traces of famous artists; it is this which makes them authentic, and hence valuable (Bown 2015).
14. By contrast, Jaś Elsner (2006), in an echo of Berger on zoo animals, suggests that Classical statues no longer returned the viewer's gaze as their Archaic predecessors did; instead, they cast the viewer as passive observer. In doing so he places Classical Greek naturalism at the start of Western naturalism.
15. Another symbolic interpretation is that this fresco represents the constellations Orion and Taurus (MacGillivray 2000a). There are no astrological texts from Bronze Age Crete, meaning that there is no contemporary evidence for this symbolism.
16. Knappett et al. (2010) argue for a broader use of containers, and rhyta such as this, as metaphors for bodies.

3 HERDING

1. http://geodata.gov.gr/dataset/apographe-georgias-ktenotrophias-1999-2000.
2. The biographies of donkeys in the Agia Marina donkey sanctuary on Crete (www.agia-marina-donkeyrescue.com) show that some animals are still used for traction. The work of the donkey sanctuary is a very different type of animal practice specific to modernity.

3. The study of animal dung at the Neolithic site of Koutroulou Magoula in mainland Greece showed how archaeological science can reveal traces of animal presence in the development of the mound (Koromila et al. 2018).
4. Isaakidou (2008) has also demonstrated that increasing proportions of cattle bones compared with other domesticates over time are not just the result of recovery bias and taphonomic processes as had been suggested (Winder 1991; Whitelaw 1992).
5. For the Final Neolithic dating of levels I–III, see Peter Tomkins (2007).
6. Other candidates are polecat, marten or ichneumon (Egyptian mongoose), all of which can be tamed. Various mustelids are found in zooarchaeological assemblages from prehistoric Greek sites (Yannouli 2003: 184).
7. This could even be true when the sign showing a cow's head is used as a syllabogram; although the relationship between sign and sound could be arbitrary and symbolic, John Younger (2013) suggests by analogy with Linear B the iconic sound value 'mu', and similarly 'ma' for the cat face. This opens the intriguing possibility of onomatopoeia, another type of iconic relationship between sign and signifier because they are linked by resemblance.
8. English surnames such as 'Fowler', 'Potter' or 'Weaver' refer originally to occupations.
9. Middle Bronze Age Old Assyrian texts from Anatolia refer to imported 'Akkadian' sheep which appear to be related to wool production (Lassen 2014: 258).
10. CHIC sign 013 as a syllabogram, 152 as an ideogram. There are three occurrences of the ideogram 152 listed in CHIC, two of them on the same document (Hh (08) 01). As with the clay bar Hh (08) 02, the ideogram is associated with the numeral 1, and both documents are from Knossos (Olivier and Godart 1996: 118–21, 423 [#065, #067]).
11. Ideogram 336, for example, clearly depicts an animal head. This has been suggested to be a dog (Godart and Tzedakis 1992: 139), but Schoep disagrees on the basis that it could equally be an equid, and identifies another ideogram as a cow (306) (Schoep 2002a: 125–7).
12. The tablets were categorised by Emmett Bennett (1951), prior to decipherment, on the basis of ideograms; the letter 'D' refers to livestock tablets. That the system is still used illustrates again that the structure of the tablets was apparent even before the syllabograms could be read.
13. The tablet shown in Figure 3.1, Dd 1171, has been changed in this way; on the bottom row the sheep ideogram has been changed from female to male (Chadwick et al. 1992: 49).
14. Faunal totals taken from the following publications where NISP given: Ayia Triada (Prepalatial NISP = 194, Neopalatial NISP = 200) (Wilkens 1996a); Debla (Prepalatial NISP = 60) (Warren and Tzedhakis 1974); Knossos (Neopalatial NISP = 219, Final Palace NISP = 2,888) (Jones 1979; Bedwin 1984); Kommos (Protopalatial NISP = 390, Neopalatial NISP = 695, Final Palace NISP = 923) (Reese 1995b; Ruscillo 2006); Mochlos island (Final Palace NISP = 1,005) (Soles 2008); Mochlos shore [Artisan's Quarter] (Neopalatial NISP = 1,039) (Soles 2003); Pseira (Neopalatial NISP = 88) (Betancourt and Davaras 1995a, 1995b; Floyd 1998; Reese 2009); Palaikastro (Final Palace NISP = 476) (Wall-Crowther 2007).
15. A fragment of a rhyton showing boxing was found in the same context as the Taureador frescoes at Knossos (Evans 1901b: 96).
16. CMS II 8 232.
17. The fragments of relief frescoes of bulls at the North Entrance are Neopalatial, although Evans reconstructed the North Bastion with the charging bull in order to materialise his theory that scenes of bulls were visible even after the destruction of the palace, inspiring the myth of the Minotaur (Evans 1921b: 1–2).
18. See, for instance, a cattle-tail fly-whisk in the collection of the Horniman Museum (www.horniman.ac.uk/object/1972.39).
19. Horn and tortoiseshell are made of the same material, keratin, and dyed horn was traditionally used as a substitute for tortoiseshell (Aikin 1838: 341). Horn cups were a traditional craft product in historic European cultures, and were sometimes composites of other materials such as glass or silver. See the collection of the Worshipful Company of Horners on loan to the Museum of Design in Plastics (www.horners.org.uk/the-company/the-horn-collection/). This shows the variability of horns; some horn products have a flecked pattern which could also be linked to designs in late Protopalatial and early Neopalatial pottery.
20. According to Günter Nobis (1993: 110), the largest diameter of a horn core from Poros was

12.4 cm, compared with the largest diameters of the Vapheio cups of 10.3 and 10.7 cm (Davis 1974: 487). The diameter of the ceramic Vapheio cup AN1938.456 shown in Figure 3.13 is 9.2 cm.

21. Fragments of a miniature fresco from Tylissos show boxers; the close similarity of these depictions to the miniature 'Grandstand' and 'Sacred grove and dance' frescoes at Knossos help support Evans's (1930: 56) original suggestion that arena sports, such as bull-leaping and boxing, were originally part of these scenes (Rethemiotakis 2020: 123–8).

4 BUTCHERING

1. Including the reports discussed below for Kommos, Mochlos, Nopigeia-Drapanias and Vronda, as well as Valasia Isaakidou's (2004, 2007b) reanalysis of historic Knossos excavations.
2. Cut marks found on the bones of children at Knossos, with implications of cannibalism, further showed how assumptions about Minoan society could be undermined by the traces of an unexpected practice (Wall et al. 1986). Although an exceptional context, this indicates that human bodies did not always have different affordances from non-human bodies.
3. 'Butchery, n., 3', OED Online, Oxford University Press, March 2021 (www.oed.com/view/Entry/25338).
4. E.g., Pausanias, *History of Greece* 5.13.8.
5. Another possibility is that the human on the platform was being treated for a hunting injury, since the spear is inscribed with a boar's head, and could have been used for boar hunting (Younger and Rehak 2010a: 170).
6. Rehak (1995c: 451) argues that they were smashed with a blow to the muzzle because 'a strong blow to the muzzle would have approximated the blow actually delivered to stun sacrificial bulls'. This is contrary to the usual method of stunning bulls, which is either a blow to the forehead, or sometimes the neck (Aldrete 2014).
7. Helène Whittaker (2016) argues that the association is instead based on the use of the axe for polling cattle horns but there is no conclusive evidence for this.
8. The bronze objects sometimes described as 'cutters' are the same form as each half of the 'votive' double axes (Boardman 1961: 23, 42–5). A sturdier version from Psychro is in the New York Metropolitan Museum of Art (26.31.474). Compare the Inuit ulu or Tibetan kartika, both curved flaying knives. Knives were also abundant at Psychro, which could have been used for filleting.
9. Nanno Marinatos (1986: 22) suggests the double axe could also be used for stunning, although sees a mace as a better weapon for this. Maria Shaw (2012: 156–7) also proposes that the double axe was used to stun cattle before butchery and consumption. She suggests this was a part of bull-leaping, although there is no clear evidence for this.

5 HUNTING

1. Spratt was employed by the British Admiralty to survey the Mediterranean, providing data for its charts of the region. For Spratt's career, including his contributions to Cretan archaeology, see Dudley Moore (2013: 85–144).
2. Support for the idea that seabirds were eaten comes from the Middle to Late Bronze Age site of Kolonna on Aegina, where butchery marks are reported on the remains of cormorants and raptors, if these are not for the removal of feathers (Forstenpointner et al. 2010).
3. Many are listed in the 2009 EU 'Birds Directive' (http://ec.europa.eu/environment/nature/conservation/wildbirds/hunting/guide_en.htm).
4. There were over 100 bones in the cooking pot and a further 32 in the same room, and another 24 in a Final Palatial context (Soles 2008: 94). Hare bones have been identified at: Archanes-Phourni (Sakellarakis and Sapouna-Sakellaraki 1997: 263); Ayia Triada (Wilkens: 1996a: 248); Chania (Hallager and Hallager 2003: 144); Knossos, Acropolis Houses (Jones 1979); Kommos (Reese 1995b: 166; Ruscillo 2006: 835); Pseira (Betancourt 2009: 142); Vronda (Preston Day et al. 2009).
5. See, for example, the feral populations of various domesticates, including cattle on Hawaii (Henke 1929: 21–5). For several generations after their introduction, cattle were left to roam the islands of Hawaii and were only controlled by hunting (Fischer 2015: 60–4). Cattle are still hunted on Hawaii.

6. Günter Nobis (1996) argued that wild cattle were part of the Pleistocene fauna of Crete but there is no evidence for this in the fossil record. Cattle were brought to Crete by humans at the start of the Neolithic period (Broodbank and Strasser 1991: 236). Conrad Keller (1909) and Nobis (1990, 1993) identified a number of specimens as aurochs but these identifications of wild versus domestic cattle on the basis of the size of the remains have been questioned (Vigne 1999: 300).
7. Kenna highlights the last of these, which 'even in its attenuated form shows feeling for nature which was once thought to be the preserve of the nineteenth century in England and France' (Kenna 1960: 44). This impression is now thought to come from a seal showing three fish (CMS II 8 159).
8. Linda Evans (2010) has demonstrated that the varied poses of animals depicted in Old Kingdom Egyptian tombs can be linked with close observation of animal behaviour.
9. Now available online (www.uni-heidelberg.de/fakultaeten/philosophie/zaw/cms/) and in a searchable database (arachne.dainst.org).
10. Some Late Bronze Age seals cannot be identified specifically as either Minoan or Mycenaean manufacture. Seals dated Late Bronze Age I, for example, by CMS are included within the Neopalatial total. This is the reason for restricting the sample to seals known to have come from Crete because Late Bronze Age seals found on Crete are likely to have been used there in the Bronze Age; conversely, stylistically Late Minoan seals found on the mainland were not necessarily ever in use on Crete. Indeed, Younger (1984) identified certain seals as having been made in Mycenae by a Cretan engraver; whether or not the identification of individual biographies is accepted, it illustrates the stylistic interchange in Late Bronze Age between Crete and the mainland.
11. The Cretan collections are the museums in Heraklion, Ayios Nikolaos, Sitia and Chania, as published in CMS II–V. The seals from Crete in the Athens National Museum are also included (CMS I). The individual collectors are Demargne, Evans and Seager, whose collections were dispersed to the Ashmolean Museum, Oxford, the British Museum, London, the Bibliothèque nationale de France, Paris, the New York Metropolitan Museum of Art and the Penn Museum, Philadelphia (CMS VI–IX, XI). British Museum seals known to have been collected on Crete by Thomas Burgon, Colonel J. S. Cameron and Augustus Franks are also included. All of these collectors were active before the Second World War, and the sample does not include seals acquired after 1970, the date of the UNESCO Convention on the Means of Prohibiting and Preventing the Illicit Import, Export and Transfer of Ownership of Cultural Property.
12. See the author's PhD thesis (Shapland 2009) for a more extensive statistical treatment of seals and other animal depictions from Bronze Age Crete.
13. Alexander MacGillivray (1984: 153) identifies the birds as partridges although there is debate about their identification (Panagiotaki 1999: 136).
14. A recent exception is Dieter Rumpel (2007), who is sceptical that the figures are harvesters and suggests that they are marines with grappling hooks.
15. Also known as the 'Captain and warriors', it was recovered with the other miniature fresco fragments reconstructed as the 'Grandstand' and 'Sacred grove and dance' frescoes, which help to date it to the Neopalatial period (Hood 2005: 63–4).
16. Roman poleaxes and hammers used for stunning cattle are very similar in appearance to those carried by the harvesters (Aldrete 2014: 34). A similar axe-hammer from the Sutton Hoo ship burial has been regarded as a tool for killing and butchering cattle (Dobat 2006). See also the third-century-CE butchery axes from the sacrificial bog at Oberdorla (Behm-Blancke 2003: 64, pl. 21).
17. The curious 'scissor' handles appear to be skeuomorphs of horn handles.
18. The boar's-tusk helmet is described in *Iliad* 10.260–5, where it features as an heirloom given to Odysseus by the Cretan hero Meriones.
19. The same criss-cross design as on the garment in Figure 5.9c appears on a stone vessel which also appears to be a skeuomorph for leather (Evans 1921b: 412).

6 FISHING

1. Adolf Furtwängler and Georg Loeschke's (1886) *Mykenische Vasen* established the idea that pottery from across the Aegean could be described

with reference to the finds at Mycenae and given the term 'Mycenaean'.

2. 'The Mycenaean ornament seems derived not so much from traditional forms as from nature herself' (Newton 1880: 284).
3. Cretaquarium (www.hcmr.gr/en/cretaquarium/).
4. The importance of sieving was pointed out in a review of the Myrtos publication (Vitelli 1974: 192).
5. These include Chania (Reese 2016a: 407), Kommos (Reese 1995c: 244–50), Pseira (Floyd 1998: 133–4; Betancourt 2009: 135–8) and Mochlos (Mylona 2020: 197).
6. The data in Figure 6.3 come from the following publications detailing more than 300 shells (only families comprising 5 per cent or more of assemblage are included in graph): limpet (*Patellidae*), top shell (*Trochidae*), purple (*Muricidae*), cockle (*Cardiidae*), whelk (*Buccinidae*), dog cockle (*Glycymerididae*), duck clam (*Mactridae*), Venus clam (*Veneridae*), triton (*Charoniidae*), lucinid clam (*Lucinidae*), cowrie (*Cypraeidae*)). Recording and collection strategies differ: Chalinomouri (Neopalatial MNI = 432) (Soles 2003); Chryssi (Neopalatial MNI = 60,030) (Mylona 2020: 190); Knossos (Neolithic NISP = 826, Neopalatial Temple Repositories NISP = 6,338) (Shackleton 1968; Panagiotaki 1999: 175); Kommos (Protopalatial NISP = 959, Neopalatial NISP = 5,357, Final Palace NISP = 31,591) (Reese 1995c; Ruscillo 2006); Mochlos island (Neopalatial MNI = 5,879) (Mylona 2020: 197); Mochlos shore (Neopalatial MNI = 564, Final Palace MNI = 4,198 (Soles 2003, 2008); Myrtos (Prepalatial NISP = 318) (Shackleton 1972); Papadiokampos (Neopalatial MNI = 19,820) (Mylona 2020: 200); Pseira (Protopalatial NISP= 953, Neopalatial NISP = 1,162) (Betancourt and Davaras 1995a, 1995b; Floyd 1998; Reese 2009).
7. For the evidence for the use of waterworn shells from the northern Aegean in the Bronze Age, see Theodoropoulou (2014a).
8. With the notable exception of Bernd Kaiser (1976).
9. Fish occasionally appear on Prepalatial light-on-dark pottery (Bosanquet and Dawkins 1923: 11). A gable seal from Platanos shows a ship with fish swimming underneath (CMS II 1 287c).
10. This vindicates Boardman's (1966: 266) minor criticism of the first CMS volume that it did not record the colour of each seal, since he suggested it might be expected to be associated with the 'device' or shape.
11. There are a small number of Talismanic hardstone three-sided prisms, although more amygdaloid in shape. Although it is not clear when in Middle Minoan III the Talismanic style started to be produced, this is one indication that the Talismanic seals were the direct replacements of the soft-stone three-sided prisms.
12. Jasink's (2016: 22) identification of one sign as either a fish or a cuttlefish is problematic.
13. See CMS II,8 025 and II,8 115.
14. Caroline Tully (2020) uses a gold ring from Pylos to argue for a set of relations between the sea, textiles and ritual.
15. It is worth noting in this context that one of the Greek myths about Theseus on Crete was his diving to recover the ring of Minos, in addition to his exploits with the Minotaur which seem to relate in some way to the animal practice of bull-leaping.

7 COLLECTING

1. Reports of a petrified monkey skull from Bronze Age Thera have been debunked (Greenlaw 2006: 66).
2. Although other details of the restoration have been challenged (Galanakis et al. 2017).
3. Michael Jarman (1996: 214) had previously identified cat remains from a mixed Bronze Age/Roman level at Knossos.
4. Older archaeological reports mention horse remains, from Prepalatial Ayia Triada (Banti 1933: 216) and Neopalatial Tylissos (Hazzidakis 1921: 76). Donkey (*Equus asinus*) remains were also reported from Tylissos. These reports are problematic because they do not give any information beyond presence.
5. CMS II 8 133.
6. Ayia Triada Tholos A contained a number of imported items (Legarra Herreo 2014: 54).
7. Donkey (*Equus asinus*) remains with pathologies indicating an old animal or one used to carrying heavy loads have been identified at Knossos. Although the stated context was Middle Minoan IA, carbon dating suggested that they belonged to the thirteenth century BCE. The animal appears to have been skinned

(Preston et al. 2013: 63–7). Donkeys were also identified in various contexts in Postpalatial Chania (Hallager and Hallager 2003: 103, 113–14, 140).

8. Using Linear B values, the Linear A ideogram spells out ma-ru, and the Linear B ideogram usually looks more like ma-re. Figure 7.4b (CMS VI 93a) has sign 31, 're', between the cat's ears in exactly the same position as the later wool ideogram. This could be the first instance of the later wool ideogram. The word entered into later Greek meaning wool, μάλλος (Younger 2005: 409). Because it is unknown whether Linear A syllabograms had the same value as Linear B syllabograms, this reading of the wool ideogram is doubted by some authors (Del Freo et al. 2010).
9. As noted above, the Petsophas figurines include a small predator. Polecats or ferrets were kept in ancient Greece as pest-controllers alongside cats (Engels 1999: 66–70).
10. For detailed discussions and catalogues of depictions of monkeys in the Aegean Bronze Age, see Cybelle Greenlaw (2011) and Marie Nicole Pareja (2017).
11. Colonel Henry Drummond-Hay noted that what he called the blue rock pigeon (*Columba livia*) 'breeds both in the caves on the sea-shore and on the highest mountains, where they were found in great numbers' (Drummond-Hay 1865: 404). Pigeons are now common in cities worldwide because they have adapted to using roofs and ledges as nesting places.
12. The terracotta head from Zakro usually identified as a cat also closely resembles these mastiff depictions, particularly the large ears and wrinkled face (Platon 1971: 252).
13. Analagous to the Egyptian 'pariah' dogs (Osborn and Osbornová 1998: 59). 'Pariah' is widely used as a term for dogs around the world which have evolved in and around human settlements in the absence of selective breeding (Coppinger and Coppinger 2016).
14. A unique depiction of parading lions on painted pottery, seemingly derived from seal depictions, also appears at Protopalatial Phaistos (Sanavia 2019).
15. Arthur Evans rejected the possible reading of 'po-lo' from the Cypro-Minoan syllabary next to a sign resembling a 'maneless' horse on one Knossos tablet because he was convinced that the language of the Linear B tablets could not be Greek (Evans 1935: 799). Instead, Evans noted that the horses' heads were examples of 'primitive pictography', which on other tablets had become 'conventionalised' into a linear form. It turns out that Evans was right on the linguistic reading and wrong on the conventional forms, some of which are now recognised as cattle ideograms. The signs resembling horses' heads are still recognised as horse ideograms and 'po-lo' is recognised as representing the Mycenaean Greek word for foal, here neatly explaining the lack of mane on the ideogram (Ventris and Chadwick 1956: 48).
16. The griffins on the Phaistos sealings have wings and crests but there is some debate about whether they can be called griffins (Crowley 1989b: 52). CMS identifies them as 'Greif?'. Morgan (2010a: 304) accepts them as griffins.
17. The rowing boats on the Theran minature fresco have lions painted on their side (Morgan 1988: 45; Shapland 2010b: 119–20), suggesting an analogy between those on board and lions as predators, realised through the animal practice of lion-hunting.

REFERENCES

Adams, E. 2017. *Cultural Identity in Minoan Crete: Social Dynamics in the Neopalatial Period*. Cambridge: Cambridge University Press.

Aikin, A. 1838. 'On horn and tortoiseshell', *Transactions of the Society, Instituted at London, for the Encouragement of Arts, Manufactures, and Commerce* 52(2), 334–49.

Alberti, B. 2002. 'Gender and the figurative art of Late Bronze Age Knossos', in Hamilakis, Y. (ed.), *Labyrinth Revisited: Rethinking 'Minoan' Archaeology*. Oxford: Oxbow, 98–117.

Alberti, B. 2016. 'Archaeologies of ontology', *Annual Review of Anthropology* 45, 163–79.

Alberti, L. 2013. 'The funerary meanings of the octopus in LM IIIC Crete', in Graziadio, G., Guglielmino, R., Lenuzza, V. and Vitale, S. (eds.), *Filiki Sinavlia: Studies in Mediterranean Archaeology for Mario Benzi* (BAR IS 2460). Oxford: Archaeopress, 69–77.

Aldrete, G. S. 2014. 'Hammers, axes, bulls, and blood: some practical aspects of Roman animal sacrifice', *Journal of Roman Studies* 104, 28–50.

Alexiou, S. 1968. *Υστερομινωικοί Τάφοι Λιμένος Κνωσού (Κατσαμπά)* (BAE 56). Athens: Archaeological Society at Athens.

Allen, H. 2001. *Mediterranean Ecogeography*. Harlow: Pearson Education.

Allsen, T. T. 2006. *The Royal Hunt in Eurasian History*. Philadelphia: University of Pennsylvania Press.

Almond, R. 2003. *Medieval Hunting*. Stroud, UK: Sutton.

Alušík, T. 2007. *Defensive Architecture of Prehistoric Crete*. Oxford: Tempus Reparatum.

Anastadiadou, M. 2011. *The Middle Minoan Three-Sided Soft Stone Prism: A Study of Style and Iconography* (CMS Beiheft 9). Mainz: von Zabern.

Anastasiadou, M. 2016. 'Drawing the line: seals, script, and regionalism in Protopalatial Crete', *American Journal of Archaeology* 120(2), 159–93.

Anderson, E. 2016. *Seals, Craft, and Community in Bronze Age Crete*. Cambridge: Cambridge University Press.

Anderson, E. 2020. 'The poetics of the Cretan lion: glyptic and oral culture in the Bronze Age Aegean', *American Journal of Archaeology* 124(3), 345–79.

Andreadaki-Vlazaki, M. 2015. 'Sacrifices in LM IIIB: early Kydonia palatial centre', *Pasiphae* 9, 27–42.

Apostolakou, S., Betancourt, P., Brogan, T. and Mylona, D. 2016. 'Chryssi and Pefka: the production and use of purple dye on Crete in the Middle and Late Bronze Age', in Ortiz, J., Alfaro, C., Turell, L. and Martínez, J. (eds.), *Textiles, Basketry and Dyes in the Ancient Mediterranean World* (Purpureae Vestes V). Valencia: Universitat de València, 199–208.

Apostolakou, V., Betancourt, P., Brogan, T., Mylona, D. and Sofianou, C. 2014. 'Tritons revisited', in Touchais, G., Laffineur, R. and Rougemont, F. (eds.), *PHYSIS. L'environnement naturel et la relation homme-milieu dans le monde égéen protohistorique. Actes de la 14e Rencontre égéenne internationale, Paris, Institut National d'Histoire de l'Art (INHA), 11–14 décembre 2012* (Aegaeum 37). Leuven: Peeters, 325–32.

Apostolakou, S., Brogan, T. and Betancourt, P. P. 2012. 'The Minoan settlement on Chryssi and its murex dye industry', in Nosch, M.-L. and Laffineur, R. (eds.), *KOSMOS. Jewellery, Adornment and Textiles in the Aegean Bronze Age. Proceedings of the 13th International Aegean Conference, University of Copenhagen, Danish*

National Research Foundation's Centre for Textile Research, 21–26 April 2010 (Aegaeum 33). Leuven: Peeters, 179–82.

Apostolakou, V., Brogan, T. and Betancourt, P. P. 2020. *Alatzomouri Pefka: A Middle Minoan IIB Workshop Making Organic Dyes* (Prehistory Monographs 62). Philadelphia, PA: INSTAP Academic Press.

Arkell, A. J. 1948. 'The shoulder ornament of Near Eastern lions', *Journal of Near Eastern Studies* 7, 52.

Aruz, J. 2008. *Marks of Distinction: Seals and Cultural Exchange between the Aegean and the Orient* (CMS Beiheft 7). Mainz: von Zabern.

Åström, P. and Reese, D. S. 1990. 'Triton shells in East Mediterranean cults', *Journal of Prehistoric Religion* 3–4, 5–14.

Atran, S. 1990. *Cognitive Foundations of Natural History: Towards an Anthropology of Science.* Cambridge: Cambridge University Press.

Atran, S. 1998. 'Folk biology and the anthropology of science: cognitive universals and cultural particulars', *Behavioural and Brain Sciences* 21, 547–609.

Baker, S. 2000. *The Postmodern Animal.* London: Reaktion.

Baker, S. 2001. *Picturing the Beast: Animals, Identity and Representation.* Urbana and Chicago: University of Illinois Press.

Baker, S. 2013. *Artist Animal.* Minneapolis: University of Minnesota Press.

Ballintijn, M. 1995. 'Lions depicted on Aegean seals – how realistic are they?', in Müller, W. (ed.), *Sceaux minoens et mycéniens. IVe symposium international, 10–12 septembre 1992, Clermont-Ferrand* (CMS Beiheft 5). Berlin: Gebr. Mann Verlag, 23–37.

Banti, L. 1933. 'La grande tombe a tholos di Haghia Triadha', *Annuario della Scuola Archeologica di Atene e delle Missioni Italiane in Oriente* 13–14, 155–251.

Barbanera, F., Guerrini, M., Khan, A. A. et al. 2009a. 'Human-mediated introgression of exotic chukar (*Alectoris chukar*, Galliformes) genes from East Asia into native Mediterranean partridges', *Biological Invasions* 11(2), 333–48.

Barbanera, F., Marchi, C., Guerrini, M. et al. 2009b. 'Genetic structure of Mediterranean chukar (*Alectoris chukar*, Galliformes) populations: conservation and management implications', *Naturwissenschaften* 96, 1203–12.

Barber, E. J. W. 1991. *Prehistoric Textiles: The Development of Cloth in the Neolithic and Bronze Ages with Special Reference to the Aegean.* Princeton, NJ: Princeton University Press.

Bar-Gal, G. K., Smith, P., Tchernov, E. et al. 2002. 'Genetic evidence for the origin of the agrimi goat (*Capra aegagrus cretica*)', *Journal of the Zoological Society of London* 256, 369–77.

Bate, D. 1905a. 'Four and a half months in Crete in search of Pleistocene mammalian remains', *The Geological Magazine* Decade V, II(V) (May 1905), 195–202.

Bate, D. 1905b. 'On the mammals of Crete', *Proceedings of the Zoological Society of London* 75 (3), 315–23.

Bate, D. 1907. 'On elephant remains from Crete, with description of *Elephas creticus*, sp. n.', *Proceedings of the Zoological Society of London* 77 (1), 238–50.

Bate, D. 1913. 'Birds in Crete', in Trevor-Battye, A. (ed.), *Camping in Crete. With Notes upon the Animal and Plant Life of the Island.* London: Witherby & Co, 257–63.

Bate, D. 1950. 'The "shoulder ornament" of Near Eastern lions', *Journal of Near Eastern Studies* 9, 43–54.

Baurain, C. and Darcque, P. 1983. 'Un triton en pierre à Malia', *Bulletin de Correspondance Hellénique* 107(1), 3–73.

Beaman, M. and Madge, S. 1998. *The Handbook of Bird Identification: For Europe and the Western Palearctic.* London: Christopher Helm.

Bedwin, O. 1984. 'Appendix 2: the animal bones', in Popham, M. R. (ed.), *The Minoan Unexplored Mansion at Knossos.* London: British School at Athens and Thames and Hudson, 307–8.

Behm-Blancke, G. 2003. *Heiligtümer der Germanen und ihrer Vorgänger in Thüringen: die Kultstätte Oberdorla; Forschungen zum alteuropäischen Religions- und Kultwesen. Vol. I: Text und Fototafeln* (Weimarer Monographien zur Ur- und Frühgeschichte 38.1). Stuttgart: Theiss.

Belon, P. 1555. *Les observations de plusieurs singularitez & choses memorables, trouvées en Grece, Asie, Iudée, Egypte, Arabie, & autres pays estranges, redigées en trois liures*. Paris: Chez Guillaume Cauellet.

Belting, H. 2005. 'Image, medium, body: a new approach to iconology', *Critical Inquiry* 31, 302–19.

Bendall, L. M. 2004. 'Fit for a king? Hierarchy, exclusion, aspiration and desire in the social structure of Mycenaean banqueting', in Halstead, P. and Barrett, J. C. (eds.), *Food, Cuisine and Society in Prehistoric Greece* (Sheffield Studies in Aegean Archaeology 5). Oxford: Oxbow, 105–35.

Bendall, L. M. 2007. *Economies of Religion in the Mycenaean World: Resources Dedicated to Religion in the Mycenaean Palatial Economy*. Oxford: Oxford University School of Archaeology.

Bennet, J. 1985. 'The structure of the Linear B administration at Knossos', *American Journal of Archaeology* 89(2), 231–49.

Bennett, E. L. 1951. *The Pylos Tablets: A Preliminary Transcription*. Princeton, NJ: Princeton University Press.

Berg, I. 2011. 'Towards a conceptualisation of the sea: artefacts, iconography and meaning', in Vavouranakis, G. (ed.), *The Seascape in Aegean Prehistory*. Athens: The Danish Institute at Athens, 119–37.

Berger, J. 1971. 'Animal world', *New Society* 18 (478), 1042–3.

Berger, J. 1980. 'Why look at animals?', in *About Looking*. London: Writers and Readers, 1–26.

Berlin, B., Breedlove, D. and Raven, P. 1973. 'General principles of classification and nomenclature in folk biology', *American Anthropologist* 75(1), 214–42.

Betancourt, P. P. 1973. 'The Polyp Workshop: a stylistic group from LM IB', *American Journal of Archaeology* 77(3), 333–4.

Betancourt, P. P. 1985. *The History of Minoan Pottery*. Princeton, NJ: Princeton University Press.

Betancourt, P. P. 2009. *Pseira X: The Excavation of Block AF* (Prehistory Monographs 28). Philadelphia, PA: INSTAP Academic Press.

Betancourt, P. P., Apostolakou, S. and Brogan, T. 2012. 'The workshop for making dyes at Pefka, Crete', in Nosch, M.-L. and Laffineur, R. (eds.), *KOSMOS. Jewellery, Adornment and Textiles in the Aegean Bronze Age. Proceedings of the 13th International Aegean Conference, University of Copenhagen, Danish National Research Foundation's Centre for Textile Research, 21–26 April 2010* (Aegaeum 33). Leuven: Peeters, 183–6.

Betancourt, P. P. and Davaras, C. 1995a. *Pseira I: The Minoan Buildings on the West Side of Area A*. Philadelphia: University of Pennsylvania Museum.

Betancourt, P. P. and Davaras, C. 1995b. *Pseira II: Building AC (The "Shrine") and Other Buildings in Area A*. Philadelphia: University of Pennsylvania Museum.

Betts, J. 1967. 'New light on Minoan bureaucracy: a re-examination of some Cretan sealings', *Kadmos* 6, 15–40.

Betts, J. 1974. 'Review of: V.E.G. Kenna, The Cretan Talismanic Stone in the Late Minoan Age', *Bibliotheca Orientalis* 31, 309–14.

Betts, J. 1981. 'The "Jasper Lion Master": some principles of establishing LM/LH workshops and artists', in Niemeier, W.-D. (ed.), *Studien zur minoischen und helladischen Glyptik* (CMS Beiheft 1). Berlin: Gebr. Mann Verlag, 1–15.

Betts, J. 1989. 'Seals of Middle Minoan III: chronology and technical revolution', in Pini, I. (ed.), *Fragen und Probleme der bronzezeitlichen ägäischen Glyptik. Beiträge zum 3. Internationalen Marburger Siegel Symposium, 5–7 September 1985* (CMS Beiheft 3). Berlin: Gebr. Mann Verlag, 1–17.

Betts, J. and Younger, J. G. 1982. 'Aegean seals of the Late Bronze Age: masters and workshops', *Kadmos* 21, 104–21.

Bevan, A. 2007. *Stone Vessels and Value in the Bronze Age Mediterranean*. Cambridge: Cambridge University Press.

Bicknell, P. 2000. 'Late Minoan IB ware, the marine environment of the Aegean, and the Bronze Age eruption of the Thera volcano', in McGuire, W. J., Griffiths, D. R., Hancock, P. L. and Stewart, I. S. (eds.), *The Archaeology of Geological Catastrophes* (Geological

Society Special Publications 171). London: The Geological Society of London, 95–103.

Bietak, M., Marinatos, N. and Palivou, C. 2007. *Taureador Scenes in Tell El-Dab'a (Avaris) and Knossos* (Denkschriften der Gesamtakademie 43). Vienna: Verlag der Österreichischen Akademie der Wissenschaften.

Binnberg, J. 2013. 'Form, Funktion und Kontext der Tritonschnecken in der minoischen Kultur', *Archäologischer Anzeiger* 2013/1, 1–30.

Binnberg, J. 2017. 'Animal identification in iconography: an interdisciplinary approach combining zoology, anthropology, and archaeology', in O'Sullivan, R., Marini, C. and Binnberg, J. (eds.), *Interaction, Integration and Division: Archaeological Approaches to Breaking Boundaries* (BAR 2869). Oxford: BAR Publishing, 279–89.

Binnberg, J. 2019a. 'Animism or analogism? Bird depictions and their significance for the reconstruction of Cretan Bronze Age ontologies', in Zebrowska, K., Ulanowska, A. and Lewartowski, K. (eds.), *Sympozjum Egejskie: Papers in Aegean Archaeology 2*. Warsaw: University of Warsaw.

Binnberg, J. 2019b. 'Like a duck to water – birds and liquids in the Aegean Bronze Age', *Annual of the British School at Athens* 114, 41–78.

Bird-David, N. 1999. '"Animism" revisited: personhood, environment and relational epistemology', *Current Anthropology* 40(S1), 567–91.

Bird-David, N. 2018. 'Size matters! The scalability of modern hunter-gatherer animism', *Quaternary International* 464, 305–14.

BirdLife International. 2021. *IUCN Red List for birds* (www.birdlife.org) [Accessed on 30 May 2021].

Birke, L. and Oma, K. A. 2013. 'Archaeology and human-animal studies', *Society and Animals* 21(2), 113–19.

Blakolmer, F. 2007. 'Die "Schnittervase" von Agia Triada. Zu Narrativität, Mimik und Prototypen in der minoischen Bildkunst', *Creta Antica* 8, 201–42.

Blakolmer, F. 2010. 'Small is beautiful: the significance of Aegean glyptic for the study of wall paintings, relief frescoes and minor relief arts', in Müller, W. (ed.), *Die Bedeutung der minoischen und mykenischen Glyptik. VI. Internationales Siegel-Symposium, Marburg, 9–12 Oktober 2008* (CMS Beiheft 8). Mainz: von Zabern, 91–108.

Blakolmer, F. 2016. 'Hierarchy and symbolism of animals and mythical creatures in the Aegean Bronze Age: a statistical and contextual approach', in Alram-Stern, E., Blakolmer, F., Deger-Jalkotzy, S., Laffineur, R. and Weilhartner, J. (eds.), *METAPHYSIS. Ritual, Myth and Symbolism in the Aegean Bronze Age. Proceedings of the 15th International Aegean Conference, Vienna, Institute for Oriental and European Archaeology, Aegean and Anatolia Department, Austrian Academy of Sciences and Institute of Classical Archaeology, University of Vienna, 22–25 April 2014* (Aegaeum 39). Leuven and Liège: Peeters, 61–8.

Blakolmer, F. 2020a. 'At the dawn of Minoan iconography: the "Archivio di cretule" at MM II B Phaistos', in Blakolmer, F. (ed.), *Current Approaches and New Perspectives in Aegean Iconography*. Louvain-la-Neuve: Presses universitaires de Louvain, 47–70.

Blakolmer, F. 2020b. 'Gab es Löwen und Affen im minoischen Kreta? Ein ikonographisches Problem', in Berger, L. , Huber, L. , Lang, F. and Weilhartner, J. (eds.), *Akten des 17. Österreichischen Archäologentages am Fachbereich Klassische Altertumswissenschaften, Klassische und Frühägäische Archäologie der Universität Salzburg vom 26 bis 28 Februar 2018*. Salzburg: Universität Salzburg, 39–49.

Blegen, C. and Rawson, M. 1966. *The Palace of Nestor at Pylos in Western Messenia*. Princeton, NJ: Princeton University Press.

Bloedow, E. 1990. 'The "Sanctuary Rhyton" from Zakros: what do the goats mean?', *Aegaeum 6*. Liège: Université de Liège, 59–78.

Bloedow, E. 1992. 'Minoan talismanic goats', *Journal of Prehistoric Religion* 6, 15–23.

Bloedow, E. 2003. 'The significance of the goat in Minoan culture', *Praehistorische Zeitschrift* 78, 1–59.

Blondel, J. and Aronson, J. 1995. 'Biodiversity and ecosystem function in the Mediterranean Basin: human and non-human determinates', in Davis, G. W. and Richardson, D. M. (eds.), *Mediterranean-type Ecosystems* (Ecological Studies 109). Berlin: Springer-Verlag, 43–119.

Blondel, J. and Aronson, J. 1999. *Biology and Wildlife of the Mediterranean Region*. Oxford: Oxford University Press.

Boardman, J. 1961. *The Cretan Collection in Oxford: The Dictaean Cave and Iron Age Crete*. Oxford: Oxford University Press.

Boardman, J. 1966. 'Review of: Corpus der minoischen und mykenischen Siegel I', *Gnomon* 38, 264–7.

Boardman, J. 2001. *Greek Gems and Finger Rings*. London: Thames and Hudson.

Borgna, E. 2004. 'Aegean feasting: a Minoan perspective', *Hesperia* 73(2), 247–79.

Bosanquet, R. C. and Dawkins, R. 1923. *The Unpublished Objects from the Palaikastro Excavations 1902–1906* (BSA Supplement 1). London: British School at Athens.

Bosanquet, R. C. 1902a. 'Excavations at Palaikastro. I', *Annual of the British School at Athens* 8, 286–316.

Bosanquet, R. C. 1902b. 'Excavations at Petras', *Annual of the British School at Athens* 8, 282–5.

Boucher, A. 2014. *La Grèce des origines: entre rêve et archéologie. Musée d'archéologie nationale, domaine national de Saint-Germain-en-Laye 5 octobre 2014–19 janvier 2015*. Paris: Réunion des musées nationaux-Grand palais.

Bown, M. 2015. 'Traces of the holy: the contemporary art work as "crypto-relic"', *The Times Literary Supplement* 5845, 14–15.

Boyd, B. 2017. 'Archaeology and human-animal relations: thinking through anthropocentrism', *Annual Review of Anthropology* 46, 299–316.

Boyd Hawes, H., Williams, B., Seager, R. B. and Hall, E. H. 1908. *Gournia, Vasiliki and Other Prehistoric Sites on the Isthmus of Hierapetra, Crete: Excavations of the Wells-Houston-Cramp Expeditions, 1901, 1903, 1904*. Philadelphia, PA: American Exploration Society.

Boyd-Dawkins, W. 1902. '114. Remains of animals found in the Dictaean Cave in 1901', *Man* 2, 162–5.

Branigan, K. 1993. *Dancing with Death: Life and Death in Southern Crete c. 3000–2000 B.C.* Amsterdam: A. M. Hakkert.

Briault, C. 2007. 'High fidelity or Chinese whispers? Cult symbols and ritual transmission in the Bronze Age Aegean', *Journal of Mediterranean Archaeology* 20(2), 239–65.

Brice, W. C. 1988. 'Notes on Linear A: III. The "livestock" tablet, Her. Mus. 1609. IV. Ligatured signs at Mallia', *Kadmos* 27(2), 155–65.

Brodie, N. 2008. 'The donkey: an appropriate technology for Early Bronze Age land transport and traction', in Brodie, N., Doole, J., Gavalas, G. and Renfrew, C. (eds.), *Horizon: A Colloquium on the Prehistory of the Cyclades*. Cambridge: McDonald Institute for Archaeological Research, 299–304.

Brogan, T. M., Betancourt, P. P. and Apostolakou, S. 2012. 'The purple dye industry of eastern Crete', in Nosch, M.-L. and Laffineur, R. (eds.), *KOSMOS. Jewellery, Adornment and Textiles in the Aegean Bronze Age. Proceedings of the 13th International Aegean Conference, University of Copenhagen, Danish National Research Foundation's Centre for Textile Research, 21–26 April 2010* (Aegaeum 33). Leuven: Peeters, 187–92.

Brogan, T. M. and Hallager, E. 2011. *LM IB Pottery: Relative Chronology and Regional Differences. Acts of a Workshop Held at the Danish Institute at Athens in Collaboration with the INSTAP Study Center for East Crete, 27–29 June 2007* (Monographs of the Danish Institute at Athens 11, 1). Athens: Danish Institute at Athens.

Brogan, T. M., Mylona, D., Apostolakou, V., Betancourt, P. P. and Sofianou, C. 2019. 'A Bronze Age fishing village on Chryssi', in Chalikias, K. and Oddo, E. (eds.), *Exploring a Terra Incognita on Crete: Recent Research on Bronze Age Habitation in the Southern Ierapetra Isthmus*. Philadelphia, PA: INSTAP Academic Press, 97–109.

Broodbank, C. 1992. 'The Neolithic Labyrinth: social change at Knossos before the Bronze Age', *Journal of Mediterranean Archaeology* 5(1), 39–75.

Broodbank, C. 2004. 'Minoanisation', *The Cambridge Classical Journal* 50, 46–91.

Broodbank, C. 2013. *The Making of the Middle Sea. A History of the Mediterranean from the*

Beginning to the Emergence of the Classical World. London: Thames and Hudson.

Broodbank, C. and Strasser, T. 1991. 'Migrant farmers and the Neolithic colonization of Crete', *Antiquity* 65, 233–45.

Brower, M. 2005. 'Trophy shots: early North American photographs of nonhuman animals and the display of masculine prowess', *Society and Animals* 13(1), 13–31.

Brown, A. (ed.) 2001. *Arthur Evans's Travels in Crete, 1894–1899* (BAR IS 1000). Oxford: Archaeopress.

Brunner, B. 2005. *The Ocean at Home: An Illustrated History of the Aquarium.* New York: Princeton Architectural Press.

Brysbaert, A. 2008. *The Power of Technology in the Bronze Age Eastern Mediterranean: The Case of the Painted Plaster* (Monographs in Mediterranean Archaeology 12). London and Oakville: Equinox.

Bryson, N. 1983. *Vision and Painting: The Logic of the Gaze.* Basingstoke: Macmillan.

Buchholz, H. 1999. *Ugarit, Zypern und Ägäis : Kulturbeziehungen im zweiten Jahrtausend v.Chr* (Alter Orient und Altes Testament 261). Münster: Ugarit-Verlag.

Burke, B. 2005. 'Materialization of Mycenaean ideology and the Ayia Triada Sarcophagus', *American Journal of Archaeology* 109(3), 403–22.

Burke, B. 2010. *From Minos to Midas: Ancient Cloth Production in the Aegean and in Anatolia.* Oxford: Oxbow.

Burkert, W. 1983. *Homo Necans: The Anthropology of Ancient Greek Sacrificial Ritual and Myth.* Berkeley: University of California Press

Burns, B. 2010. *Mycenaean Greece, Mediterranean Commerce, and the Formation of Identity.* Cambridge: Cambridge University Press.

Burt, J. 2005. 'John Berger's "Why look at animals?": a close reading', *Worldviews: Environment Culture Religion* 9(2), 203–18.

Bynum, C. W. 2011. *Christian Materiality: An Essay on Religion in Late Medieval Europe.* Cambridge, MA and London: The MIT Press.

Cadogan, G. 2000. 'The pioneers: 1900–1914', in Huxley, D. (ed.), *Cretan Quests: British Explorers, Excavators and Historians* (BSA Studies 18). London: British School at Athens, 15–27.

Cain, D. 2001. 'Dancing in the dark: deconstructing a narrative epiphany on the Isopata Ring', *American Journal of Archaeology* 105(1), 27–49.

Calarco, M. 2008. *Zoographies: The Question of the Animal from Heidegger to Derrida.* New York: Columbia University Press.

Caloi, I. 2017. 'Preserving memory in Minoan Crete: filled-in bench and platform deposits from the First Palace of Phaistos', *Journal of Greek Archaeology* 2, 33–52.

Cameron, M. A. S. 1967. 'Unpublished fresco fragments of a chariot composition from Knossos', *Archäologischer Anzeiger* 3, 330–4.

Cameron, M. A. S. 1968. 'Unpublished paintings from the "House of the Frescoes" at Knossos', *Annual of the British School at Athens* 63, 1–31.

Carabott, P. 2006. 'A country in a "state of destitution" labouring under an "unfortunate regime": Crete at the turn of the 20th century (1898–1906)', in Hamilakis, Y. and Momigliano, N. (eds.), *Archaeology and European Modernity: Producing and Consuming the 'Minoans'* (Creta Antica 7). Padua: Bottega d'Erasmo, 39–53.

Carinci, F. and La Rosa, V. 2009. 'Revisioni Festie II', *Creta Antica* 10/I, 147–300.

Cassidy, R. and Mullin, M. (eds.) 2007. *Where the Wild Things Are Now: Domestication Reconsidered.* Oxford: Berg.

Cauvin, J. 2000. *The Birth of the Gods and the Origins of Agriculture.* Cambridge: Cambridge University Press.

Chadwick, J., Godart, L., Killen, J. T. et al. 1992. *Corpus of Mycenaean Inscriptions from Knossos, II* (Incunabula Graeca 88). Rome and Cambridge: Edizione dell'Ateneo and Cambridge University Press.

Chalikias, K. 2013. 'Searching for the missing palace: Protopalatial and Neopalatial settlement dynamics in the Southern Ierapetra Isthmus. The recent evidence', *Aegean Archaeology* 10, 33–46.

Chapin, A. P. 2004. 'Power, privilege, and landscape in Minoan art', in Chapin, A. P. (ed.), *ΧΑΡΙΣ: Essays in Honor of Sara A. Immerwahr*

(Hesperia Supplement 33). Athens: American School of Classical Studies at Athens, 47–64.

Chapin, A. P. and Shaw, M. C. 2006. 'The frescoes from the House of the Frescoes at Knossos: a reconsideration of their architectural context and a new reconstruction of the crocus panel', *Annual of the British School at Athens* 101, 57–88.

Chapouthier, F. and Charbonneaux, J. 1928. *Fouilles exécutées à Mallia. Premier rapport (1922–1924)* (Études crétoises 1). Paris: P. Geuthner.

Cherry, J. F. 1988. 'Pastoralism and the role of animals in the pre- and protohistoric economies of the Aegean', in Whittaker, C. R. (ed.), *Pastoral Economies in Classical Antiquity*. Cambridge: Cambridge Philological Society.

Cheylan, G. 1991. 'Patterns of Pleistocene turnover, current distribution and speciation among Mediterranean mammals', in Groves, R. H. and Di Castri, F. (eds.), *Biogeography of Mediterranean Invasions*. Cambridge: Cambridge University Press, 227–62.

Chlouveraki, S., Betancourt, P. P., Davaras, C. et al. 2008. 'Excavations in the Hagios Charalambos Cave: a preliminary report', *Hesperia* 77(4), 539–605.

Christakis, K. 2020. 'Palatial Crete: recent discoveries & research, 2014–2019', *Archaeological Reports* 66, 83–115.

Claassen, C. 1998. *Shells*. Cambridge: Cambridge University Press.

Cline, E. H. 1994. *Sailing the Wine-Dark Sea: International Trade and the Late Bronze Age Aegean* (BAR IS 591). Oxford: Tempus Reparatum.

Colley, A. C. 2014. *Wild Animal Skins in Victorian Britain: Zoos, Collections, Portraits and Maps*. Farnham: Ashgate.

Collins, B. J. 2003. 'On the trail of the deer: Hittite *kūrala-*', in Beckman, G., Beal, R. H. and McMahon, G. (eds.), *Hittite Studies in Honor of Harry A. Hoffner Jr. on the Occasion of His 65th Birthday*. Winona Lake, IN: Eisenbrauns, 73–82.

Collon, D. 1994. 'Bull-leaping in Syria', *Ägypten und Levante* 4, 81–8.

Cook, A. B. 1925. *Zeus: A Study in Ancient Religion, II*. Cambridge: Cambridge University Press.

Coppinger, R. and Coppinger, L. 2016. *What Is a Dog?* Chicago, IL: University of Chicago Press.

Cosmopoulos, M. and Ruscillo, D. 2014. 'Mycenaean burnt animal sacrifice at Eleusis', *Oxford Journal of Archaeology* 33(3), 257–73.

Coulomb, J. 1979. 'Le "Prince aux Lis" de Knossos reconsidéré', *Bulletin de Correspondance Hellénique* 103, 29–50.

Cromarty, R. J. 2008. *Burning Bulls, Broken Bones: Sacrificial Ritual in Context of Palace Period Minoan Religion* (BAR IS 1792). Oxford: Archaeopress.

Crooks, S., Tully, C. J. and Hitchcock, L. A. 2016. 'Numinous tree and stone: re-animating the Minoan landscape', in Alram-Stern, E., Blakolmer, F., Deger-Jalkotzy, S., Laffineur, R. and Weilhartner, J. (eds.), *METAPHYSIS. Ritual, Myth and Symbolism in the Aegean Bronze Age. Proceedings of the 15th International Aegean Conference, Vienna, Institute for Oriental and European Archaeology, Aegean and Anatolia Department, Austrian Academy of Sciences and Institute of Classical Archaeology, University of Vienna, 22–25 April 2014* (Aegaeum 39). Leuven and Liège: Peeters, 157–64.

Crouwel, J. 1981. *Chariots and Other Means of Land Transport in Bronze Age Greece*. Amsterdam: Allard Pierson Series.

Crowley, J. 1989a. 'Subject matter in Aegean art: the crucial changes', in Laffineur, R. (ed.), *TRANSITION. Le monde égéen du Bronze moyen au Bronze récent. Actes de la 2e Rencontre égéenne internationale de l'Université de Liège, 18–20 avril 1988* (Aegaeum 3). Liège: Université de Liège, 203–14.

Crowley, J. 1989b. *The Aegean and the East: An Investigation into the Transference of Artistic Motifs between the Aegean, Egypt, and the Near East in the Bronze Age* (SIMA Pocket Book 51). Partille: Åström.

Crowley, J. 2013. *The Iconography of Aegean Seals* (Aegaeum 34). Leuven: Peeters.

Cultraro, M. 2004. 'Exercise in dominance: boar-hunting in Mycenaean religion and Hittite royal rituals', in Hutter, M. and Hutter-Braunsar, S. (eds.), *Offizielle Religion, lokale Kulte und individuelle Religiosität*. Münster: Ugarit-Verlag, 117–35.

Cutting, J. E. 1982. 'Two ecological perspectives: Gibson vs. Shaw and Turvey', *American Journal of Psychology* 95(2), 199–222.

Cutting, J. E. 1993. 'Perceptual artifacts and phenomena: Gibson's role in the 20th century', *Advances in Psychology* 99, 231–60.

Czernohaus, K. 1988. *Delphindarstellungen von der minoischen bis zur geometrischen Zeit* (SIMA Pocket Book 67). Gothenburg: Paul Åströms Förlag.

D'Agata, A. L. 2000. 'Ritual and rubbish in Dark Age Crete: the settlement of Thronos/Kephala (ancient Sybrita) and the pre-classical roots of a Greek city', *Aegean Archaeology* 4, 45–59.

Damiani Indelicato, S. 1988. 'Were Cretan girls playing at bull-leaping?', *Cretan Studies* 1, 39–47.

Davis, E. N. 1974. 'The Vapheio Cups: one Minoan and one Mycenean?', *The Art Bulletin* 56(4), 472–87.

Dawkins, R. and Currelly, C. T. 1904. 'Excavations at Palaikastro. III', *Annual of the British School at Athens* 10, 192–231.

Day, J. 2011a. 'Crocuses in context: a diachronic survey of the crocus motif in the Aegean Bronze Age', *Hesperia* 80(3), 337–79.

Day, J. 2011b. 'Counting threads: saffron in Aegean Bronze Age writing and society', *Oxford Journal of Archaeology* 30(4), 369–91.

Day, P. M. and Wilson, D. E. 2002. 'Landscapes of memory, craft and power in Prepalatial and Protopalatial Knossos', in Hamilakis, Y. (ed.), *Labyrinth Revisited: Rethinking 'Minoan' Archaeology*. Oxford: Oxbow, 143–66.

Del Freo, M. 2020. 'Wool working at Hagia Triada: the HT 24 tablet and the 45 noduli from the Quartiere Sudovest', in Davis, B. and Laffineur, R. (eds.), *ΝΕΩΤΕΡΟΣ. Studies in Bronze Age Aegean Art and Archaeology in Honor of Professor John G. Younger on the Occasion of his Retirement* (Aegaeum 44). Leuven: Peeters, 55–66.

Del Freo, M., Nosch, M.-L. and Rougemont, F. 2010. 'The terminology of textiles in the Linear B Tablets, including some considerations on Linear A logograms and abbreviations', in Michel, C. and Nosch, M.-L. (eds.), *Textile Terminologies in the Ancient Near East and Mediterranean from the Third to the First Millennia BC*. Oxford: Oxbow, 338–73.

DeMello, M. 2012. *Animals and Society: An Introduction to Human-Animal Studies*. New York: Columbia University Press.

Dermargne, J. 1902. 'Antiquités de Praesos et de l'Antre Dictéen', *Bulletin de Correspondance Hellénique* 26, 571–83.

Derrida, J. 2002. 'The animal that therefore I am (more to follow)', *Critical Inquiry* 28(2), 369–418.

Descola, P. 1996a. 'Constructing natures: symbolic ecology and social practice', in Descola, P. and Palsson, G. (eds.), *Nature and Society: Anthropological Perspectives*. London: Routledge, 82–103.

Descola, P. 1996b. *The Spears of Twilight: Life and Death in the Amazon Jungle*. London: HarperCollins.

Descola, P. 2005. *Par-delà nature et culture*. Paris: Gallimard.

Descola, P. 2006. 'Beyond nature and culture', *Proceedings of the British Academy* 139, 137–55.

Descola, P. 2009. 'Human natures', *Social Anthropology* 17(2), 145–57.

Descola, P. 2010a. 'Cognition, perception and worlding', *Interdisciplinary Science Reviews* 35 (3–4), 334–40.

Descola, P. 2010b. *La Fabrique des images*. Paris: Musée du quai Branly and Somogy.

Descola, P. 2013a. *Beyond Nature and Culture*. Chicago, IL: University of Chicago Press.

Descola, P. 2013b. *The Ecology of Others: Anthropology and the Question of Nature*. Chicago, IL: Prickly Paradigm Press.

Descola, P. and Pálsson, G. 1996. 'Introduction', in Descola, P. and Pálsson, G. (eds.), *Nature and Society: Anthropological Perspectives*. London: Routledge, 1–21.

Detournay, B., Poursat, J.-C. and Vandenabeele, F. 1980. *Fouilles exécutées à Mallia. Le Quartier Mu II. Vases de pierre et de métal, vannerie, figurines et reliefs d'applique, éléments de parure et de décoration, armes, sceaux et empreintes* (Études crétoises 26). Paris: École française d'Athènes.

Dimopoulou, N. 1999a. 'The Marine Style Ewer from Poros', in Betancourt, P. P., Karageorghis, V., Laffineur, R. and Niemeier, W.-D. (eds.), *MELETEMATA. Studies in Aegean Archaeology Presented to Malcolm H. Wiener as he Enters His 65th Year, I* (Aegaeum 20). Liège and Austin: Université de Liège and University of Texas at Austin, 217–26.

Dimopoulou, N. 1999b. 'The Neo-palatial cemetery of the Knossian harbour-town at Poros: mortuary behaviour and social ranking', in Kilian-Dirlmeier, I. and Egg, M. (eds.), *Eliten in der Bronzezeit* (Monographien des Romisch Germanischen Zentralmuseums 43). Mainz: Verlag des Römisch-Germanischen Zentralmuseums, 27–36.

Dimopoulou, N. 2010. 'A gold discoid from Poros, Herakleion: the guard dog and the garden', in Krzyszkowska, O. (ed.), *Cretan Offerings: Studies in Honour of Peter Warren* (BSA Studies 18). London: British School at Athens, 89–100.

Dobat, A. S. 2006. 'The king and his cult: the axe-hammer from Sutton Hoo and its implications for the concept of sacral leadership in early medieval Europe', *Antiquity* 80(310), 880–93.

Driessen, J. 2000. *The Scribes of the Room of the Chariot Tablets at Knossos: Interdisciplinary Approach to the Study of a Linear B Deposit* (Suplementos a Minos 15). Salamanca: Ediciones Universidad de Salamanca.

Driessen, J. 2002. '"The King Must Die": some observations on the use of Minoan court compounds', in Driessen, J., Schoep, I. and Laffineur, R. (eds.), *Monuments of Minos: Rethinking the Minoan Palaces* (Aegaeum 23). Liège and Austin: Université de Liège and University of Texas at Austin, 1–14.

Driessen, J. 2004. 'The central court of the palace at Knossos', in Cadogan, G., Hatzaki, E. and Vasilakis, A. (eds.), *Knossos: Palace, City, State* (BSA Studies 12). London: British School at Athens, 75–82.

Driessen, J. 2008. 'Chronology of the Linear B texts', in Duhoux, Y. and Morpurgo Davies, A. (eds.), *A Companion to Linear B: Mycenaean Greek Texts and Their World*. Louvain: Peeters, 69–79.

Driessen, J. 2010. 'Spirit of place: Minoan houses as major actors', in Pullen, D. J. (ed.), *Political Economies of the Aegean Bronze Age*. Oxford: Oxbow, 35–65.

Driessen, J. 2018. 'Beyond the collective: the Minoan palace in action', in Relaki, M. and Papadatos, Y. (eds.), *From the Foundations to the Legacy of Minoan Archaeology: Studies in Honour of Professor Keith Branigan* (Sheffield Studies in Aegean Archaeology 12). Oxford: Oxbow, 291–313.

Driessen, J. and Macdonald, C. F. 1997. *The Troubled Island: Minoan Crete Before and after the Santorini Eruption* (Aegaeum 17). Liège and Austin: Université de Liège and University of Texas at Austin.

Drummond-Hay, H. M. 1865. 'Birds noticed in the island of Crete, during a stay of nearly two months', in Spratt, T. A. B. (ed.), *Travels and Researches in Crete, II*. London: John van Voorst, 397–407.

Duke, P. 2007. *The Tourists Gaze, the Cretans Glance*. Walnut Creek, CA: Left Coast Press.

Eccles, E. 1943. 'Unpublished objects from Palaikastro and Praisos: the seals and sealings', *Annual of the British School at Athens* 40, 43–9.

Economidis, P. S. 2000. 'The "Little Fisherman" and the fish he holds', in Sherratt, S. (ed.), *The Wall Paintings of Thera, II*. Athens: Thera Foundation, 555–60.

Elder, G., Wolch, J. and Emel, J. 1998. 'Race, place and the bounds of humanity', *Society and Animals* 6(2), 183–202.

Elsner, J. 2006. 'Reflections on the "Greek Revolution" in art: from changes in viewing to the transformation of subjectivity', in Goldhill, S. and Osborne, R. (eds.), *Rethinking*

Revolutions through Ancient Greece. Cambridge: Cambridge University Press.

Engels, D. 1999. *Classical Cats: The Rise and Fall of the Sacred Cat*. London: Routledge.

EU 2016. *Facts and Figures on the Common Fisheries Policy*. Luxembourg: Publications Office of the European Union.

Evans, A. J. 1894. 'Primitive pictographs and a Prae-Phoenician script, from Crete and the Peloponnese', *Journal of Hellenic Studies* 14, 270–372.

Evans, A. J. 1897. 'Further discoveries of Cretan and Aegean script: with Libyan and Proto-Egyptian comparisons', *Journal of Hellenic Studies* 17, 327–95.

Evans, A. J. 1900. 'Knossos: summary report of the excavations in 1900', *Annual of the British School at Athens* 6, 3–70.

Evans, A. J. 1901a. 'Mycenaean tree and pillar cult and Its Mediterranean relations', *Journal of Hellenic Studies* 21, 99–204.

Evans, A. J. 1901b. 'The Palace of Knossos: provisional report of the excavations for the year 1901', *Annual of the British School at Athens* 7, 1–120.

Evans, A. J. 1903. 'The Palace of Knossos: provisional report for the year 1903', *Annual of the British School at Athens* 9, 1–153.

Evans, A. J. 1906a. *Essai de classification des époques de la civilisation minoenne: résumé d'un discours fait au Congrès d'Archéologie à Athènes*. London: Quaritch.

Evans, A. J. 1906b. *The Prehistoric Tombs of Knossos*. London: Quaritch.

Evans, A. J. 1909. *Scripta Minoa*. Oxford: Clarendon Press.

Evans, A. J. 1914. 'The "Tomb of the Double Axes" and associated group, and the pillar rooms and ritual vessels of the "Little Palace" at Knossos', *Archaeologia* 65, 1–94.

Evans, A. J. 1921a. 'On a Minoan bronze group of a galloping bull and acrobatic figure from Crete. With glyptic comparisons and a note on the Oxford relief showing the taurokathapsia', *Journal of Hellenic Studies* 41 (2), 247–59.

Evans, A. J. 1921b. *The Palace of Minos at Knossos, I*. London: Macmillan.

Evans, A. J. 1928. *The Palace of Minos at Knossos, II*. London: Macmillan.

Evans, A. J. 1930. *The Palace of Minos at Knossos, III*. London: Macmillan.

Evans, A. J. 1935. *The Palace of Minos at Knossos, IV*. London: Macmillan.

Evans, A. J., Cameron, M. A. S. and Hood, M. S. F. 1967. *Knossos Fresco Atlas*. Farnborough: Gregg.

Evans, A. J. and Myres, J. L. 1895. 'A Mycenaean military road in Crete', *Academy* 1204 (June 1), 469.

Evans, J. D. 1964. 'Excavations in the Neolithic settlement at Knossos, 1957–60: part I', *Annual of the British School at Athens* 59, 132–240.

Evans, L. 2010. *Animal Behaviour in Egyptian Art: Representations of the Natural World in Memphite Tomb Scenes*. Oxford: Aris and Phillips.

Evely, D. 1993. *Minoan Crafts: Tools and Techniques, an Introduction*, I (SIMA 92). Gothenburg: Paul Åströms Förlag.

Evely, D. 1996. 'The Neo-palatial Minoan warrior: fact or fiction?', in Evely, D., Lemos, I. S. and Sherratt, S. (eds.), *Minotaur and Centaur: Studies in the Archaeology of Crete and Euboea Presented to Mervyn Popham* (BAR IS 638). Oxford: Tempus Reparatum.

Evely, D. 1999. *Fresco: A Passport into the Past. Minoan Crete Through the Eyes of Mark Cameron*. Athens: British School at Athens and N.P. Goulandris Foundation.

Farmar, H. 1952. 'The protection of the agrimi', *Oryx* 1(7), 327–37.

Feldman, M. H. 2002. 'Luxurious forms: redefining a Mediterranean "International Style", 1400–1200 B.C.E.', *The Art Bulletin* 84(1), 6–29.

Feldman, M. H. 2006. *Diplomacy by Design: Luxury Arts and an 'International Style' in the Ancient Near East, 1400–1200 BCE*. Chicago, IL: University of Chicago Press.

Ferrara, S. 2015. 'The beginnings of writing on Crete: theory and context', *Annual of the British School at Athens* 110, 27–49.

Ferrara, S., Weingarten, J. and Cadogan, G. 2016. 'Cretan Hieroglyphic at Myrtos-Pyrgos', *Studi Micenei ed Egeo-Anatolici NS* 2, 81–99.

Fielding, X. 1953. *The Stronghold*. London: Secker and Warburg.

Fischer, J. R. 2015. *Cattle Colonialism: An Environmental History of the Conquest of California and Hawai'i*. Chapel Hill: University of North Carolina Press.

Floyd, C. 1998. *Pseira III: The Plateia Building*. Philadelphia: University of Pennsylvania Museum.

Forsdyke, J. 1931. *Minoan Art* (Proceedings of the British Academy 15). London: Humphrey Milford.

Forsdyke, J. 1952. 'Minos of Crete', *Journal of the Warburg and Courtauld Institutes* 15(1/2), 13–19.

Forsdyke, J. 1954. 'The "Harvester" Vase of Hagia Triada', *Journal of the Warburg and Courtauld Institutes* 17(1/2), 1–9.

Forstenpointner, G., Galik, A., Weissengruber, G. E. et al. 2010. 'Subsistence and more in MBA Aegina Kolonna: patterns of husbandry, hunting and agriculture', in Touchais, G. (ed.), *Mesohelladika. The Greek Mainland in the Middle Bronze Age. International Conference, Athens, 8–12 March 2006*. (BCH Supplément 52). Athens: École française d'Athènes, 733–42.

Foster, K. P. 1979. *Aegean Faience of the Bronze Age*. New Haven, CT and London: Yale University Press.

Foster, K. P. 1982. *Minoan Ceramic Relief* (SIMA 64). Gothenburg: Paul Åströms Förlag.

Frankfort, H. 1937. 'Notes on the Cretan Griffin', *Annual of the British School at Athens* 37, 106–22.

Franklin, A. 1999. *Animals and Modern Cultures: A Sociology of Human–Animal Relations in Modernity*. London: Sage.

Freedberg, D. 1989. *The Power of Images: Studies in the History and Theories of Response*. Chicago, IL and London: University of Chicago Press.

Friedman, J. B. 2000. *The Monstrous Races in Medieval Art and Thought*. New York: Syracuse University Press.

Fudge, E. 2002a. *Animal*. London: Reaktion.

Fudge, E. 2002b. *Perceiving Animals: Humans and Beasts in Early Modern English Culture*. Champaign: University of Illinois Press.

Fudge, E. 2012. 'Renaissance animal things', in Landes, J. B., Lee, P. Y. and Youngquist, P. (eds.), *Gorgeous Beasts: Animal Bodies in Historical Perspective*. University Park: Pennsylvania State University Press, 41–56.

Furtwängler, A. 1900. *Die antiken Gemmen: Geschichte der Steinschneidekunst im klassischen Altertum*. Leipzig and Berlin: Giesecke & Devrient.

Furtwängler, A. and Loeschcke, G. 1886. *Mykenische Vasen*. Berlin: Asher.

Furumark, A. 1941. *The Mycenaean Pottery: Analysis and Classification*. Stockholm: Kungl. Vitterhets Historie och Antikvitets Akademien.

Galanakis, Y., Tsitsa, E. and Günkel-Maschek, U. 2017. 'The power of images: re-examining the wall paintings from the Throne Room at Knossos', *Annual of the British School at Athens* 112, 47–98.

Gell, A. 1998. *Art and Agency*. Oxford: Clarendon Press.

Gere, C. 2009. *Knossos and the Prophets of Modernism*. Chicago, IL: University of Chicago Press.

German, S. 2005. *Performance, Power and the Art of the Aegean Bronze Age* (BAR IS 1347). Oxford: Archaeopress.

Gesell, G. C. 1985. *Town, Palace and House Cult in Minoan Crete* (SIMA 67). Gothenburg: Paul Åströms Förlag.

Gibson, J. J. 1979. *The Ecological Approach to Visual Perception*. Boston, MA: Houghton Mifflin Company.

Gill, M. A. V. 1963. 'The Minoan Dragon', *Bulletin of the Institute of Classical Studies* 10, 1–12.

Gill, M. A. V. 1964. 'The Minoan "Genius"', *Mitteilungen des Deutschen Archäologischen Instituts, Athenische Abteilung* 79, 1–21.

Gill, M. A. V. 1985. 'Some observations on representations of marine animals in Minoan art, and their identification', in Darcque, P. and Poursat, J.-C. (eds.), *L'iconographie minoenne: Actes de la Table Ronde d'Athènes (21–22 avril 1983)* (BCH Supplément 11). Athens: École française d'Athènes, 63–81.

Girella, L. 2008. 'Feasts in "transition"? An overview of feasting', in Hitchcock, L. A., Laffineur, R. and Crowley, J. (eds.), *DAIS. The Aegean Feast. Proceedings of the 12th International Aegean Conference, University of Melbourne, Centre for Classics and Archaeology, 25–29 March 2008* (Aegaeum 29). Liège and Austin: Université de Liège and University of Texas, 167–78.

Godart, L. 1999. 'Les sacrifices d'animaux dans les textes mycéniens', in Deger-Jalkotzy, S. , Hiller, S. and Panagl, O. (eds.), *Floreant Studia Mycenaea I–II.* Vienna: Verlag der Österreichischen Akademie der Wissenschaften, 249–54.

Godart, L. and Olivier, J.-P. 1976–85. *Recueil des Inscriptions en Linéaire A, I–V* (Études crétoises 21). Paris: L'École française D'Athènes.

Godart, L. and Tzedakis, Y. 1992. *Témoignages archéologiques et épigraphiques en Crète occidentale du Néolithique au Minoen Récent III B.* Rome: Gruppo Editoriale Internazionale.

Gombrich, E. H. 1977. *Art and Illusion*, 5th edn. London: Phaidon.

Gombrich, E. H. 1982. 'Visual discovery through art', in Gombrich, E. H. (ed.), *The Image and the Eye.* London: Phaidon, 11–39.

Goodison, L. 1989. *Death, Women and the Sun: Symbolism of Regeneration in Early Aegean Religion* (BICS Supplement 53). London: Institute of Classical Studies.

Goodman, N. 1976. *Languages of Art: An Approach to a Theory of Symbols.* Indianapolis, IN: Hackett.

Gordon, A. H. and Schwabe, C. W. 2004. *The Quick and the Dead: Biomedical Theory in Ancient Egypt* (Egyptological Memoirs 4). Leiden and Boston, MA: Brill.

Goren, Y. and Panagiotopoulos, D. 2009. 'The "Lords of the Rings": an analytical approach to the riddle of the "Knossian replica rings"', *Bulletin of the Institute of Classical Studies* 52(1), 257–8.

Gorogianni, E., Pavúk, P. and Girella, L. (eds.) 2016. *Beyond Thalassocracies: Understanding Processes of Minoanisation and Mycenaeanisation in the Aegean.* Oxford: Oxbow.

Graham, J. W. 1957. 'The central court as the Minoan bull-ring', *American Journal of Archaeology* 61(3), 255–62.

Greenlaw, C. 2006. 'Monkeying around the Mediterranean: a fresh perspective on ancient primates', in Day, J. (ed.), *SOMA 2004: Symposium on Mediterranean Archaeology* (BAR IS 1514). Oxford: Archaeopress, 63–7.

Greenlaw, C. 2011. *The Representation of Monkeys in the Art and Thought of Mediterranean Cultures: A New Perspective on Ancient Primates* (BAR IS 2192). Oxford: Archaeopress.

Groenewegen-Frankfort, H. 1951. *Arrest and Movement.* London: Faber and Faber.

Grove, A. T. and Rackham, O. 2001. *The Nature of Mediterranean Europe: An Ecological History.* New Haven, CT and London: Yale University Press.

Günkel-Maschek, U. 2012a. 'Spirals, bulls, and sacred landscapes: the meaningful appearance of pictorial objects within their spatial and social contexts', in Panagiotopoulos, D. and Günkel-Maschek, U. (eds.), *Minoan Realities: Approaches to Images, Architecture, and Society in the Aegean Bronze Age* (Aegis 5). Louvain: UCL Presses Universitaires de Louvain, 115–39.

Günkel-Maschek, U. 2012b. 'Unterwegs im Labyrinth. Zur Gestaltung des ‚Prozessionskorridors' im spätminoischen Palast von Knossos', in Reinholdt, C. and Wohlmayr, W. (eds.), *Akten des 13. Österreichischen Archäologentages. Klassische und Frühägäische Archäologie Paris-Lodron-Universität Salzburg vom 25. bis 27. Februar 2010.* Vienna: Phoibos, 39–46.

Hägg, R. (ed.) 1997. *The Function of the 'Minoan Villa'. Proceedings of the Eighth International Symposium at the Swedish Institute at Athens, 6–8 June 1992.* Stockholm: Svenska institutet i Athen.

Haggis, D. 1999. 'Staple finance, peak sanctuaries, and economic complexity in Late Prepalatial Crete', in Chaniotis, A. (ed.), *From Minoan Farmers to Roman Traders: Sidelights on the Economy of Ancient Crete.* Stuttgart: Franz Steiner Verlag, 53–86.

Haggis, D. 2002. 'What future for the "Minoan" past? Rethinking Minoan archaeology', in Hamilakis, Y. (ed.), *Labyrinth Revisited: Rethinking 'Minoan' Archaeology*. Oxford: Oxbow, 120–42.

Haggis, D. 2007. 'Stylistic diversity and diacritical feasting at protopalatial Petras: a preliminary analysis of the Lakkos deposit', *American Journal of Archaeology* 111(4), 715–75.

Halbherr, F. 1903. 'Resti dell' Età Micenea scoperti ad Haghia Triada presso Phaestos. Rapporto delle ricerche del 1902', *Monumenti Antichi* 13, 5–74.

Halbherr, F. and Orsi, P. 1888. 'Scoperte nell'Antro di Psychrò', *Museo italiano di antichità classica* 2, 905–12.

Hallager, E. 1996a. *The Minoan Roundel and Other Sealed Documents in the Neopalatial Linear A Administration, I* (Aegaeum 14). Liège and Austin: Université de Liège and University of Texas at Austin.

Hallager, E. 1996b. *The Minoan Roundel and Other Sealed Documents in the Neopalatial Linear A Administration, II* (Aegaeum 14). Liège and Austin: Université de Liège and University of Texas at Austin.

Hallager, E. 2001. 'A waste deposit from a LBA-shrine in Khania (?)', in Laffineur, R. and Hägg, R. (eds.), *POTNIA. Deities and Religion in the Aegean Bronze Age. Proceedings of the 8th International Aegean Conference, Göteborg, Göteborg University, 12–15 April 2000* (Aegaeum 22). Liège and Austin: Université de Liège and University of Texas at Austin, 175–9.

Hallager, B. P. and Hallager, E. 1995. 'The Knossian bull: political propaganda in Neo-Palatial Crete?', in Laffineur, R. and Niemeier, W.-D. (eds.), *POLITEIA. Society and State in the Aegean Bronze Age. Proceedings of the 5th International Aegean Conference, University of Heidelberg, Archäologisches Institut, 10–13 April 1994* (Aegaeum 12). Liège and Austin: Université de Liège and University of Texas at Austin, 547–56.

Hallager, E. and Hallager, B. P. (eds.) 2000. *The Greek-Swedish Excavations at the Agia Aikaterini Square Kastelli, Khania 1970–1987, II.* Stockholm: Paul Åströms Förlag.

Hallager, E. and Hallager, B. P. (eds.) 2003. *The Greek-Swedish Excavations at the Agia Aikaterini Square Kastelli, Khania 1970–1987 and 2001, III: 1*. Stockholm: Paul Åströms Förlag.

Halstead, P. 1987. 'Traditional and ancient rural economy in Mediterranean Europe: plus ça change?', *Journal of Hellenic Studies* 107, 77–87.

Halstead, P. 1992. 'The Mycenaean palatial economy: making the most of the gaps in the evidence', *Proceedings of the Cambridge Philological Society* 38, 57–86.

Halstead, P. 1993a. 'Banking on livestock: indirect storage in Greek agriculture', *Bulletin on Sumerian Agriculture* 7, 63–75.

Halstead, P. 1993b. 'Lost sheep? On the Linear B evidence for breeding flocks at Mycenaean Knossos and Pylos', *Minos* 25–26, 343–65.

Halstead, P. 1998. 'Linear B evidence for the management of sheep breeding at Knossos: production targets and deficits in the KN Dl (1) and Do sets', *Minos* 31–32, 187–99.

Halstead, P. 1999a. 'Missing sheep: on the meaning and wider significance of O in Knossos sheep records', *Annual of the British School at Athens* 94, 145–66.

Halstead, P. 1999b. 'Surplus and share-croppers: the grain production strategies of Mycenaean Palaces', in Betancourt, P. P., Karageorghis, V., Laffineur, R. and Niemeier, W.-D. (eds.), *MELETEMATA. Studies in Aegean Archaeology Presented to Malcolm H. Wiener as he Enters his 65th Year, I* (Aegaeum 20). Liège and Austin: Université de Liège and University of Texas at Austin, 319–26.

Halstead, P. 2001. 'Mycenaean wheat, flax and sheep: palatial intervention in farming and its implications for rural society', in Voutsaki, S. and Killen, J. T. (eds.), *Economy and Politics in the Mycenaean Palace States* (Proceedings of the Cambridge Philological Society Supplementary Volume 27). Cambridge: Cambridge Philological Society, 38–50.

Halstead, P. 2002. 'Texts, bones and herders: approaches to animal husbandry in Late Bronze Age Greece', in Bennet, J. and Driessen, J. (eds.), *A-NA-QO-TA: Studies Presented to John T. Killen* (Minos 33–34).

Salamanca: Ediciones Universidad de Salamanca, 149–89.

Halstead, P. 2003. 'Texts and bones: contrasting Linear B and archaeozoological evidence for animal exploitation in Mycenaean Southern Greece', in Kotjabopoulou, E., Hamilakis, Y., Halstead, P., Gamble, C. and Elefanti, P. (eds.), *Zooarchaeology in Greece: Recent Advances* (BSA Studies 9). London: British School at Athens, 257–61.

Halstead, P. 2014. *Two Oxen Ahead: Pre-Mechanized Farming in the Mediterranean*. Chichester: John Wiley & Sons.

Halstead, P. and Isaakidou, V. 2004. 'Faunal evidence for feasting: burnt offerings from the Palace of Nestor at Pylos', in Halstead, P. and Barrett, J. C. (eds.), *Food, Cuisine and Society in Prehistoric Greece* (Sheffield Studies in Aegean Archaeology 5). Oxford: Oxbow, 136–54.

Halstead, P. and Isaakidou, V. 2011. 'A pig fed by hand is worth two in the bush: ethnoarchaeology of pig husbandry in Greece and its archaeological implications', in Albarella, U. and Trentacoste, A. (eds.), *Ethnozooarchaeology: The Present and Past of Human-Animal Relationships*. Oxford: Oxbow, 160–74.

Halstead, P. and Isaakidou, V. 2017. 'Sheep, sacrifices, and symbols: animals in Later Bronze Age Greece', in Umbarella, U. (ed.), *The Oxford Handbook of Zooarchaeology*. Oxford: Oxford University Press, 113–26.

Hamilakis, Y. 1998. 'Eating the dead: mortuary feasting and the politics of memory in the Aegean Bronze Age societies', in Branigan, K. (ed.), *Cemetery and Society in the Aegean Bronze Age*. Sheffield: Sheffield Academic Press, 115–32.

Hamilakis, Y. 2002a. 'Too many chiefs? Factional competition in Neopalatial Crete', in Driessen, J., Schoep, I. and Laffineur, R. (eds.), *Monuments of Minos: Rethinking the Minoan Palaces* (Aegaeum 23). Liège and Austin: Université de Liège and University of Texas at Austin, 179–99.

Hamilakis, Y. 2002b. 'What future for the "Minoan" past? Rethinking Minoan archaeology', in Hamilakis, Y. (ed.), *Labyrinth Revisited: Rethinking 'Minoan' Archaeology*. Oxford: Oxbow, 2–28.

Hamilakis, Y. 2003. 'The sacred geography of hunting: wild animals, social power and gender in early farming societies', in Kotjabopoulou, E., Hamilakis, Y., Halstead, P., Gamble, C. and Elefanti, P. (eds.), *Zooarchaeology in Greece: Recent Advances* (BSA Studies 9). London: British School at Athens, 239–47.

Hamilakis, Y. and Harris, K. 2011. 'The social zooarchaeology of feasting: the evidence from the "ritual" deposit at Nopigeia-Drapanias', in *Proceedings of the 10th International Congress of Cretan Studies, A*. Chania: The Philological Association 'Chrysostomos', 199–218.

Hamilakis, Y. and Konsolaki, E. 2004. 'Pigs for the gods: burnt animal sacrifices as embodied rituals at a Mycenaean sanctuary', *Oxford Journal of Archaeology* 23(2), 135–51.

Haraway, D. 2008. *When Species Meet*. Minneapolis: University of Minnesota Press.

Harris, K. 2014a. 'Hunting, performance and incorporation: human-deer encounter in Late Bronze Age Crete', in Baker, K., Carden, R. F. and Madgwick, R. (eds.), *Deer and People*. Oxford: Windgather Press, 48–58.

Harris, K. 2014b. *The Social Role of Hunting and Wild Animals in Late Bronze Age Crete: A Social Zooarchaeological Analysis*. Unpublished PhD thesis, University of Southampton.

Harris, K. and Hamilakis, Y. 2014. 'Beyond the wild, the feral, and the domestic: lessons from prehistoric Crete', in Marvin, G. and McHugh, S. (eds.), *Routledge Handbook of Human-Animal Studies*. London and New York: Routledge, 97–113.

Haskell, H. W. 2011. 'Trade', in Haskell, H. W., Jones, R. E., Day, P. M. and Killen, J. T. (eds.), *Transport Stirrup Jars of the Bronze Age Aegean and East Mediterranean* (Prehistory Monographs 33). Philadelphia, PA: INSTAP Academic Press, 125–31.

Hatzaki, E. 2007. 'Neopalatial (MM IIIB-LM IB): KS 178, Gypsades Well (Upper Deposit) and SEX North House Groups', in Momigliano, N. (ed.), *Knossos Pottery Handbook: Neolithic and Bronze Age* (BSA

Studies 14). London: British School at Athens, 151–96.

Hatzaki, E. 2009. 'Structured deposition as ritual action at Knossos', in D'Agata, A. L. and Van de Moortel, A. (eds.), *Archaeologies of Cult: Essays on Ritual and Cult in Crete in Honor of Geraldine C. Gesell* (Hesperia Supplement 42). Athens: American School of Classical Studies at Athens, 19–30.

Haysom, M. 2010. 'The double-axe: a contextual approach to the understanding of a Cretan symbol in the Neopalatial period', *Oxford Journal of Archaeology* 29(1), 35–55.

Haysom, M. 2011. 'Fish and ships: Neopalatial seascapes in context', in Vavouranakis, G. (ed.), *The Seascape in Aegean Prehistory*. Athens: The Danish Institute at Athens, 139–60.

Haysom, M. 2015. 'Recent research into Minoan extra-urban sanctuaries', *Archaeological Reports* 61, 94–103.

Haysom, M. 2018. 'Mass and elite in Minoan peak sanctuaries', in Vavouranakis, G., Kopanias, K. and Kanellopoulos, C. (eds.), *Popular Religion and Ritual in Prehistoric and Ancient Greece and the Eastern Mediterranean*. Oxford: Archaeopress, 19–28.

Hazzidakis, J. 1915. 'Άνασκαφαί εν Κρήτη παρά τον χοριόν Μάλια. Γ΄. Μινωικόν Άνάκτορον', *Praktika tes en Athenais Archaiologikes Hetaireias* 70, 115–30.

Hazzidakis, J. 1921. *Tylissos à l'époque minoenne, suivi d'une note sur les larnax de Tylissos*. Paris: P. Geuthner.

Helms, M. 1993. *Craft and the Kingly Ideal: Art, Trade and Power*. Austin: University of Texas Press.

Helms, M. 1999. 'Why Maya lords sat on jaguar thrones', in Robb, J. (ed.), *Material Symbols*. Carbondale: Southern Illinois University, 55–66.

Henke, L. A. 1929. *A Survey of Livestock in Hawaii*. Honolulu: University of Hawaii.

Herva, V.-P. 2005. 'The life of buildings: Minoan building deposits in an ecological perspective', *Oxford Journal of Archaeology* 24 (3), 215–27.

Herva, V.-P. 2006a. 'Flower lovers, after all? Rethinking religion and human-environment relations in Minoan Crete', *World Archaeology* 38(4), 586–98.

Herva, V.-P. 2006b. 'Marvels of the system: art, perception and engagement with the environment in Minoan Crete', *Archaeological Dialogues* 13(2), 221–40.

Higgins, R. 1967. *Minoan and Mycenaean Art*. London: Thames & Hudson.

Hill, J. D. 1996. 'The identification of ritual deposits of animals: a general perspective from a specific study of "special animal deposits" from the Southern English Iron Age', in Anderson, S. and Boyle, K. (eds.), *Ritual Treatment of Human and Animal Remains: Proceedings of the First Meeting of the Osteoarchaeological Research Group, Cambridge, 1994*. Oxford: Oxbow, 17–32.

Hirsch, E. 1977. *Painted Decorations on the Floors of Bronze Age Structures* (SIMA 53). Gothenburg: Paul Åströms Förlag.

Hodder, I. 1990. *The Domestication of Europe*. Oxford: Basil Blackwell.

Hogarth, D. 1900. 'The Dictaean Cave', *Annual of the British School at Athens* 6, 94–116.

Hogarth, D. 1901. 'Excavations at Zakro, Crete', *Annual of the British School at Athens* 7, 121–49.

Hogarth, D. 1902. 'The Zakro sealings', *Journal of Hellenic Studies* 22, 76–93.

Hood, M. S. F. 1978. *The Arts in Prehistoric Greece*. Harmondsworth: Penguin.

Hood, M. S. F. 1985. 'Pigs or pulse? The pens at Knossos', *American Journal of Archaeology* 89(2), 308–13.

Hood, M. S. F. 2000. 'Cretan fresco dates', in Sherratt, S. (ed.), *The Wall Paintings of Thera, I*. Athens: Thera Foundation, 191–208.

Hood, M. S. F. 2005. 'Dating the Knossos frescoes', in Morgan, L. (ed.), *Aegean Wall Painting: A Tribute to Mark Cameron* (BSA Studies 13). London: British School at Athens, 45–81.

Hood, S., Huxley, G., Sandars, N. and Werner, A. E. 1959. 'A Minoan cemetery on Upper Gypsades (Knossos Survey 156)', *Annual of the British School at Athens* 54, 194–262.

Hooker, J. T. 1980. *Linear B: An Introduction*. London: Bristol Classical Press.

Horden, P. and Purcell, N. 2000. *The Corrupting Sea*. Oxford: Blackwell.

Horwitz, L. K. and Bar-Gal, G. K. 2006. 'The origin and genetic status of insular caprines in the Eastern Mediterranean: a case study of free-ranging goats (*Capra aegagrus cretica*) on Crete', *Human Evolution* 21, 123–38.

Hughes, J. D. 2003. 'Europe as consumer of exotic biodiversity: Greek and Roman times', *Landscape Research* 28(1), 21–31.

Hull, D. B. 1964. *Hounds and Hunting in Ancient Greece*. Chicago, IL: University of Chicago Press.

Immerwahr, S. A. 1985. 'A possible influence of Egyptian Art in the creation of Minoan wall painting', in Darcque, P. and Poursat, J.-C. (eds.), *L'iconographie minoenne: Actes de la Table Ronde d'Athènes (21–22 avril 1983)* (BCH Supplément 11). Athens: École française d'Athènes, 41–50.

Immerwahr, S. A. 1990. *Aegean Painting in the Bronze Age*. University Park and London: Pennsylvania State University Press.

Ingold, T. 1988. 'Introduction', in Ingold, T. (ed.), *What Is an Animal?* London: Unwin Hyman, 1–16.

Ingold, T. 1994. 'From trust to domination: an alternative history of human-animal relations', in Manning, A. and Serpell, J. (eds.), *Animals and Human Society: Changing Perspectives*. London: Routledge, 1–22.

Ingold, T. 2000. *The Perception of the Environment: Essays on Livelihood Dwelling and Skill*. London: Routledge.

Isaakidou, V. 2004. *Bones from the Labyrinth: Faunal Evidence for Management and Consumption of Animals at Neolithic and Bronze Age Knossos, Crete*. Unpublished PhD thesis, UCL Institute of Archaeology.

Isaakidou, V. 2006. 'Ploughing with cows: Knossos and the secondary products revolution', in Serjeantson, D. and Field, D. (eds.), *Animals in the Neolithic of Britain and Europe*. Oxford: Oxbow, 95–112.

Isaakidou, V. 2007a. 'Bioarchaeological analysis', in Macdonald, C. F. and Knappett, C. (eds.), *Knossos: Protopalatial Deposits in Early Magazine A and the South-West Houses* (BSA Supplementary Volume 41). London: British School at Athens, 139–42.

Isaakidou, V. 2007b. 'Cooking in the Labyrinth: exploring "cuisine" at Bronze Age Knossos', in Mee, C. and Renard, J. (eds.), *Cooking Up the Past: Food and Culinary Practices in the Neolithic and Bronze Age Aegean*. Oxford: Oxbow, 5–24.

Isaakidou, V. 2008. 'The fauna and economy of Neolithic Knossos', in Isaakidou, V. and Tomkins, P. (eds.), *Escaping the Labyrinth: The Cretan Neolithic in Context* (Sheffield Studies in Aegean Archaeology 8). Oxford: Oxbow, 90–114.

Isaakidou, V. 2019. 'The animal remains from Tholos A and B', in Girella, L. and Caloi, I. (eds.), *Kamilari. Una necropoli di tombe a tholos nella Messarà (Creta)* (Monographie della Scuola Archeologica di Atene e delle Missioni Italiane in Oriente 29). Athens: Scuola Archeologica di Atene, 627–31.

Isaakidou, V. and Halstead, P. 2013. 'Bones and the body politic? A diachronic analysis of structured deposition in the Neolithic–Early Iron Age Aegean', in Ekroth, G. and Wallensten, J. (eds.), *Bones, Behaviour and Belief: The Zooarchaeological Evidence as a Source for Ritual Practice in Ancient Greece and Beyond* (Skrifter utgivna av Svenska institutet i Athen, 40, 55). Stockholm: Svenska institutet i Athen, 87–99.

Isaakidou, V., Halstead, P., Davis, J. and Stocker, S. 2002. 'Burnt animal sacrifice at the Mycenaean "Palace of Nestor", Pylos', *Antiquity* 76(291), 86–92.

Jackson, C. E. 2012. *Fish in Art*. London: Reaktion.

Jarman, M. R. 1972. 'Appendix VI: the fauna', in Warren, P. (ed.), *Myrtos: An Early Bronze Age Settlement in Crete* (BSA Supplementary Volume 7). London: British School at Athens and Thames and Hudson, 318–20.

Jarman, M. R. 1996. 'Human influence in the development of the Cretan mammalian fauna', in Reese, D. S. (ed.), *Pleistocene and Holocene Fauna of Crete and Its First Settlers*

(Monographs in World Archaeology 28). Madison, WI: Prehistory Press, 211–29.

Jarman, M. R. and Jarman, H. N. 1968. 'The fauna and economy of Early Neolithic Knossos', *Annual of the British School at Athens* 63, 241–64.

Jasink, A. 2009. *Cretan Hieroglyphic Seals: A New Classification of Symbols and Ornamental/Filling Motifs* (Biblioteca di "Pasiphae" 8). Pisa: F. Serra.

Jasink, A. 2016. 'From "Hieroglyphic" to "Talismanic" Symbols', *Pasiphae* 10, 19–28.

Jasink, A., Bombadieri, L., Carminati, F. and Faralli, C. 2008. 'The Database of Hieroglyphic and Related Seals of Middle Minoan Glyptic' (www.sagas.unifi.it/vp-394-dbas-chs-cretan-hieroglyphic-seals.html) [Accessed on 7 June 2021].

Jones, G. 1979. 'Appendix A: The Animal Remains. In H. Catling et al. Knossos 1975: Middle Minoan III and Late Minoan I Houses by the Acropolis', *Annual of the British School at Athens* 74, 78–9.

Kaiser, B. 1976. *Untersuchungen zum minoischen Relief.* Bonn: Habelt.

Kanta, A. 2019. 'The religious centre of the city of Knossos and religious worship through time', in Stampolidis, N. C., Papadopoulou, E., Lourentzatou, I. G. and Fappas, I. D. (eds.), *Crete: Emerging Cities. Aptera, Eleutherna, Knossos.* Athens: Museum of Cycladic Art, 244–55.

Kantor, H. 1947a. 'The Aegean and the Orient in the second millennium B.C.', *American Journal of Archaeology* 51(1), 1–103.

Kantor, H. 1947b. 'The shoulder ornament of Near Eastern lions', *Journal of Near Eastern Studies* 6(4), 250–74.

Kantor, H. 1956. 'Syro-Palestinian ivories', *Journal of Near Eastern Studies* 15(3), 153–74.

Karali, L. 1999. *Shells in Aegean Prehistory* (BAR IS 761). Oxford: Archaeopress.

Karetsou, A. 1981. 'The peak sanctuary of Mt. Juktas', in Hägg, R. and Marinatos, N. (eds.), *Sanctuaries and Cults in the Aegean Bronze Age. Proceedings of the First International Symposium at the Swedish Institute in Athens, 12–13 May 1980* (Skrifter utgivna av Svenska institutet i Athen, 40, 28). Stockholm: Svenska institutet i Athen, 137–53.

Karetsou, A. 2013. 'The Middle Minoan III building at Alonaki, Juktas', in Macdonald, C. F. and Knappett, C. (eds.), *INTERMEZZO: Intermediacy and Regeneration in Middle Minoan III Palatial Crete* (BSA Studies 21). London: British School at Athens, 71–91.

Karetsou, A. and Andreadaki-Vlazaki, M. 2001. *Crete–Egypt: Three Thousand Years of Cultural Links.* Heraklion: Hellenic Ministry of Culture.

Karetsou, A. and Koehl, R. 2013. 'Cult object–image–emblem: a life-sized bull's head from the Juktas peak sanctuary', in Koehl, R. (ed.), *Amilla: The Quest for Excellence. Studies Presented to Guenter Kopcke in Celebration of His 75th Birthday* (Prehistory Monographs 43). Philadelphia, PA: INSTAP Academic Press, 135–43.

Karetsou, A. and Koehl, R. B. 2014. 'The Minoan mastiffs of Juktas', in Touchais, G., Laffineur, R. and Rougemont, F. (eds.), *PHYSIS. L'environnement naturel et la relation homme-milieu dans le monde égéen protohistorique. Actes de la 14e Rencontre égéenne internationale, Paris, Institut National d'Histoire de l'Art (INHA), 11–14 décembre 2012* (Aegaeum 37). Leuven: Peeters, 333–40.

Karo, G. 1930. *Die Schachtgräber von Mykenai.* Munich: Bruckmann.

Keane, W. 2005. 'Signs are not the garb of meaning: on the social analysis of material things', in Miller, D. (ed.), *Materiality.* Durham, NC: Duke University Press, 182–205.

Keller, C. 1909. 'Die ausgestorbene Fauna von Kreta und ihre Beziehungen zur Minotaurus-Sage', *Vierteljahrsschrift der Naturforschenden Gesellschaft in Zürich* 54(1–2), 424–35.

Kemp, B. and Merrillees, R. 1980. *Minoan Pottery In Second Millennium Egypt.* Mainz: von Zabern.

Kenna, V. 1960. *Cretan Seals, with a Catalogue of the Minoan Gems in the Ashmolean Museum.* Oxford: Clarendon Press.

Kenna, V. 1968. 'Studies of birds on seals of the Aegean and Eastern Mediterranean in the Bronze Age', *Opuscula Atheniensia* 8, 23–38.

Kenna, V. 1969. *The Cretan Talismanic Stone in the Late Minoan Age* (SIMA 24). Lund: Paul Åströms Förlag.

Killen, J. T. 1962. 'Mycenaean po-ka: a suggested interpretation', *La Parola del Passato* 17, 26–31.

Killen, J. T. 1963. 'Some adjuncts to the sheep ideogram on the Knossos tablets', *Eranos* 61, 69–93.

Killen, J. T. 1964. 'The wool industry of Crete in the Late Bronze Age', *Annual of the British School at Athens* 59, 1–15.

Killen, J. T. 1984. 'The textile industries at Pylos and Knossos', in Shelmerdine, C. W. and Palaima, T. G. (eds.), *Pylos Comes Alive*. New York: Fordham University, 49–63.

Killen, J. T. 1993. 'The oxen's names on the Knossos Ch tablets', *Minos* 27–28, 101–7.

Killen, J. T. 1994. 'Thebes sealings, Knossos tablets and Mycenaean state banquets', *Bulletin of the Institute of Classical Studies* 39, 67–84.

Killen, J. T. 2008a. 'A note on the Knossos Mc tablets', *Faventia* 30(1–2), 47–52.

Killen, J. T. 2008b. 'Mycenaean economy', in Duhoux, Y. and Morpurgo Davies, A. (eds.), *A Companion to Linear B: Mycenaean Greek Texts and Their World*. Louvain: Peeters, 159–200.

Klein, R. G. and Cruz-Uribe, K. 1984. *The Analysis of Animal Bones from Archeological Sites*. Chicago, IL and London: University of Chicago Press.

Klingender, F. 1971. *Animals in Art and Thought*. London: Routledge.

Klippel, W. and Snyder, L. 1991. 'Dark-Age fauna from Kavousi, Crete: the vertebrates from the 1987 and 1988 excavations', *Hesperia* 60(2), 179–86.

Knappett, C. 1999a. 'Assessing a polity in Protopalatial Crete: the Malia-Lasithi state', *American Journal of Archaeology* 103(4), 615–39.

Knappett, C. 1999b. 'Tradition and innovation in pottery forming technology: wheel-throwing at Middle Minoan Knossos', *Annual of the British School at Athens* 94, 101–29.

Knappett, C. 2002a. 'Agency and object photographs, skeuomorphs and marionettes: some thoughts on mind, agency and object', *Journal of Material Culture* 7, 97–117.

Knappett, C. 2002b. 'Mind the gap: between pots and politics in Minoan studies', in Hamilakis, Y. (ed.), *Labyrinth Revisited: Rethinking 'Minoan' Archaeology*. Oxford: Oxbow, 167–88.

Knappett, C. 2005a. 'The affordances of things: a post-Gibsonian perspective on the relationality of mind and matter', in DeMarrais, E., Gosden, C. and Renfrew, C. (eds.), *Rethinking Materiality: The Engagement of Mind with the Material World*. Cambridge: McDonald Institute, 43–51.

Knappett, C. 2005b. *Thinking Through Material Culture*. Philadelphia: University of Pennsylvania Press.

Knappett, C. 2007. 'Materials with materiality?', *Archaeological Dialogues* 14(1), 20–3.

Knappett, C. 2008. 'The neglected networks of material agency: artefacts, pictures and texts', in Knappett, C. and Malafouris, L. (eds.), *Material Agency: Towards a Non-Anthropocentric Approach*. New York: Springer, 139–56.

Knappett, C. 2016. 'Minoanisation and Mycenaeanisation: a commentary', in Gorogianni, E., Pavúk, P. and Girella, L. (eds.), *Beyond Thalassocracies: Understanding Processes of Minoanisation and Mycenaeanisation in the Aegean*. Oxford: Oxbow, 202–6.

Knappett, C. 2018. 'From network connectivity to human mobility: models for Minoanization', *Journal of Archaeological Method and Theory* 25(4), 974–95.

Knappett, C. 2020. *Aegean Bronze Age Art: Meaning in the Making*. Cambridge: Cambridge University Press.

Knappett, C., Malafouris, L. and Tomkins, P. 2010. 'Ceramics (as containers)', in Hicks, D. and Beaudry, M. (eds.), *The Oxford Handbook of Material Culture Studies*. Oxford: Oxford University Press, 588–612.

Knappett, C. and Nikolakopoulou, I. 2005. 'MBA interaction networks: Crete, the Cyclades and the South-East Aegean', in Laffineur, R. and Greco, E. (eds.),

EMPORIA. Aegeans in Central and Eastern Mediterranean. Proceedings of the 10th International Aegean Conference, Athens, Italian School of Archaeology, 14–18 April 2004 (Aegaeum 25). Liège and Austin: Université de Liège and University of Texas at Austin, 175–8.

Knight, J. 2005. 'Introduction', in Knight, J. (ed.), *Animals in Person: Cultural Perspectives on Human-Animal Intimacies*. Oxford: Berg, 1–13.

Kockelman, P. 2006. 'Residence in the world: affordances, instruments, actions, roles, and identities', *Semiotica* 162, 19–71.

Koehl, R. B. 1986a. 'A marinescape floor from the Palace at Knossos', *American Journal of Archaeology* 90(4), 407–17.

Koehl, R. B. 1986b. 'The Chieftain Cup and a Minoan rite of passage', *Journal of Hellenic Studies* 106, 99–110.

Koehl, R. B. 2006. *Aegean Bronze Age Rhyta* (Prehistory Monographs 19). Philadelphia, PA: INSTAP Academic Press.

Koehl, R. B. and Karetsou, A. 2011. 'An enigmatic piece of gold-work from the Juktas peak sanctuary', in Carinci, F., Cucuzza, N., Militello, P. and Palio, O. (eds.), *Κρήτης Μινωίδος. Tradizione e identità Minoica tra Producione Artigianale, Pratiche Cerimoniali e Memoria del Passato*. Padova: Bottega d'Erasmo, 207–22.

Koh, A. J., Betancourt, P. P., Pareja, M. N., Brogan, T. M. and Apostolakou, V. 2016. 'Organic residue analysis of pottery from the dye workshop at Alatsomouri-Pefka, Crete', *Journal of Archaeological Science: Reports* 7, 536–8.

Kopaka, K. 1995. 'Ο Μίνως Καλοκαιρινός και οι πρώτες ανασκαφές στην Κνωσό', in *Proceedings of the 7th International Congress of Cretan Studies, A1*. Rethymnon: Municipality of Rethymnon, 501–11.

Kopaka, K. 2015. 'Minos Kalokairinos and his early excavation at Knossos: an overview, a portrait and a return to the Kefala pithoi', in Macdonald, C. F., Hatzaki, E. and Andreou, S. (eds.), *The Great Islands: Studies of Crete and Cyprus Presented to Gerald Cadogan*. Athens: Kapon Editions, 143–51.

Koromila, G., Karkanas, P., Hamilakis, Y. et al. 2018. 'The Neolithic tell as a multi-species monument: human, animal, and plant relationships through a micro-contextual study of animal dung remains at Koutroulou Magoula, central Greece', *Journal of Archaeological Science: Reports* 19, 753–68.

Kourou, N. and Karetsou, A. 1997. 'Terracotta wheelmade bull figurines from Central Crete: types, fabrics, technique and tradition', in Laffineur, R. and Betancourt, P. P. (eds.), *TEXNH. Craftsmen, Craftswomen and Craftsmanship in the Aegean Bronze Age. Proceedings of the 6th International Aegean Conference, Philadelphia, Temple University, 18–21 April 1996* (Aegaeum 16). Liège and Austin: Université de Liège and University of Texas at Austin, 107–16.

Krzyszkowska, O. 1983. 'Wealth and prosperity in Pre-Palatial Crete: the case of ivory', in Krzyszkowska, O. and Nixon, L. (eds.), *Minoan Society: Proceedings of the Cambridge Colloquium 1981*. Bristol: Bristol Classical Press, 11–21.

Krzyszkowska, O. 1988. 'Ivory in the Aegean Bronze Age: elephant tusk or hippopotamus ivory?', *Annual of the British School at Athens* 83, 209–34.

Krzyszkowska, O. 1989. 'Early Cretan seals: new evidence for the use of bone, ivory and boar's tusk', in Pini, I. (ed.), *Fragen und Probleme der bronzezeitlichen ägäischen Glyptik. Beiträge zum 3. Internationalen Marburger Siegel Symposium, 5–7 September 1985* (CMS Beiheft 3). Berlin: Gebr. Mann Verlag, 111–26.

Krzyszkowska, O. 2005. *Aegean Seals: An Introduction* (BICS Supplement 85). London: Institute of Classical Studies.

Krzyszkowska, O. 2010. 'Impressions of the natural world: landscape in Aegean glyptic', in Krzyszkowska, O. (ed.), *Cretan Offerings: Studies in Honour of Peter Warren* (BSA Studies 18). London: British School at Athens, 169–88.

Krzyszkowska, O. 2012. 'Seals from the Petras cemetery: a preliminary overview', in Tsipopoulou, M. (ed.), *Petras, Siteia – 25 years of excavations and studies. Acts of a Two-Day Conference Held at the Danish Institute at Athens, 9–10 October 2010* (Monographs of the Danish Institute at Athens 16). Athens: Danish Institute of Athens, 145–60.

Krzyszkowska, O. 2014. 'Cutting to the chase: hunting in Minoan Crete', in Touchais, G., Laffineur, R. and Rougemont, F. (eds.), *PHYSIS. L'environnement naturel et la relation homme-milieu dans le monde égéen protohistorique. Actes de la 14e Rencontre égéenne internationale, Paris, Institut National d'Histoire de l'Art (INHA), 11–14 décembre 2012* (Aegaeum 37). Leuven: Peeters, 341–7.

Krzyszkowska, O. 2015. 'Why were cats different? Script and imagery at Middle Minoan Pyrgos', in Macdonald, C. F., Hatzaki, E. and Andreou, S. (eds.), *The Great Islands: Studies of Crete and Cyprus Presented to Gerald Cadogan*. Athens: Kapon, 100–6.

Kyriakidis, E. 2005. *Ritual in the Bronze Age Aegean: The Minoan Peak Sanctuaries*. London: Duckworth.

La Rosa, V. 2012. 'Ayia Triada', in Cline, E. H. (ed.), *The Oxford Handbook of the Bronze Age Aegean*. Oxford: Oxford University Press, 496–508.

Laffineur, R. 1987. 'Weitere Beiträge zur Symbolik im mykenischen Bestattungsritual', in Schiering, W. (ed.), *Kolloquium zur ägäischen Vorgeschichte. Mannheim, 20–22.2.1986*. Mannheim: Vorstand des Deutschen Archäologen-Verbandes, 125–32.

Laffineur, R. 1992. 'Iconography as evidence of social and political status in Mycenaean Greece', in Laffineur, R. and Crowley, J. (eds.), *EIKON. Aegean Bronze Age Iconography: Shaping a Methodology. Proceedings of the 4th International Aegean Conference, University of Tasmania, Hobart, Australia, 6–9 April 1992* (Aegaeum 8). Liège: Université de Liège, 105–12.

Lapatin, K. 2002. *Mysteries of the Snake Goddess: Art, Desire, and the Forging of History*. Cambridge, MA: Da Capo.

Lassen, A. W. 2014. 'Wool in Anatolia in the Old Assyrian period', in Breniquet, C. and Michel, C. (eds.), *Wool Economy in the Ancient Near East and the Aegean: From the Beginnings of Sheep Husbandry to Institutional Textile Industry*. Oxford: Oxbow.

Latour, B. 1993. *We Have Never Been Modern*. London: Harvester Wheatsheaf.

Latour, B. 2004. *Politics of Nature*. Cambridge, MA: Harvard University Press.

Latour, B. 2005. *Reassembling the Social: An Introduction to Actor-Network-Theory*. Oxford: Oxford University Press.

Latour, B. 2009. 'Perspectivism: "type" or "bomb"?', *Anthropology Today* 25(2), 1–2.

Latour, B. 2013. *An Inquiry into Modes of Existence: An Anthropology of the Moderns*. Cambridge, MA: Harvard University Press.

Latour, B. 2017. *Facing Gaia: Eight Lectures on the New Climatic Regime*. Cambridge: Polity Press.

Lawrence, E. A. 1982. *Rodeo: An Anthropologist Looks at the Wild and the Tame*. Chicago, IL: University of Chicago Press.

Lax, E. and Strasser, T. 1992. 'Early Holocene extinctions on Crete: the search for the cause', *Journal of Mediterranean Archaeology* 5(2), 205–28.

Layton, R. 1977. 'Naturalism and cultural relativity in art', in Ucko, P. (ed.), *Form in Indigenous Art: Schematisation in the Art of Aboriginal Australia and Prehistoric Europe*. Canberra: Australian Institute of Aboriginal Studies, 33–43.

Lebessi, A. and Muhly, P. 1990. 'Aspects of Minoan cult: sacred enclosures. the evidence from the Syme Sanctuary (Crete)', *Archäologischer Anzeiger* 3, 315–36.

Legarra Herrero, B. 2014. *Mortuary Behaviour and Social Trajectories in Pre- and Protopalatial Crete* (Prehistory Monographs 44). Philadelphia, PA: INSTAP Academic Press.

Legarra Herrero, B. 2016. 'An elite infested sea: interaction and change in Mediterranean paradigms', in Molloy, B. (ed.), *Of Odysseys and Oddities: Scales and Modes of Interaction between Prehistoric Aegean Societies and their Neighbours*. Oxford: Oxbow, 25–52.

Legendart, A. 2020. 'Une classification animale crétoise à l'âge du bronze? Le point de vue des Minoens par le prisme de l'iconographie', *Anthropozoologica* 55(17), 247–55.

Legendart, A. forthcoming. 'Some jellyfish in Minoan Art?', in Theodoropoulou, T. and Gallant, T. (eds.), *Harvesting the Gifts of the Sea: Aegean Societies and Marine Life through Time*. New York: Springer.

Lévi-Strauss, C. 1963. *Totemism*. Boston, MA: Beacon Press.

Lévi-Strauss, C. 1966. *The Savage Mind*. London: Weidenfeld & Nicolson.

Lewis, B. 2009. *Hunting in Britain: From the Ice Age to the Present*. Stroud, UK: The History Press.

Lloret, J. and Font, T. 2013. 'A comparative analysis between recreational and artisanal fisheries in a Mediterranean coastal area', *Fisheries Management and Ecology* 20, 148–60.

Lloyd, G. E. R. 2004. *Ancient Worlds, Modern Reflections: Philosophical Perspectives on Greek and Chinese Science and Culture*. Oxford: Oxford University Press.

Lloyd, G. E. R. 2007. *Cognitive Variations: Reflections on the Unity and Diversity of the Human Mind*. Oxford: Oxford University Press.

Lloyd, G. E. R. 2010. 'History and human nature: cross-cultural universals and cultural relativities', *Interdisciplinary Science Reviews* 35 (3–4), 201–14.

Logue, W. 2004. 'Set in stone: the role of relief-carved stone vessels in Neopalatial Minoan elite propaganda', *Annual of the British School at Athens* 99, 149–72.

Long, C. R. 1974. *The Ayia Triadha Sarcophagus: a Study of Late Minoan and Mycenaean Funerary Practices and Beliefs* (SIMA 41). Gothenburg: Paul Åströms Förlag.

Long, C. R. 1978. 'The Lasithi Dagger', *American Journal of Archaeology* 82(1), 35–46.

Loughlin, E. 2000. *Representations of the Cow and Calf in Minoan Art*. Unpublished PhD thesis, University of Edinburgh.

Loughlin, E. 2004a. 'Grasping the bull by the horns: Minoan bull sport', in Bell, S. and Davies, G. (eds.), *Games and Festivals in Classical Antiquity. Proceedings of the Conference held in Edinburgh 10–12 July 2000* (BAR IS 1220). Oxford: Archaeopress, 1–8.

Loughlin, E. 2004b. 'The calf in Bronze Age Cretan art and society', in Santillo Frizell, B. (ed.), *PECUS. Man and Animal in Antiquity. Proceedings of the Conference at the Swedish Institute in Rome, 9–12 September 2002*. Rome: The Swedish Institute in Rome, 183–9.

Lowe Fri, M. 2011. *The Minoan Double Axe: An Experimental Study of Production and Use* (BAR IS 2304). Oxford: BAR Publishing.

Macdonald, C. F. and Knappett, C. 2013. *INTERMEZZO: Intermediacy and Regeneration in Middle Minoan III Palatial Crete* (BSA Studies 21). London: British School at Athens.

MacGillivray, J. A. 1984. 'Cycladic jars from MM III contexts at Knossos', in Hägg, R. and Marinatos, N. (eds.), *The Minoan Thalassocracy: Myth and Reality. Proceedings of the Third International Symposium at the Swedish Institute in Athens, 31 May–5 June 1982* (Skrifter utgivna av Svenska institutet i Athen, 40, 32). Stockholm: Svenska institutet i Athen, 153–7.

MacGillivray, J. A. 1998. *Knossos: Pottery Groups of the Old Palace Period* (BSA Studies 5). London: British School at Athens.

MacGillivray, J. A. 2000a. 'Labyrinths and bull-leapers', *Archaeology* 53(6), 53–5.

MacGillivray, J. A. 2000b. *Minotaur: Sir Arthur Evans and the Archaeology of the Minoan Myth*. London: Jonathan Cape.

MacGillivray, J. A., Driessen, J. and Sackett, L. H. 2000. *The Palaikastro Kouros: A Minoan Chryselephantine Statuette and its Aegean Bronze Age Context* (BSA Studies 6). London: British School at Athens.

MacKenzie, J. M. 1988. *The Empire of Nature*. Manchester: Manchester University Press.

Madge, S. and McGowan, P. 2002. *Pheasants, Partridges, and Grouse: A Guide to the Pheasants, Partridges, Quails, Grouse, Guineafowl, Buttonquails, and Sandgrouse of the World*. Princeton, NJ: Princeton University Press.

Maran, J. 2011. 'Lost in translation: the emergence of Mycenaean culture as a phenomenon of glocalization', in Wilkinson, T., Sherratt, S. and Bennet, J. (eds.), *Interweaving Worlds: SystemicInteractions in Eurasia, 7th to the 1st Millennia BC*. Oxford: Oxbow, 282–94.

Marinatos, N. 1986. *Minoan Sacrificial Ritual: Cult Practice and Symbolism*. Stockholm: Paul Åströms Forlag.

Marinatos, N. 1989. 'The bull as an adversary: some observations on bull-hunting and bull-leaping', *Ariadne* 5, 23–32.

Marinatos, N. 1993. *Minoan Religion: Ritual, Image, and Symbol*. Columbia: University of South Carolina Press.

Marinatos, N. 1994. 'The "export" significance of Minoan bull hunting and bull leaping scenes', *Ägypten und Levante* 4, 89–93.

Marinatos, N. 1999. 'Bull hides as dadoes and emblems of prestige in Creto-Mycenaean palaces', in Betancourt, P. P., Karageorghis, V., Laffineur, R. and Niemeier, W.-D. (eds.), *MELETEMATA. Studies in Aegean Archaeology Presented to Malcolm H. Wiener as he Enters his 65th Year, II* (Aegaeum 20). Liège and Austin: Université de Liège and University of Texas at Austin, 481–4.

Marinatos, N. 2007. 'Bull-leaping and royal ideology', in Bietak, M., Marinatos, N. and Palivou, C. (eds.), *Taureador Scenes in Tell El-Dab'a (Avaris) and Knossos* (Denkschriften der Gesamtakademie 43). Vienna: Verlag der Österreichischen Akademie der Wissenschaften, 127–32.

Marinatos, N. 2010. 'Lions from Tell el Dab'a', *Ägypten und Levante* 20, 325–55.

Marinatos, N. and Morgan, L. 2005. 'The dog pursuit scenes from Tell el Dab'a and Kea', in Morgan, L. (ed.), *Aegean Wall Painting: A Tribute to Mark Cameron* (BSA Studies 13). London: British School at Athens, 119–22.

Marinatos, S. and Hirmer, M. 1960. *Crete and Mycenae*. London: Thames and Hudson.

Marshall, F. H. 1907. *Catalogue of the Finger Rings, Greek, Etruscan, and Roman, in the Departments of Antiquities, British Museum*. London: Trustees of the British Museum.

Martinez, L., Underhill, P. A., Zhivotovsky, L. A. et al. 2007. 'Paleolithic Y-haplogroup heritage predominates in a Cretan highland plateau', *European Journal of Human Genetics* 15(4), 485–93.

Marvin, G. 1988. *Bullfight*. Champaign: University of Illinois Press.

Marvin, G. 2005. 'Seeing, looking, watching, observing nonhuman animals', *Society and Animals* 13(1), 1–11.

Masala, S. 2009. 'I resti faunistici provenienti dal saggio sotto il vano XIX a Festòs', *Creta Antica* 10/1, 97–103.

Mason, J. 2007. 'Animals: from souls and the sacred in prehistoric times to symbols and slaves in antiquity', in Kalof, L. (ed.), *A Cultural History of Animals in Antiquity*. London: Berg, 17–45.

Masseti, M. 1997. 'Representations of birds in Minoan art', *International Journal of Osteoarchaeology* 7(4), 354–63.

Masseti, M. 2009. 'A possible approach to the "conservation" of the mammalian populations of ancient anthropochorous origin of the Mediterranean islands', *Folia Zoologica* 58(3), 303–8.

Masseti, M. 2012. *Atlas of Terrestrial Mammals of the Ionian and Aegean Islands*. Berlin: de Gruyter.

Mathioudaki, I. 2018. 'The pottery deposit from the Houses of the Fallen Blocks and the Sacrificed Oxen at the south-eastern corner of the palace of Knossos', *Annual of the British School at Athens* 113, 19–73.

Maturana, H. and Varela, F. 1980. *Autopoiesis and Cognition: The Realization of the Living*. Dordrecht: D. Reidel.

Matz, F. 1928. *Die Frühkretischen Siegel. Eine Untersuchung über das Werden des minoischen Stiles*. Berlin: de Gruyter.

Matz, F. 1951. *Torsion: eine formenkundliche Untersuchung zur agaiischen Vorgeschichte*. Mainz: Akadamie der Wissenschaften und der Literatur.

Mavriyannaki, C. 1978. 'Double axe-tool with an engraved bucranium from the district of Amari (nome of Rethymno)', *Athens Annals of Archaeology* 11, 198–208.

Mayr, E. 1942. *Systematics and the Origin of Species*. New York: Columbia University Press.

Mayr, E. 1949. 'The species concept: semantics versus semantics', *Evolution* 3, 371–2.

Mazarakis Ainian, A. 1997. *From Rulers' Dwellings to Temples: Architecture, Religion and Society in Early Iron Age Crete (1100–700 B.C.)* (SIMA 121). Jonsered: Paul Åströms Förlag.

McGowan, E. 2018. 'Cryptic glyptic: multivalency in Minoan glyptic imagery', in Ameri, M., Kielt Costello, S., Jamison, G. and Jarmer Scott, S. (eds.), *Seals and Sealing in the Ancient World*. Cambridge: Cambridge University Press, 368–86.

McInerney, J. 2010. *The Cattle of the Sun: Cows and Culture in the World of the Ancient Greeks*. Princeton, NJ: Princeton University Press.

McLeod, W. 1970. *Composite Bows from the Tomb of Tut'ankhamūn* (Tut'ankhamūn's tomb series 3). Oxford: Griffith Institute.

McNeal, R. A. 1973. 'The legacy of Arthur Evans', *California Studies in Classical Antiquity* 6, 205–20.

Merrillees, R. 2003. 'The first appearances of Kamares Ware in the Levant', *Ägypten und Levante* 13, 127–42.

Mezzalira, F. 2001. *Beasts and Bestiaries: The Representation of Animals from Prehistory to the Renaissance*. Turin: Alberto Allemandi.

Midgley, M. 1978. *Beast and Man: The Roots of Human Nature*. Hassocks: Harvester Press.

Milchhoefer, A. 1883. *Die Anfänge der Kunst in Griechenland*. Leipzig: Brockhaus.

Militello, P. 1998. *Haghia Triadha I. Gli Affreschi* (Monografie della Scuola Archeologica di Atene e delle Missione Italiane in Oriente 9). Padova: Bottega D'Erasmo.

Militello, P. 2000. 'Nilotic models and local reelaboration: the Hagia Triada example', in Karetsou, A. (ed.), *Κρήτη – Αίγυπτος: πολιτιστικοί δεσμοί τριών χιλιετιών*. Athens: Ministry of Culture, 78–85.

Militello, P. 2012. 'Emerging authority: a functional analysis of the MM II settlement of Festòs', in Schoep, I., Tomkins, P. and Driessen, J. (eds.), *Back to the Beginning: Reassessing Social and Political Complexity on Crete during the Early and Middle Bronze Age*. Oxford: Oxbow, 236–72.

Militello, P. 2014. 'Wool production in Neolithic and Early Bronze Age Aegean', in Touchais, G., Laffineur, R. and Rougemont, F. (eds.), *PHYSIS. L'environnement naturel et la relation homme-milieu dans le monde égéen protohistorique. Actes de la 14e Rencontre égéenne internationale, Paris, Institut National d'Histoire de l'Art (INHA), 11–14 décembre 2012* (Aegaeum 37). Leuven: Peeters, 317–24.

Miller, E. B. 1984. *Zoomorphic Vases in the Bronze Age Aegean*. Unpublished PhD thesis, New York University, Institute of Fine Arts.

Mitchell, W. J. T. 1986. *Iconology: Image, Text, Ideology*. Chicago, IL and London: University of Chicago Press.

Mittman, A. S. 2006. *Maps and Monsters in Medieval England*. New York and London: Routledge.

Molloy, B. P. C. 2012. 'Martial Minoans? War as social process, practice and event in Bronze Age Crete', *Annual of the British School at Athens* 107, 87–142.

Momigliano, N. 2007. 'Late Prepalatial (EM III–MM IA): South Front House Foundation Trench, Upper East Well and House C / Royal Road South Fill Groups', in Momigliano, N. (ed.), *Knossos Pottery Handbook. Neolithic and Bronze Age* (BSA Studies 14). London: British School at Athens, 79–103.

Momigliano, N. 2020. *In Search of the Labyrinth: The Cultural Legacy of Minoan Crete*. London: Bloomsbury.

Momigliano, N. and Karadimas, N. 2004. 'On the term "Minoan" before Evans's work in Crete (1894)', *Studi Micenei ed Egeo-Anatolici* 46(2), 243–58.

Moore, D. 2013. *Dawn of Discovery: The Early British Travellers to Crete*. Newcastle upon Tyne: Cambridge Scholars Publishing.

Morgan, L. 1985. 'Idea, idiom and iconography', in Darcque, P. and Poursat, J.-C. (eds.), *L'Iconographie minoenne. Actes de la table ronde d'Athènes (21–22 avril 1983)* (BCH Supplément 11). Athens: Ecole française d'Athènes, 5–19.

Morgan, L. 1988. *The Miniature Wall Paintings of Thera: A Study in Aegean Culture and Iconography*. Cambridge: Cambridge University Press.

Morgan, L. 1990. 'Island iconography: Thera, Kea, Milos', in Hardy, D. A. (ed.), *Thera and the Aegean World III, I*. London: The Thera Foundation, 252–66.

Morgan, L. 1995a. 'Frontal face and the symbolism of death in Aegean glyptic', in Müller, W.

(ed.), *Sceaux minoens et mycéniens. IVe symposium international, 10–12 septembre 1992, Clermont-Ferrand* (CMS Beiheft 5). Berlin: Gebr. Mann Verlag, 135–49.

Morgan, L. 1995b. 'Of animals and men: the symbolic parallel', in Morris, C. E. (ed.), *Klados: Essays in Honour of J.N. Coldstream* (BICS Supplement 63). London: Institute of Classical Studies, 171–84.

Morgan, L. 2000. 'Form and meaning in figurative painting', in Sherratt, S. (ed.), *The Wall Paintings of Thera, II*. Athens: Thera Foundation, 925–46.

Morgan, L. 2005. 'New discoveries and new ideas in Aegean wall painting', in Morgan, L. (ed.), *Aegean Wall Painting: A Tribute to Mark Cameron* (BSA Studies 13). London: British School at Athens, 21–44.

Morgan, L. 2010a. 'An Aegean griffin in Egypt: the Hunt Frieze at Tell el-Dab'a', *Ägypten und Levante* 20, 303–23.

Morgan, L. 2010b. 'A pride of leopards: a unique aspect of the Hunt Frieze from Tell el-Dab'a', *Ägypten und Levante* 20, 263–301.

Morgan, L. 2020. *Keos XI. Wall Paintings and Social Context. The Northeast Bastion at Ayia Irini*. Philadelphia, PA: INSTAP Academic Press.

Morris, B. 1998. *The Power of Animals: An Ethnography*. Oxford: Berg.

Morris, B. 2000. *Animals and Ancestors: An Ethnography*. Oxford: Berg.

Morris, C. and Peatfield, A. 2004. 'Experiencing ritual: shamanic elements in Minoan religion', in Wedde, M. (ed.), *Celebrations: Sanctuaries and the Vestiges of Cult Activity* (Papers from the Norwegian Institute of Athens 6). Bergen: Norwegian Institute at Athens, 35–59.

Morris, C. E. 1990. 'In pursuit of the white tusked boar: aspects of hunting in Mycenaean society', in Hägg, R. and Nordquist, G. C. (eds.), *Celebrations of Death and Divinity in the Bronze Age Argolid. Proceedings of the Sixth International Symposium at the Swedish Institute at Athens, 11–13 June 1988* (Skrifter utgivna av Svenska institutet i Athen, 40, 40). Stockholm: Svenska institutet i Athen, 149–55.

Morris, C. E. 1995. 'Fishy tales from Knossos: a Late Minoan larnax and vase-painter', in Morris, C. E. (ed.), *Klados: Essays in Honour of J.N. Coldstream* (BICS Supplement 63). London: Institute of Classical Studies, 185–93.

Morris, C. E. 2001. 'The language of gesture in Minoan religion', in Laffineur, R. and Hägg, R. (eds.), *POTNIA. Deities and Religion in the Aegean Bronze Age. Proceedings of the 8th International Aegean Conference, Göteborg, Göteborg University, 12–15 April 2000* (Aegaeum 22). Liège and Austin: Université de Liège and University of Texas at Austin, 245–51.

Morris, J. 2008. 'Associated bone groups: one archaeologist's rubbish is another's ritual deposition', in Davis, O. , Sharples, N. and Waddington, K. (eds.), *Changing Perspectives on the First Millennium BC*. Oxford: Oxbow, 83–98.

Mountjoy, P. A. 1972. 'A Late Minoan IB Marine Style stirrup-jar', *Annual of the British School at Athens* 67, 125–8.

Mountjoy, P. A. 1984. 'The Marine Style pottery of LMIB/LHIIA: towards a corpus', *Annual of the British School at Athens* 79, 161–219.

Mountjoy, P. A. 1985. 'Ritual associations for LM I B Marine Style vases', in Darcque, P. and Poursat, J.-C. (eds.), *L'iconographie minoenne: Actes de la Table Ronde d'Athènes (21–22 avril 1983)* (BCH Supplément 11). Athens: École française d'Athènes, 231–42.

Mountjoy, P. A., Jones, R. E. and Cherry, J. F. 1978. 'Provenance studies of the LMIB/LHIIA Marine Style', *Annual of the British School at Athens* 73, 143–71.

Mrva-Montoya, A. 2015. 'Cycles of anthropocenic interdependencies on the island of Cyprus', in Human Animal Research Network Editorial Collective (ed.), *Animals in the Anthropocene: Critical Perspectives on Non-Human Futures*. Sydney: Sydney University Press, 33–62.

Muhly, P. 1992. *Μινωικός λαξευτός τάφος στον Πόρο Ηρακλείου (Άνασκαφης 1967)* (BAE 129). Athens: Archaeological Society at Athens.

Müller, W. 1997. *Kretische Tongefässe mit Meeresdekor: Entwicklung und Stellung innerhalb der feinen Keramik von Spätminoisch IB auf Kreta* (Archäologische Forschungen 19). Berlin: Gebr. Mann Verlag.

Murock Hussein, A. 2015. 'Minoan goat hunting: social status and the economics of war', in Duistermaat, K. and Regulski, I. (eds.), *Intercultural Contacts in the Ancient Mediterranean*. Leuven: Peeters, 557–75.

Murock Hussein, A. 2018. 'The magic and the mundane: the function of "talismanic-class" stones in Minoan Crete', in Ameri, M., Kielt Costello, S., Jamison, G. and Jarmer Scott, S. (eds.), *Seals and Sealing in the Ancient World*. Cambridge: Cambridge University Press 387–99.

Murray, A. S. 1890. 'A vase of the Mykenai type in New York', *The American Journal of Archaeology and of the History of the Fine Arts* 6, 437–44.

Murray, A. S. 1892. *Handbook of Greek Archaeology: Vases, Bronzes, Gems, Sculpture, Terra-Cottas, Mural Paintings, Architecture &c.* London: John Murray.

Mylona, D. 2004. 'The fish remains', in Soles, J. and Davaras, C. (eds.), *Mochlos IC: Period III. Neopalatial Settlement on the Coast: The Artisans' Quarter and the Farmhouse at Chalinomouri. The Small Finds* (Prehistory Monographs 9). Philadelphia, PA: INSTAP Academic Press, 121–5.

Mylona, D. 2007. 'The fish remains', in MacGillivray, J. A., Sackett, L. H. and Driessen, J. M. (eds.), *Palaikastro: Two Late Minoan Wells* (BSA Supplementary Volume 43). London: British School at Athens, 217–20.

Mylona, D. 2010. 'The bones', in Tsipopoulou, M. and Hallager, E. (eds.), *The Hieroglyphic Archive at Petras, Siteia* (Monographs of the Danish Institute at Athens 9). Athens: Danish Institute at Athens, 221–31.

Mylona, D. 2015. 'Sacrifices in LM IIIB: Early Kydonia Palatial Centre. The animal remains', *Pasiphae* 9, 53–8.

Mylona, D. 2016. 'Το Ιερό Κορυφής του Βρύσινα (IV): Βοσκοί, γεωργοί και ψαράδες προσκυνητές – η συμβολή της ζωοαρχαιολογίας', *Archaiologia* 121, 48–53.

Mylona, D. 2020. 'Marine resources and coastal communities in the Late Bronze Age southern Aegean: a seascape approach', *American Journal of Archaeology* 124(2), 179–213.

Mylona, D. forthcoming. 'The Bronze Age birds in Greece. A zooarchaeological perspective', *Quaternary International*.

Nanoglou, S. 2009. 'Animal bodies and ontological discourse in the Greek Neolithic', *Journal of Archaeological Method and Theory* 16 (3), 184–204.

Neri, J. 2011. *The Insect and the Image: Visualizing Nature in Early Modern Europe, 1500–1700*. Minneapolis: University of Minnesota Press.

Newton, C. T. 1880. *Essays on Art and Archaeology*. London: Macmillan.

Niemeier, W.-D. 1987. 'Das Stuckrelief der "Prinzen mit der Federkrone" aus Knossos und minoische Götterdarstellungen', *Mitteilungen des Deutschen Archaeologischen Instituts, Athenische Abteilung* 102, 65–98.

Niemeier, W.-D. 1988. 'The "Priest-King" fresco from Knossos: a new reconstruction and interpretation', in French, E. B. and Wardle, K. A. (eds.), *Problems in Greek Prehistory*. Bristol: Bristol Classical Press, 235–52.

Niemeier, W.-D. and Niemeier, B. 1998. 'Minoan frescoes in the Eastern Mediterranean', in Cline, E. H. and Harris-Cline, D. (eds.), *The Aegean and the Orient in the Second Millennium* (Aegaeum 18). Liège and Austin: Université de Liège and University of Texas at Austin, 69–98.

Niemeier, W.-D. and Niemeier, B. 2000. 'Aegean frescoes in Syria-Palestine: Alalakh and Tel Kabri', in Sherratt, S. (ed.), *The Wall Paintings of Thera: Proceedings of the First International Symposium*. Athens: Thera Foundation, 763–800.

Nikouloudis, S. 2001. 'Animal sacrifice in the Mycenaean world', *Journal of Prehistoric Religion* 15, 11–31.

Nilsson, M. P. 1950. *The Minoan-Mycenaean Religion and Its Survival in Greek Religion*. 2nd edn. Lund: Gleerup.

Nobis, G. 1990. 'Der "Minotauros" von Knossos auf Kreta – im Lichte moderner archäozoologischer Forschung', *Tier und Museum* 2(1), 15–19.

Nobis, G. 1993. 'Zur antiken Wild- und Haustierfauna Kretas – nach Studien an Tierresten aus den archäologischen Grabungen Poros bei Iraklion und Eléftherna bei Arkadhi', *Tier und Museum* 3(4), 109–20.

Nobis, G. 1996. 'Der Aurochse oder Ur (Bos primigenius) auf Kreta', in Reese, D. S. (ed.), *Pleistocene and Holocene Fauna of Crete and Its First Settlers* (Monographs in World Archaeology 28). Madison, WI: Prehistory Press, 263–72.

Noble, W. 1981. 'Gibsonian theory and the pragmatist perspective', *Journal for the Theory of Social Behaviour* 11(1), 65–85.

Nowicki, K. 1994. 'Some remarks on the Pre- and Protopalatial peak sanctuaries in Crete', *Aegean Archaeology* 1, 31–48.

Olivier, J.-P. 1967a. 'La série Dn de Cnossos', *Studi Micenei ed Egeo-Anatolici* 2, 71–93.

Olivier, J.-P. 1967b. *Les scribes de Cnossos. Essai de classement des archives d'un palais mycénien* (Incunabula Graeca 27). Rome: Edizioni dell'Ateneo.

Olivier, J.-P. 1981. 'Les sceaux avec des signes hieroglyphiques. Que lire? Une question de definition', in Niemeier, W.-D. (ed.), *Studien zur minoischen und helladischen Glyptik* (CMS Beiheft 1). Berlin: Gebr. Mann Verlag, 105–15.

Olivier, J.-P. 1988. 'KN: Da-Dg', in Olivier, J.-P. and Palaima, T. (eds.), *Texts, Tablets and Scribes: Studies in Mycenaean Epigraphy and Economy Offered to Emmett Bennett, Jr.* (Suplementos a Minos 10). Salamanca: Ediciones Universidad de Salamanca, 219–67.

Olivier, J.-P. and Godart, L. 1996. *Corpus Hieroglyphicarum Inscriptionum Cretae* (Études crétoises 31). Paris: École Française d'Athènes.

Olsen, B. 2010. *In Defense of Things: Archaeology and the Ontology of Objects*. Lanham, MD: Altamira.

Oma, K. A. 2010. 'Between trust and domination: social contracts between humans and animals', *World Archaeology* 42(2), 175–87.

Oma, K. A. and Hedeager, L. 2010. 'Introduction: humans and animals', *World Archaeology* 42(2), 155–6.

Onassoglou, A. 1985. *Die ›talismanischen‹ Siegel* (CMS Beiheft 2). Berlin: Gebr. Mann Verlag.

Onassoglou, A. 1995. '"Data Morgana" sur les sceaux 'Talismaniques"', in Müller, W. (ed.), *Sceaux minoens et mycéniens. IVe symposium international, 10–12 septembre 1992, Clermont-Ferrand* (CMS Beiheft 5). Berlin: Gebr. Mann Verlag, 183–7.

O'Regan, H., Turner, A. and Sabin, R. 2006. 'Medieval big cat remains from the Royal Menagerie at the Tower of London', *International Journal of Osteoarchaeology* 16(5), 385–94.

Orton, D. 2010. 'Both subject and object: herding, inalienability and sentient property in prehistory', *World Archaeology* 42(2), 188–200.

Osborn, D. and Osbornová, J. 1998. *The Mammals of Ancient Egypt*. Warminster: Aris & Phillips.

Ottoni, C., Van Neer, W., De Cupere, B. et al. 2017. 'The palaeogenetics of cat dispersal in the ancient world', *Nature Ecology & Evolution* 1(7), 0139.

Oulié, M. 1926. *Les animaux dans la peinture de la Crète Préhellénique*. Paris: Félix Alcan.

Palaima, T. 1989. 'Perspectives on the Pylos Oxen tablets: textual (and archaeological) evidence for the use and management of oxen in Late Bronze Age Messenia (and Crete)', in Palaima, T., Shelmerdine, C. W. and Ilievski, P. H. (eds.), *Studia Mycenaea 1988* (Ziva Antika Monograph 7). Skopje: Univerzitet vo Skopje, 85–124.

Palaima, T. 1992. 'The Knossos oxen dossier: the use of oxen in Mycenaean Crete. Part I: general background and Scribe 107', in Olivier, J.-P. (ed.), *Mykenaïka. Actes du IXe Colloque international sur les textes mycéniens et égéens* (BCH Supplément 25). Athens: École française d'Athènes, 463–74.

Palaima, T. 2004. 'Sacrificial feasting in the Linear B documents', *Hesperia* 73(2), 217–46.

Palaima, T. and Blackwell, N. 2020. 'Pylos Ta 716 and Mycenaean ritual paraphernalia: a reconsideration', *Studi Micenei ed Egeo-Anatolici NS* 6, 67–96.

Palmer, L. R. and Boardman, J. 1963. *On the Knossos Tablets*. Oxford: Oxford University Press.

Palmer, R. 1995. 'Linear A commodities: a comparison of resources', in Laffineur, R. and Niemeier, W.-D. (eds.), *POLITEIA. Society and State in the Aegean Bronze Age. Proceedings of the 5th International Aegean Conference, University of Heidelberg, Archäologisches Institut, 10–13 April, 1994* (Aegaeum 12). Liège and Austin: Université de Liège and University of Texas at Austin, 133–55.

Palyvou, C. 1987. 'Circulatory patterns in Minoan architecture', in Hägg, R. and Marinatos, N. (eds.), *The Function of the Minoan Palaces. Proceedings of the Fourth International Symposium at the Swedish Institute in Athens, 10–16 June 1984* (Skrifter utgivna av Svenska institutet i Athen, 40, 35). Stockholm: Svenska institutet i Athen, 195–203.

Panagiotaki, M. 1999. *The Central Palace Sanctuary at Knossos* (BSA Supplementary Volume 31). London: British School at Athens.

Panagiotakopulu, E. 2000. *Archaeology and Entomology in the Eastern Mediterranean* (BAR IS 836). Oxford: Archaeopress.

Panagiotopoulos, D. 2018. '14. Κουμάσα Κρήτης', *To Ergon tes En Athenais Archaiologikes Hetaireias* 64, 49–52.

Papageorgiou, I. 2014. 'The practice of bird hunting in the Aegean of the second millennium BC: an investigation', *Annual of the British School at Athens* 109, 111–28.

Pareja, M. N. 2017. *Monkey and Ape Iconography in Aegean Art* (SIMA Pocket Book 186). Uppsala: Astrom Editions.

Pareja, M. N., McKinney, T., Mayhew, J. A. et al. 2020. 'A new identification of the monkeys depicted in a Bronze Age wall painting from Akrotiri, Thera', *Primates* 61(2), 159–68.

Paribeni, R. 1903. 'Lavori eseguiti dalla Missione Archeologica Italiana nel Palazzo e nella Necropoli di Haghia Triada dal 23 Febbraio al 15 Luglio 1903', *Rendiconti della Reale Accademia dei Lincei, Classe di scienze morali, storiche e filologiche* 12, 317–51.

Parmentier, R. 1997. 'The pragmatic semiotics of cultures', *Semiotica* 116(1), 1–115.

Paschalidis, C. 2009. *The LMIII Cemetery at Tourloti, Siteia. The 'Xanthoudidis Master' and the Octopus Style in East Crete* (BAR IS 1917). Oxford: Archaeopress.

Pashley, R. 1837a. *Travels in Crete, I.* London: John Murray.

Pashley, R. 1837b. *Travels in Crete, II.* London: John Murray.

Peatfield, A. 1987. 'Palace and peak: the political and religious relationship between palaces and peak sanctuaries', in Hägg, R. and Marinatos, N. (eds.), *The Function of the Minoan Palaces. Proceedings of the Fourth International Symposium at the Swedish Institute in Athens, 10–16 June 1984* (Skrifter utgivna av Svenska institutet i Athen, 40, 35). Stockholm: Svenska institutet i Athen, 89–93.

Peatfield, A. 1992. 'Rural ritual in Bronze Age Crete: the peak sanctuary at Atsipadhes', *Cambridge Archaeological Journal* 2(1), 59–87.

Peatfield, A. 2000. 'Minoan religion for ordinary people', in *Proceedings of the 8th International Congress of Cretan Studies, A3*. Heraklion: Society of Cretan Historical Studies, 9–17.

Peirce, C. S. 1998. 'Nomenclature and divisions of triadic relations, as far as they are determined [MS 540]', in The Peirce Edition Project (ed.), *The Essential Peirce: Selected Philosophical Writings. Volume 2 (1893–1913).* Bloomington: Indiana University Press, 289–99.

Pendlebury, J. D. S. 1930. *Aegyptiaca: A Catalogue of Egyptian Objects in the Aegean Area.* Cambridge: Cambridge University Press.

Pernier, L. and Banti, L. 1935. *Il palazzo minoico di Festòs: scavi e studi della Missione archeologica italiana a Creta dal 1900 al 1934, I.* Rome: Libreria Dello Stato.

Pernier, L. and Banti, L. 1951. *Il palazzo minoico di Festòs: scavi e studi della Missione archeologica italiana a Creta dal 1900 a 1950, II*. Rome: Libreria Dello Stato.

Phillips, A. A. and Willcock, M. M. 1999. *Xenophon and Arrian on Hunting*. Warminster: Aris & Phillips.

Phillips, J. 1995. 'False analogies: the Minoan origin of some so-called "Egyptianizing" features', in *Proceedings of the 7th International Congress of Cretan Studies, A2*. Rethymnon: Municipality of Rethymnon, 757–65.

Phillips, J. 2004. 'The odd man out: Minoan scarabs and scaraboids', in Bietak, M. and Czerny, E. (eds.), *Scarabs of the Second Millennium BC from Egypt, Nubia, Crete and the Levant: Chronological and Historical Implications. Papers of a Symposium, Vienna, 10–13 January 2002*. Vienna: Verlag der Österreichischen Akademie der Wissenschaften.

Phillips, J. 2008. *Aegyptiaca on the Island of Crete in their Chronological Context: A Critical Review, I*. Vienna: Verlag der Österreichischen Akademie der Wissenschaften.

Philo, C. and Wilbert, C. 2000. 'Animal spaces, beastly places: an introduction', in Philo, C. and Wilbert, C. (eds.), *Animal Spaces, Beastly Places: New Geographies of Human-Animal Relations*. London and New York: Routledge, 1–34.

Philo, C. and Wolch, J. 1998. 'Through the geographical looking glass: space, place and society-animal relations', *Society and Animals* 6(2), 103–18.

Pilali-Papasteriou, A. 1985. *Die bronzenen Tierfiguren aus Kreta* (Prähistorische Bronzefunde 1,3). Munich: Beck'sche.

Pini, I. 1985. 'Das Motiv des Löwenüberfalls in der spätminoischen und mykenischen Glyptyk', in Darcque, P. and Poursat, J.-C. (eds.), *L'iconographie minoenne: Actes de la Table Ronde d'Athènes (21–22 avril 1983)* (BCH Supplément 11). Athens: École française d'Athènes, 153–66.

Pini, I. 1990. 'Eine Frühkretische Siegelwerkstatt?', in *Proceedings of the 6th International Congress of Cretan Studies, A2*. Chania: The Philological Association 'Chrysostomos', 115–27.

Pini, I. 1992. 'Towards a standardization of terminology: problems of description and identification', in Laffineur, R. and Crowley, J. (eds.), *EIKON. Aegean Bronze Age Iconography: Shaping a Methodology. Proceedings of the 4th International Aegean Conference, University of Tasmania, Hobart, Australia, 6–9 April 1992* (Aegaeum 8). Liège: Université de Liège, 11–19.

Pini, I. 1995. 'Bemurkungen zur Datierung von Löwendarstellungen der spätminoischen Weichsteinglyptik', in Müller, W. (ed.), *Sceaux minoens et mycéniens. IVe symposium international, 10–12 septembre 1992, Clermont-Ferrand* (CMS Beiheft 5). Berlin: Gebr. Mann Verlag, 193–207.

Pini, I. 2000a. 'Der Cut Style in der spätbronzezeitlichen ägäischen Glyptik', in Mattern, T. (ed.), *Munus: Festschrift für Hans Wiegartz*. Münster: Scriptorium, 209–20.

Pini, I. 2000b. 'Eleven early Cretan scarabs', in Karetsou, A. (ed.), *Κρήτη – Αίγυπτος: πολιτιστικοί δεσμοί τριών χιλιετιών*. Athens: Ministry of Culture, 107–13.

Pini, I. 2010. 'Soft stone versus hard stone seals in Aegean glyptic: some observations on style and iconography', in Müller, W. (ed.), *Die Bedeutung der minoischen und mykenischen Glyptik. VI. Internationales Siegel-Symposium, Marburg, 9–12 Oktober 2008* (CMS Beiheft 8). Mainz: von Zabern, 325–39.

Platon, L. 2012. 'Kato Zakros', in Cline, E. H. (ed.), *The Oxford Handbook of the Bronze Age Aegean*. Oxford: Oxford University Press, 509–17.

Platon, N. 1956. 'La chronologie minoenne', in Zervos, C. (ed.), *L'art de la Crète néolithique et minoenne*. Paris: Éditions Cahiers d'art, 509–12.

Platon, N. 1971. *Zakros: The Discovery of a Lost Palace of Ancient Crete*. New York: Scribner.

Platt, V. 2006. 'Making an impression: replication and the ontology of the Graeco-Roman sealstone', *Art History* 29(2), 233–57.

Poliquin, R. 2012. *The Breathless Zoo: Taxidermy and the Cultures of Longing*. University Park, PA: Penn State University Press.

Pollard, J. 1977. *Birds in Greek Life and Myth*. London: Thames and Hudson.

Popham, M. R. 1984. *The Minoan Unexplored Mansion at Knossos* (BSA Supplementary Volume 17). London: BSA and Thames and Hudson.

Poursat, J. 1977. *Les ivoires mycéniens: Essai sur la formation d'un art mycénien* (Bibliothèque des écoles françaises d'Athènes et de Rome 230). Athens: École française d'Athènes.

Poursat, J.-C. 2008. *L'art égéen. Vol. 1. Grèce, Cyclades, Crète jusqu'au milieu du II millénaire av. J.-C.* Paris: Picard.

Poursat, J.-C. 2014. *L'art égéen. Vol. 2. Mycènes et le monde mycénien*. Paris: Picard.

Poursat, J.-C. and Knappett, C. 2005. *La poterie du Minoen Moyen II: production et utilisation* (Études crétoises 33). Athens: École française d'Athènes.

Powell, J. 1992. 'Archaeological and pictorial evidence for fishing in the Bronze Age: issues of identification and interpretation', in Laffineur, R. and Crowley, J. (eds.), *EIKON. Aegean Bronze Age Iconography: Shaping a Methodology. Proceedings of the 4th International Aegean Conference, University of Tasmania, Hobart, Australia, 6–9 April 1992* (Aegaeum 8). Liège: Université de Liège, 23–37.

Powell, J. 1996. *Fishing in the Prehistoric Aegean* (SIMA Pocket Book 137). Jonsered: Paul Åströms Förlag.

Prent, M. 2005. *Cretan Sanctuaries and Cults: Continuity and Change from Late Minoan IIIC to the Archaic Period* (Religions in the Graeco-Roman World 154). Leiden and Boston, MA: Brill.

Preston Day, L. 1984. 'Dog burials in the Greek world', *American Journal of Archaeology* 88(1), 21–32.

Preston Day, L. , Klein, N. L. and Turner, L. A. 2009. *Kavousi IIA. The Late Minoan IIIC Settlement at Vronda: The Buildings on the Summit* (Prehistory Monographs 26). Philadelphia, PA: INSTAP Academic Press.

Preston, L. 2004. 'Final Palatial Knossos and Postpalatial Crete: a mortuary perspective on political dynamics', in Cadogan, G., Hatzaki, E. and Vasilakis, A. (eds.), *Knossos: Palace, City, State* (BSA Studies 12). London: British School at Athens, 137–45.

Preston, L. 2008. 'Late Minoan II to IIIB Crete', in Shelmerdine, C. (ed.), *The Cambridge Companion to the Aegean Bronze Age*. Cambridge: Cambridge University Press, 310–26.

Preston, L. 2013. 'The Middle Minoan III funerary landscape at Knossos', in MacDonald, C. F. and Knappett, C. (eds.), *INTERMEZZO: Intermediacy and Regeneration in Middle Minoan III Palatial Crete* (BSA Studies 21). London: British School at Athens, 57–70.

Preston, L., Evely, D., Gowland, R., Isaakidou, V. and Krzyszkowska, O. 2013. *Knossos Monastiriako Kephali Tomb and 'Deposit'* (BSA Studies 22). London: British School at Athens.

Preucel, R. 2006. *Archaeological Semiotics*. Oxford: Blackwell.

Pulak, C. 1998. 'The Uluburun shipwreck: an overview', *The International Journal of Nautical Archaeology* 27(3), 188–224.

Rackham, O. and Moody, J. 1996. *The Making of the Cretan Landscape*. Manchester: Manchester University Press.

Randi, E., Meriggi, A., Lorenzini, R., Fusco, G. and Alkon, P. U. 1992. 'Biochemical analysis of relationships of Mediterranean *Alectoris* partridges', *The Auk* 109(2), 358–67.

Randi, E., Tabarroni, C., Rimondi, S., Lucchini, V. and Sfougaris, A. 2003. 'Phylogeography of the rock partridge (*Alectoris graeca*)', *Molecular Ecology* 12(8), 2201–14.

Rathjen, W. F. 1991. 'Cephalopod capture methods: an overview', *Bulletin of Marine Science* 49, 494–505.

Raulin, V. 1869. *Description physique de l'ile de Crète, I*. Paris: A. Bertrand.

Recht, L. 2011. *Sacrifice in the Bronze Age Aegean and Near East. A Poststructuralist Approach*. Unpublished PhD thesis, Trinity College Dublin.

Recht, L. and Morris, C. E. 2021. 'Chariot kraters and horse–human relations in Late Bronze Age Greece and Cyprus', *Annual of the British School at Athens* 116, 95–132.

Reed, E. 1988a. *James J. Gibson and the Psychology of Perception*. New Haven, CT and London: Yale University Press.

Reed, E. 1988b. 'The affordances of the animate environment: social science from an ecological point of view', in Ingold, T. (ed.), *What Is an Animal?* London: Unwin Hyman, 110–26.

Reese, D. S. 1984. 'Topshell rings in the Aegean Bronze Age', *Annual of the British School at Athens* 79, 237–8.

Reese, D. S. 1985. 'Appendix VIII: Shells, ostrich eggshells and other exotic faunal remains from Kition', in Karageorghis, V. (ed.), *Excavations at Kition V(II)*. Nicosia: The Department of Antiquities, Cyprus, 340–415.

Reese, D. S. 1987. 'The EM IIA shells from Knossos, with comments on Neolithic to EM III shell utilization', *Annual of the British School at Athens* 82, 207–11.

Reese, D. S. 1991. 'The trade of Indo-Pacific shells into the Mediterranean Basin and Europe', *Oxford Journal of Archaeology* 10(2), 159–96.

Reese, D. S. 1992. 'Fauna from the Poros tomb', in Muhly, P. (ed.), *Μινωικός λαξευτός τάφος στον Πόρο Ηρακλείου (Άνασκαφης 1967)*. Athens: Archaeological Society at Athens, 180–1.

Reese, D. S. 1994. 'Recent work in Greek zooarchaeology', in Kardulias, P. N. (ed.), *Beyond the Site: Regional Studies in the Aegean Area*. Lanham, MD: University Press of America, 191–221.

Reese, D. S. 1995a. 'The bird remains', in Shaw, J. W. and Shaw, M. C. (eds.), *Kommos I, 1: The Kommos Region, Ecology, and Minoan Industries*. Princeton, NJ: Princeton University Press, 194–204.

Reese, D. S. 1995b. 'The larger mammals', in Shaw, J. W. and Shaw, M. C. (eds.), *Kommos I, 1: The Kommos Region, Ecology, and Minoan Industries*. Princeton, NJ: Princeton University Press, 165–94.

Reese, D. S. 1995c. 'The marine invertebrates', in Shaw, J. W. and Shaw, M. C. (eds.), *Kommos I, 1: The Kommos Region, Ecology, and Minoan Industries*. Princeton, NJ: Princeton University Press, 240–73.

Reese, D. S. 2000. 'The Iron Age fauna', in Shaw, J. W. and Shaw, M. C. (eds.), *Kommos IV: The Greek Sanctuary, Part I*. Princeton, NJ: Princeton University Press, 415–64.

Reese, D. S. 2004. 'The fauna', in Soles, J. and Davaras, C. (eds.), *Mochlos IC: Period III. Neopalatial Settlement on the Coast: The Artisans' Quarter and the Farmhouse at Chalinomouri. The Small Finds* (Prehistory Monographs 9). Philadelphia, PA: INSTAP Academic Press, 118–21.

Reese, D. S. 2009. 'Faunal remains from Block AF', in Betancourt, P. P. and Davaras, C. (eds.), *Pseira X: The Excavation of Block AF* (Prehistory Monographs 28). Philadelphia, PA: INSTAP Academic Press, 131–42.

Reese, D. S. 2016a. 'The Invertebrates', in Hallager, E. and Hallager, B. (eds.), *The Greek-Swedish Excavations at the Agia Aikaterini Square Kastelli, Khania 1970–1987, 2001, 2005 and 2008, V,1* (Skrifter utgivna av Svenska institutet i Athen, 40, 47). Stockholm: Paul Åströms Förlag, 396–417.

Reese, D. S. 2016b. 'Wild Cattle at Chania', in Hallager, E. and Hallager, B. (eds.), *The Greek-Swedish Excavations at the Agia Aikaterini Square Kastelli, Khania 1970–1987, 2001, 2005 and 2008, V,1* (Skrifter utgivna av Svenska institutet i Athen, 40, 47). Stockholm: Paul Åströms Förlag, 393–5.

Rehak, P. 1995a. 'The "Genius" in Late Bronze Age glyptic: the later evolution of an Aegean cult figure', in Müller, W. (ed.), *Sceaux minoens et mycéniens. IVe symposium international, 10–12 septembre 1992, Clermont-Ferrand* (CMS Beiheft 5). Berlin: Gebr. Mann Verlag, 215–31.

Rehak, P. (ed.) 1995b. *The Role of the Ruler in the Prehistoric Aegean*. Liège: Université de Liège.

Rehak, P. 1995c. 'The use and destruction of Minoan stone bull's head rhyta', in Laffineur, R. and Niemeier, W.-D. (eds.), *POLITEIA. Society and State in the Aegean Bronze Age. Proceedings of the 5th International Aegean Conference, University of Heidelberg, Archäologisches Institut, 10–13 April 1994* (Aegaeum 12). Liège and Austin: Université de Liège and University of Texas at Austin, 435–60.

Rehak, P. 1998. 'The Mycenaean warrior goddess revisited', in Laffineur, R. (ed.), *POLEMOS. Le contexte guerrier en Égée a l'Âge du Bronze. Actes de la 7è Rencontre égéenne international (Liège, 14–17 avril 1998)* (Aegaeum 19). Liège and Austin: Université de Liège and University of Texas at Austin, 227–39.

Rehak, P. and Younger, J. G. 1998. 'Review of Aegean prehistory VII: Neopalatial, Final Palatial, and Postpalatial Crete', *American Journal of Archaeology* 102(1), 91–173.

Relaki, M. 2012. 'The social arenas of tradition: investigating corporate and individual social strategies in Prepalatial and Protopalatial Mesara', in Schoep, I., Tomkins, P. and Driessen, J. (eds.), *Back to the Beginning: Reassessing Social and Political Complexity on Crete during the Early and Middle Bronze Age*. Oxford: Oxbow, 290–324.

Relaki, M. and Driessen, J. (eds.) 2020. *OIKOS: Archaeological Approaches to House Societies in Aegean Prehistory*. Louvain-la-Neuve: Presses universitaires de Louvain.

Renfrew, C. 1972. *The Emergence of Civilisation*. London: Methuen.

Rethemiotakis, G. 2014. 'Images and semiotics in space: the case of the anthropomorphic figurines from Kophinas', *Cretica Chronica* 34, 147–62.

Rethemiotakis, G. 2020. 'Tylissians and Knossians: an interactive relationship new evidence on the miniature wall paintings from Tylissos', in Blakolmer, F. (ed.), *Current Approaches and New Perspectives in Aegean Iconography*. Louvain-la-Neuve: Presses universitaires de Louvain, 117–45.

Reusch, H. 1958. 'Zum Wandschmuck des Thronsaales in Knossos', in Grumach, E. (ed.), *Minoica, Festschrift zum 80. Geburtstag von Johannes Sundwall*. Berlin: Akadamie-Verlag, 334–58.

Rhyne, N. A. 1970. *The Aegean Animal Style: A Study of the Lion, Griffin, and Sphinx*. Unpublished PhD thesis, University of North Carolina.

Rice, M. 2006. *Swifter than the Arrow: The Golden Hunting Hounds of Ancient Egypt*. London and New York: I.B.Tauris.

Riegner, M. 1993. 'Towards a holistic understanding of place: reading a landscape through its flora and fauna', in Seamon, D. (ed.), *Dwelling, Seeing, and Designing: Toward a Phenomenological Ecology*. Albany: State University of New York Press, 181–215.

Ritvo, H. 1987. *The Animal Estate: The English and Other Creatures in the Victorian Age*. Cambridge, MA: Harvard University Press.

Robertson Smith, W. 1889. *Lectures on the Religion of the Semites*. New York: D. Appleton and Co.

Rose, M. J. 1994. *With Line and Glittering Bronze Hook: Fishing in the Aegean Bronze Age*. Unpublished PhD thesis, Indiana University.

Rose, M. J. 1995. 'The fish remains', in Shaw, J. W. and Shaw, M. C. (eds.), *Kommos I, 1: The Kommos Region, Ecology, and Minoan Industries*. Princeton, NJ: Princeton University Press, 204–39.

Rose, M. J. 2009. 'Fish remains from Block AF', in Betancourt, P. P. (ed.), *Pseira X: The Excavation of Block AF* (Prehistory Monographs 28). Philadelphia, PA: INSTAP Academic Press, 143–50.

Roussaki, M. 2014. 'New evidence in Minoan pictorial wall painting: the Swallows Fresco from the Knossos area', in Touchais, G., Laffineur, R. and Rougemont, F. (eds.), *PHYSIS. L'environnement naturel et la relation homme-milieu dans le monde égéen protohistorique. Actes de la 14e Rencontre égéenne internationale, Paris, Institut National d'Histoire de l'Art (INHA), 11–14 décembre 2012* (Aegaeum 37). Leuven: Peeters, 539–42.

Rumpel, D. 2007. 'The "Harvester" Vase Revised', *Anistoriton Journal: In Situ* 10(3), online.

Ruscillo, D. 2005. 'Reconstructing *Murex* royal purple and biblical blue in the Aegean', in Bar-Yosef, D. (ed.), *Archaeomalacology: Molluscs in Former Environments of Human Behaviour*. Oxford: Oxbow, 96–106.

Ruscillo, D. 2006. 'Faunal remains and *Murex* dye production', in Shaw, J. W. and

Shaw, M. C. (eds.), *Kommos V: The Monumental Minoan Buildings at Kommos*. Princeton, NJ: Princeton University Press, 776–844.

Ruskin, J. 1853. *The Stones of Venice, II*. London: Smith, Elder and Co.

Russell, N. 2002. 'The wild side of animal domestication', *Society and Animals* 10(3), 285–302.

Russell, N. 2012. *Social Zooarchaeology: Humans and Animals in Prehistory*. Cambridge: Cambridge University Press.

Rutkowski, B. 1986. *The Cult Places of the Aegean*. New Haven, CT and London: Yale University Press.

Rutkowski, B. 1991. *Petsophas: A Minoan Peak Sanctuary*. Warsaw: Art and Archaeology.

Ruuskanen, J.-P. 1992. *Birds on Aegean Bronze Age Seals* (Studia Archaeologica Septentrionalia 2). Rovaniemi: Pohjois-Suomen Historiallinen Ydistys.

Sakellarakis, Y. 1970. 'Das Kuppelgrab A von Archanes und das kretisch-mykenische Tieropferritual', *Praehistorische Zeitschrift* 45 (2), 135–219.

Sakellarakis, Y. 1990. 'The fashioning of ostrich-egg shell rhya in the Creto-Mycenaean Aegean', in Hardy, D. A., Doumas, C., Sakellarakis, Y. and Warren, P. (eds.), *Thera and the Aegean World III, I*. Athens: Thera Foundation, 285–308.

Sakellarakis, Y. and Sapouna-Sakellaraki, E. 1997. *Archanes. Minoan Crete in a New Light*. Athens: Ammos Publications.

Sanavia, A. 2014. 'How to improve on nature: some Middle Minoan triton shells from Phaistos (Crete)', in Touchais, G., Laffineur, R. and Rougemont, F. (eds.), *PHYSIS. L'environnement naturel et la relation homme-milieu dans le monde égéen protohistorique. Actes de la 14e Rencontre égéenne internationale, Paris, Institut National d'Histoire de l'Art (INHA), 11–14 décembre 2012* (Aegaeum 37). Leuven: Peeters, 543–6.

Sanavia, A. 2019. 'Painted parading lions on an MM IB ceremonial basin: a case of symbolic transference and remembrance of an emblem in early Protopalatial Phaistos', in Borgna, E., Caloi, I., Carinci, F. M. and Laffineur, R. (eds.), *MNHMH/MNEME: Past and Memory in the Aegean Bronze Age. Proceedings of the 17th International Aegean Conference, University of Udine, Department of Humanities and Cultural Heritage, Ca' Foscari University of Venice, Department of Humanities, 17–21 April 2018* (Aegaeum 43). Leuven: Peeters, 743–8.

Savignoni, L. 1903. 'Il vaso di Haghia Triada', *Monumenti Antichi* 13, 77–131.

Sbonias, K. 1995. *Frühkretische Siegel: Ansätze für eine Interpretation der sozial-politischen Entwicklung auf Kreta während der Frühbronzezeit* (BAR IS 620). Oxford: Tempus Reparatum.

Schiering, W. 1998. *Minoische Töpferkunst: Die bemalten Tongefäße der Insel des Minos* (Kulturgeschichte der antiken Welt 73). Mainz: von Zabern.

Schinz, H. R. 1838. 'Bemerkungen über die Arten der wilden Ziegen', *Neue Denkschriften der Allgemeinen Schweizerischen Gesellschaft für die Gesammten Naturwissenschaften* 2, 1–25.

Schliemann, H. 1878. *Mycenae: A Narrative of Researches and Discoveries at Mycenae and Tiryns*. London: John Murray.

Schliemann, H. 1880. *Ilios: The City and Country of the Trojans*. London: John Murray.

Schoep, I. 1994. 'Ritual, politics and script on Minoan Crete', *Aegean Archaeology* 1, 7–25.

Schoep, I. 1999. 'The origins of writing and administration on Crete', *Oxford Journal of Archaeology* 18(3), 265–76.

Schoep, I. 2001. 'Some notes on the "Hieroglyphic" deposit from Knossos', *Studi Micenei ed Egeo-Anatolici* 43, 143–58.

Schoep, I. 2002a. *The Administration of Neopalatial Crete: A Critical Assessment of the Linear A Tablets and their Role in the Administrative Process* (Supplementos a Minos 17). Salamanca: Ediciones Universidad de Salamanca.

Schoep, I. 2002b. 'The state of the Minoan palaces or the Minoan palace-state?', in Driessen, J., Schoep, I. and Laffineur, R. (eds.), *Monuments of Minos: Rethinking the Minoan Palaces* (Aegaeum 23). Liège and Austin: Université de Liège and University of Texas at Austin, 15–33.

Schoep, I. and Knappett, C. 2004. 'Dual emergence: evolving heterarchy, exploding hierarchy', in Barrett, J. C. and Halstead, P. (eds.), *The Emergence of Civilisation Revisited*. Oxford: Oxbow, 21–37.

Schofield, E. 1996. 'Wash and brush up at the "travellers' rest": the Caravanserai reconsidered', in Evely, D., Lemos, I. and Sherratt, S. (eds.), *Minotaur and Centaur: Studies in the Archaeology of Crete and Euboea Presented to Mervyn Popham* (BAR IS 638). Oxford: Archaeopress, 27–33.

Seetah, K. 2019. *Humans, Animals and the Craft of Slaughter in Archaeo-Historic Societies*. Cambridge: Cambridge University Press.

Serpell, J. 1986. *In the Company of Animals: A Study of Human-Animal Relationships*. Oxford: Basil Blackwell.

Serres, M. [1980] 2007. *The Parasite*. Minneapolis: University of Minnesota Press.

Shackleton, D. M. 1997. *Wild Sheep and Goats and their Relatives: Status Survey and Conservation Action Plan for Caprinae*. Cambridge: International Union for Conservation of Nature.

Shackleton, N. J. 1968. 'Knossos marine mollusca (Neolithic)', *Annual of the British School at Athens* 63, 264–6.

Shackleton, N. J. 1972. 'Appendix VII: the shells', in Warren, P. (ed.), *Myrtos: An Early Bronze Age Settlement in Crete* (BSA Supplementary Volume 7). London: British School at Athens and Thames and Hudson, 321–5.

Shapland, A. J. 2009. *Over the Horizon: Human-Animal Relations in Bronze Age Crete*. Unpublished PhD thesis, UCL Institute of Archaeology.

Shapland, A. J. 2010a. 'The Minoan lion: presence and absence on Bronze Age Crete', *World Archaeology* 42(2), 273–89.

Shapland, A. J. 2010b. 'Wild nature? human-animal relations in Neopalatial Crete', *Cambridge Archaeological Journal* 20(1), 109–27.

Shapland, A. J. 2012. 'Jumping to conclusions: bull-leaping in Minoan Crete', *Society and Animals* 21(2), 194–207.

Shapland, A. J. 2013. 'Shifting horizons and emerging ontologies in the Bronze Age Aegean', in Watts, C. M. (ed.), *Relational Archaeologies: Humans, Animals, Things*. London: Routledge, 190–208.

Shapland, A. J. 2014. 'After naturalism: human-animal relations in LMII-III Crete', in Touchais, G., Laffineur, R. and Rougemont, F. (eds.), *PHYSIS. L'environnement naturel et la relation homme-milieu dans le monde égéen protohistorique. Actes de la 14e Rencontre égéenne internationale, Paris, Institut National d'Histoire de l'Art (INHA), 11–14 décembre 2012* (Aegaeum 37). Leuven: Peeters, 555–8.

Shapland, A. J. 2021. 'Sacrificial Relics or Trophies? Animal Heads in Bronze Age Crete', in Laffineur, R. and Palaima, T. (eds.), *ZOIA: Animal Connections in the Aegean Middle and Late Bronze Age*. Leuven: Peeters, 307–16.

Shapland, A. J. forthcoming. 'Marine styles in Bronze Age Aegean material culture', in Theodoropoulou, T. and Gallant, T. (eds.), *Harvesting the Gifts of the Sea: Aegean Societies and Marine Life through Time*. New York: Springer.

Shaw, J. W. 1978a. 'Evidence for the Minoan tripartite shrine', *American Journal of Archaeology* 82(4), 429–48.

Shaw, J. W. 1978b. 'Sliding panels at Knossos', *Annual of the British School at Athens* 73, 235–48.

Shaw, J. W. 2000. 'Ritual and development in the Greek sanctuary', in Shaw, J. W. and Shaw, M. C. (eds.), *Kommos IV: The Greek Sanctuary, Part I*. Princeton, NJ: Princeton University Press, 669–731.

Shaw, M. C. 1993. 'The Aegean garden', *American Journal of Archaeology* 97(4), 661–85.

Shaw, M. C. 1995. 'Bull leaping frescoes at Knossos and their influence on the Tell el Dab^{c}a murals', *Ägypten und Levante* 5, 91–120.

Shaw, M. C. 1996. 'Terra-cotta sculpture', in Shaw, J. W. and Shaw, M. C. (eds.), *Kommos I,2. The Houses of the Minoan Town*. Princeton, NJ: Princeton University Press, 282–302.

Shaw, M. C. 2004. 'The "Priest-King" fresco from Knossos: man, woman, priest, king, or someone else?', in Chapin, A. P. (ed.), *ΧΑΡΙΣ: Essays in Honor of Sara A. Immerwahr* (Hesperia Supplement 33). Athens: American School of Classical Studies at Athens, 65–84.

Shaw, M. C. 2005. 'The painted pavilion of the "Caravanserai" at Knossos', in Morgan, L. (ed.), *Aegean Wall Painting: A Tribute to Mark Cameron* (BSA Studies 13). London: British School at Athens, 91–111.

Shaw, M. C. 2012. 'New light on the Labyrinth Fresco from the palace at Knossos', *Annual of the British School at Athens* 107, 143–59.

Sherratt, A. 1993. 'What would a Bronze-Age world system look like? Relations between temperate Europe and the Mediterranean in later prehistory', *Journal of European Archaeology* 1(2), 1–58.

Sherratt, A. and Sherratt, S. 1991. 'From luxuries to commodities: the nature of Mediterranean Bronze Age trading systems', in Gale, N. H. (ed.), *Bronze Age Trade in the Mediterranean* (SIMA 90). Jonsered: Paul Åströms Förlag, 351–86.

Sherratt, S. 2000. *Arthur Evans, Knossos and the Priest-King*. Oxford: Ashmolean Museum.

Shindler, K. 2005. *Discovering Dorothea: The Life of the Pioneering Fossil-Hunter Dorothea Bate*. London: HarperCollins.

Simandiraki, A. 2006. 'The agrimi in the Minoan ceramic relief', in *Proceedings of the 9th International Congress of Cretan Studies, A3*. Heraklion: Society of Cretan Historical Studies, 95–108.

Simandiraki-Grimshaw, A. 2010. 'Minoan animal-human hybridity', in Counts, D. B. and Arnold, B. (eds.), *The Master of Animals in Old World Iconography*. Budapest: Archaeolingua Alapítvány, 93–106.

Simonelli, V. 1897. *Candia – Ricordi di escursione*. Parma: L. Battei

Simonelli, V. 1908. 'Mammiferi quaternari dell'Isola di Candia. Memoria Seconda', *Memorie della Reale Accademia delle Scienze dell'Istituto di Bologna, Classe di Scienze Fisiche* 6(5), 397–406.

Sipahi, T. 2001. 'New evidence from Anatolia regarding bull-leaping scenes in the art of the Aegean and the Near East', *Anatolica* 27, 107–25.

Smets, A. and van den Abeele, B. 2007. 'Medieval hunting', in Resl, B. (ed.), *A Cultural History of Animals in the Medieval Age*. Oxford: Berg.

Smith, B. D. and Zeder, M. A. 2013. 'The onset of the Anthropocene', *Anthropocene* 4, 8–13.

Snyder, L. and Klippel, W. 2000. 'Dark Age subsistence at the Kastro site, East Crete: exploring subsistence change and continuity during the Late Bronze Age-Early Iron Age transition', in Vaughan, S. J. and Coulson, W. D. E. (eds.), *Palaeodiet in the Aegean*. Oxford: Oxbow, 65–83.

Snyder, L. and Klippel, W. 2003. 'From Lerna to Kastro: further thoughts on dogs as food in Ancient Greece: perceptions, prejudices and reinvestigations', in Kotjabopoulou, E., Hamilakis, Y., Halstead, P., Gamble, C. and Elefanti, P. (eds.), *Zooarchaeology in Greece: Recent Advances* (BSA Studies 9). London: British School at Athens, 221–31.

Soar, K. 2009. 'Old bulls, new tricks: the reinvention of a Minoan tradition', in Georgiadis, M. and Gallou, C. (eds.), *The Past in the Past: The Significance of Memory and Tradition in the Transmission of Culture* (IS 1925). Oxford: Archaeopress, 16–27.

Soles, J. 1995. 'The functions of a cosmological centre: Knossos in palatial Crete', in Laffineur, R. and Niemeier, W.-D. (eds.), *POLITEIA. Society and State in the Aegean Bronze Age. Proceedings of the 5th International Aegean Conference, University of Heidelberg, Archäologisches Institut, 10–13 April 1994* (Aegaeum 12). Liège and Austin: Université de Liège and University of Texas at Austin, 405–14.

Soles, J. 2003. *Mochlos IA: Period III. Neopalatial Settlement on the Coast: The Artisans' Quarter and the Farmhouse at Chalinomouri. The Sites.* (Prehistory Monographs 8). Philadelphia, PA: INSTAP Academic Press.

Soles, J. 2008. *Mochlos IIA. Period IV. The Mycenaean Settlement and Cemetery. The Sites.* Philadelphia, PA: INSTAP Academic Press.

Soles, J. 2016. 'Hero, goddess, priestess: new evidence for Minoan religion and social organization', in Alram-Stern, E., Blakolmer, F., Deger-Jalkotzy, S., Laffineur, R. and Weilhartner, J. (eds.), *METAPHYSIS. Ritual, Myth and Symbolism in the Aegean Bronze Age. Proceedings of the 15th International Aegean Conference, Vienna, Institute for Oriental and European Archaeology, Aegean and Anatolia Department, Austrian Academy of Sciences and Institute of Classical Archaeology, University of Vienna, 22–25 April 2014* (Aegaeum 39). Leuven and Liège: Peeters, 247–53.

Sonesson, G. 1989. *Pictorial Concepts: Inquiries into the Semiotic Heritage and its Relevance for the Analysis of the Visual World.* Lund: Lund University Press.

Sonesson, G. 1994. 'Prolegomena to the semiotic analysis of prehistoric visual displays', *Semiotica* 100(2–4), 267–331.

Spanos, I., Platis, P., Meliadis, I. and Tsiontis, A. 2008. 'A review on the ecology and management of the Samaria Gorge, a Greek biosphere reserve', *Journal of Geography and Regional Planning* 1(2), 19–33.

Speciale, M.-S. 1999. 'La tavoletta PY Ta 716 e i sacrifici di animali', in La Rosa, V., Palermo, D. and Vagnetti, L. (eds.), *Epi ponton plazomenoi. Simposio italiano di Studi Egei, Dedicato a Luigi Bernabò Brea e Giovanni Pugliese Carratelli: Roma, 18–20 Febbraio 1998.* Rome: Scuola Archeologia Italiana di Atene.

Spratt, T. A. B. 1865. *Travels and Researches in Crete, I.* London: John van Voorst.

Starr, K. 1984. 'Minoan Flower Lovers', in Hägg, R. and Marinatos, N. (eds.), *The Minoan Thalassocracy: Myth and Reality. Proceedings of the Third International Symposium at the Swedish Institute in Athens, 31 May–5 June 1982* (Skrifter utgivna av Svenska institutet i Athen, 40, 32). Stockholm: Svenska institutet i Athen, 9–12.

Sterry, P. 2004. *Birds of the Mediterranean: A Photographic Guide.* London: Christopher Helm.

Stieglitz, R. R. 1994. 'The Minoan origin of Tyrian purple', *Biblical Archaeologist* 57(1), 46–54.

Stocker, R. S. and Davis, J. L. 2004. 'Animal sacrifice, archives, and feasting at the Palace of Nestor', *Hesperia* 73(2), 179–95.

Strasser, T., Runnels, C., Wegmann, K. et al. 2011. 'Dating Palaeolithic sites in southwestern Crete, Greece', *Journal of Quaternary Science* 26(5), 553–60.

Summers, D. 2003. *Real Spaces: World Art History and the Rise of Western Modernism.* London and New York: Phaidon Press.

Sundwall, J. 1936. 'Altkretische Urkundenstudien', *Acta Academiae Aboensis Humaniora* 10(2), 1–45.

Sykes, N., Carden, R. F. and Harris, K. 2013. 'Changes in the size and shape of fallow deer – evidence for the movement and management of a species', *International Journal of Osteoarchaeology* 23(1), 55–68.

Tanner, J. 2001. 'Nature, culture and the body in Classical Greek religious art', *World Archaeology* 33(2), 257–76.

Tapper, R. L. 1989. 'Animality, humanity, morality, society', in Ingold, T. (ed.), *What is an Animal?* London and New York: Routledge.

Theodoropoulou, T. 2012. 'Neolithic and Minoan marine exploitation at Petras: diachronic trends and cultural shifts', in Tsipopoulou, M. (ed.), *Petras, Siteia – 25 years of Excavations and Studies. Acts of a Two-Day Conference Held at the Danish Institute at Athens, 9–10 October 2010* (Monographs of the Danish Institute at Athens 16). Athens: Danish Institute of Athens, 89–103.

Theodoropoulou, T. 2014a. 'Dead from the sea: sea-worn shells in Aegean prehistory', in Szabó, K., Dupont, C., Dimitrijević, V., Gómez Gastélum, L. and Serrand, N. (eds.), *Archaeomalacology: Shells in the Archaeological Record* (BAR IS 2666). Oxford: Archaeopress, 77–90.

Theodoropoulou, T. 2014b. 'Excavating the sea: recent advances in marine zooarchaeology of the prehistoric Aegean', in Touchais, G., Laffineur, R. and Rougemont, F. (eds.), *PHYSIS. L'environnement naturel et la relation homme-milieu dans le monde égéen protohistorique. Actes de la 14e Rencontre égéenne internationale, Paris, Institut National d'Histoire de l'Art*

(INHA), 11–14 décembre 2012 (Aegaeum 37). Leuven: Peeters, 359–72.

Theodoropoulou, T. 2016. 'Marine faunal remains', in Tsipopoulou, M. (ed.), *Petras, Siteia I. A Minoan Palatial Settlement in Eastern Crete: Excavation of Houses I.1 and I.2* (Prehistory Monographs 53). Philadelphia, PA: INSTAP Academic Press, 225–34.

Thomas, K. 1983. *Man and the Natural World: Changing Attitudes in England 1500–1800*. London: Allen Lane.

Thomas, N. 2004. 'The early Mycenaean lion up to date', in Chapin, A. P. (ed.), *ΧΑΡΙΣ: Essays in Honor of Sara A. Immerwahr* (Hesperia Supplement 33). Athens: American School of Classical Studies at Athens, 161–206.

Thomas, N. 2014. 'A lion's eye view of the Greek Bronze Age', in Touchais, G., Laffineur, R. and Rougemont, F. (eds.), *PHYSIS. L'environnement naturel et la relation homme-milieu dans le monde égéen protohistorique. Actes de la 14e Rencontre égéenne internationale, Paris, Institut National d'Histoire de l'Art (INHA), 11–14 décembre 2012* (Aegaeum 37). Leuven: Peeters.

Todaro, S. and Di Tonto, S. 2008. 'The Neolithic settlement of Phaistos revisited: evidence for ceremonial activity on the eve of the Bronze Age', in Isaakidou, V. and Tomkins, P. (eds.) *Escaping the Labyrinth: The Cretan Neolithic in Context* (Sheffield Studies in Aegean Archaeology 8). Oxford: Oxbow, 177–90.

Tomkins, P. 2007. 'Neolithic: Strata IX–VII, VII–VIB, VIA–V, IV, IIIB, IIIA, IIB, IIA and IC Groups', in Momigliano, N. (ed.), *Knossos Pottery Handbook: Neolithic and Bronze Age* (BSA Studies 14). London: British School at Athens, 9–48.

Trantalidou, K. 2006. 'Companions from the oldest times: dogs in Ancient Greek literature, iconography and osteological testimony', in Snyder, L. and Moore, E. A. (eds.), *Dogs and People in Social, Working, Economic or Symbolic Interaction*. Oxford: Oxbow, 96–120.

Trevor-Battye, A. 1913. *Camping in Crete. With Notes upon the Animal and Plant Life of the Island*. London: Witherby & Co.

Tsipopoulou, M. 2012. 'Introduction: 25 years of excavations and studies at Petras', in Tsipopoulou, M. (ed.), *Petras, Siteia – 25 Years of Excavations and Studies* (Monographs of the Danish Institute at Athens 16). Athens: Danish Institute at Athens, 45–66.

Tsipopoulou, M. and Vagnetti, L. 1997. 'Workshop attributions for some Late Minoan III East Cretan larnakes', in Laffineur, R. and Betancourt, P. P. (eds.), *TEXNH. Craftsmen, Craftswomen and Craftsmanship in the Aegean Bronze Age. Proceedings of the 6th International Aegean Conference, Philadelphia, Temple University, 18–21 April 1996* (Aegaeum 16). Liège and Austin: Université de Liège and University of Texas at Austin, 473–9.

Tuchelt, K. 1962. *Tiergefässe in Kopf- und Protomengestalt: Untersuchungen zur Formengeschichte tierföemiger Giessgefässe* (Istanbuler Forschungen 22). Berlin: Gebr. Mann Verlag.

Tully, C. J. 2018. *The Cultic Life of Trees in the Prehistoric Aegean, Levant, Egypt and Cyprus* (Aegaeum 42). Leuven: Peeters.

Tully, C. J. 2020. 'Cockles, mussels, fishing nets and finery: the relationship between cult, textiles and the sea depicted on a Minoan-style gold ring from Pylos', *Journal of Eastern Mediterranean Archaeology and Heritage Studies* 8(3–4), 365–78.

Tully, C. J. and Crooks, S. 2015. 'Dropping ecstasy? Minoan cult and the tropes of shamanism', *Time and Mind* 8(2), 1–30.

Tully, C. J. and Crooks, S. 2019. 'Power ranges: identity and terrain in Minoan Crete', *Journal for the Study of Religion, Nature, and Culture* 13 (2), 130–56.

Turner, P. 2005. 'Affordance as context', *Interacting with Computers* 17, 787–800.

Tzachili, I. 2011. *Μινωικά Εικαστικά Τοπία. Τα αγγεία με τις επίθετες πλαστικές μορφές από το Ιερό Κορυφής του Βρύσινα και η αναζήτηση του βάθους* (Vrysinas I). Athens: Ta Pragmata.

Tzachili, I. 2012. 'Some particular figurines from the peak sanctuary of Vrysinas, near Rethymnon, Crete', in Mantzourani, E. and

Betancourt, P. P. (eds.), *PHILISTOR: Studies in Honor of Costis Davaras* (Prehistory Monographs 36). Philadelphia, PA: INSTAP Academic Press, 233–8.

Tzedakis, Y. 1971. 'Λάρνακες Ύστερομινωικου νεκροταφείου 'Αρμένων Ρεθύμνης', *Archaiologika Analekta ex Athenon* 4, 216–22.

Tzedakis, Y., Chryssoulaki, S., Venieri, Y. and Avgouli, M. 1990. 'Les Routes Minoennes: Le poste de Choiromandres et les controle des communications', *Bulletin de Correspondance Hellénique* 114(1), 43–65.

Tzedakis, Y., Martlew, H. and Arnott, R. 2018. *The Late Minoan III Necropolis of Armenoi, I: Introduction and Background* (Prehistory Monographs 60). Philadelphia, PA: INSTAP Academic Press.

van der Geer, A., Dermitzakis, M. and De Vos, J. 2006. 'Crete before the Cretans: the reign of dwarfs', *Pharos* 13, 121–32.

van der Geer, A., Lyras, G., de Vos, J. and Dermitzakis, M. 2010. *Evolution of Island Mammals: Adaptation and Extinction of Placental Mammals on Islands.* Chichester: Wiley-Blackwell.

van Effenterre, H. and van Effenterre, M. 1974. 'Vers une grammaire de la glyptique crétomycénienne', *Die kretische-mykenische Glyptik unde ihre gegenwärtigen Probleme.* Boppard: Harald Boldt, 22–9.

Vandenabeele, F. and Olivier, J.-P. 1979. *Les idéogrammes archéologiques du linéaire B* (Études crétoises 24). Paris: École française d'Athènes.

Vanschoonwinkel, J. 1996. 'Les animaux dans l'art minoen', in Reese, D. S. (ed.), *Pleistocene and Holocene Fauna of Crete and Its First Settlers* (Monographs in World Archaeology 28). Madison, WI: Prehistory Press, 351–412.

Varouchakis, V. 2017. 'Indigenous archaeologies of Crete, 1878–1913', *Public Archaeology* 16(1), 42–66.

Ventris, M. and Chadwick, J. 1956. *Documents in Mycenaean Greek.* Cambridge: Cambridge University Press.

Vigne, J.-D. 1999. 'The large "true" Mediterranean islands as a model for the Holocene impact on the European vertebrate fauna? Recent data and new reflections', in Benecke, N. (ed.), *The Holocene History of the European Vertebrate Fauna.* Rahden: Marie Leidorf, 295–322.

Vitelli, K. 1974. 'Review of: *Myrtos: An Early Bronze Age Site in Crete* by Peter Warren', *American Journal of Archaeology* 78(2), 192–3.

Viveiros de Castro, E. 1998. 'Cosmological deixis and Amerindian perspectivism', *Journal of the Royal Anthropological Institute* 4(3), 469–88.

Viveiros de Castro, E. 2004. 'Exchanging perspectives: the transformation of objects into subjects in Amerindian ontologies', *Common Knowledge* 10(3), 463–84.

Vlasaki, M. and Hallager, E. 1995. 'Evidence for seal use in Prepalatial Western Crete', in Müller, W. (ed.), *Sceaux minoens et mycéniens. IVe symposium international, 10–12 September 1992, Clermont-Ferrand* (CMS Beiheft 5). Berlin: Gebr. Mann Verlag, 251–70.

Vokotopoulos, L. 2015. 'Χοιρόμανδρες Ζάκρου: ο βορειοδυτικός τομέας', in Karanastasi, P., Tzigounaki, A. and Tsigonaki, C. (eds.), *Archaeological Work in Crete 3: Proceedings of the 3rd Meeting, Rethymnon, 5–8 December 2013.* Rethymnon: University of Crete and Ephorate of Antiquities of Rethymnon, 511–23.

von Rüden, C. 2014. 'Beyond the east-west dichotomy in Syrian and Levantine wall paintings', in Brown, B. A. and Feldman, M. H. (eds.), *Critical Approaches to Ancient Near Eastern Art.* Boston, MA: De Gruyter, 55–78.

von Uexküll, J. 1982. 'The theory of meaning', *Semiotica* 42(1), 25–82.

Wachsmann, S. 1987. *Aegeans in the Theban Tombs* (Orientalia Lovaniensia Analecta 20). Leuven: Peeters.

Walberg, G. 1983. *Provincial Middle Minoan Pottery.* Mainz: von Zabern.

Walberg, G. 1987. *Kamares: A Study of the Character of Palatial Middle Minoan Pottery* (SIMA Pocket Book 49). Gothenburg: Paul Åströms Förlag.

Waldau, P. 2013. *Animal Studies: An Introduction.* Oxford: Oxford University Press.

Wall, S. M., Musgrave, J. H. and Warren, P. M. 1986. 'Human bones from a Late Minoan IB house at Knossos', *Annual of the British School at Athens* 81, 333–88.

Wall-Crowther, S. 2007. 'The animal bones', in MacGillivray, J. A., Sackett, L. H. and Driessen, J. M. (eds.), *Palaikastro: Two Late Minoan Wells* (BSA Supplementary Volume 43). London: British School at Athens, 181–209.

Warren, P. M. 1969. *Minoan Stone Vases.* Cambridge: Cambridge University Press.

Warren, P. M. 1972. *Myrtos: An Early Bronze Age Settlement in Crete* (BSA Supplementary Volume 7). London: British School at Athens and Thames and Hudson.

Warren, P. M. 1981. 'Knossos: Stratigraphical Museum excavations, 1978–1980. Part I', *Archaeological Reports* 27, 73–92.

Warren, P. M. 1995. 'Minoan Crete and Pharaonic Egypt', in Davies, W. V. and Schofield, L. (eds.), *Egypt, the Aegean and the Levant: Interconnections in the Second Millennium BC.* London: British Museum, 1–18.

Warren, P. M. 2000. 'From naturalism to essentialism in Theran and Minoan art', in Sherratt, S. (ed.), *The Wall Paintings of Thera, I.* Piraeus: Thera Foundation, 364–380.

Warren, P. M. and Tzedhakis, J. 1974. 'Debla: an Early Minoan settlement in Western Crete', *Annual of the British School at Athens* 69, 299–342.

Watrous, L. V. 1991. 'The origin and iconography of the Late Minoan painted larnax', *Hesperia* 60(3), 285–307.

Watrous, L. V. 1996. *The Cave Sanctuary of Zeus at Psychro: A Study of Extra-Urban Sanctuaries in Minoan and Early Iron Age Crete* (Aegaeum 15). Liège and Austin: Université de Liège and University of Texas at Austin.

Watrous, L. V. 2004. 'New pottery from the Psychro Cave and its implications for Minoan Crete', *Annual of the British School at Athens* 99, 129–47.

Watts, C. M. 2008. 'On mediation and material agency in the Peircean semeiotic', in Knappett, C. and Malafouris, L. (eds.), *Material Agency: Towards a Non-Anthropocentric Approach.* New York: Springer, 187–207.

Watts, C. M. (ed.) 2013. *Relational Archaeologies: Humans, Animals, Things.* London: Routledge.

Weilhartner, J. 2008. 'Some observations on the commodities in the Linear B tablets referring to sacrificial banquets', in Hitchcock, L. A., Laffineur, R. and Crowley, J. (eds.), *DAIS. The Aegean Feast. Proceedings of the 12th International Aegean Conference, University of Melbourne, Centre for Classics and Archaeology, 25–29 March 2008* (Aegaeum 29). Liège and Austin: Université de Liège and University of Texas, 411–26.

Weinberg, P. and Ambarli, H. 2020. *Capra aegagrus.* The IUCN Red List of Threatened Species 2008: e.T3786A10076632 (https://dx.doi.org/10.2305/IUCN.UK.2020-2.RLTS.T3786A22145942.en) [Accessed on 7 June 2021].

Weingarten, J. 1983. *The Zakro Master and His Place in Prehistory* (SIMA Pocket Book 26). Gothenburg: Paul Åströms Förlag.

Weingarten, J. 1988. 'Seal use at Late Minoan IB Ayia Triada: a Minoan elite in action. II. Aesthetic considerations', *Kadmos* 27(2), 89–114.

Weingarten, J. 1989. 'Formulaic implications of some Late Bronze Age three-sided prisms', in Pini, I. (ed.), *Fragen und Probleme der bronzezeitlichen ägäischen Glyptik. Beiträge zum 3. Internationalen Marburger Siegel Symposium, 5–7 September 1985* (CMS Beiheft 3). Berlin: Gebr. Mann Verlag, 219–313.

Weingarten, J. 1991. *The Transformation of Egyptian Taweret into the Minoan Genius: A Study of Cultural Transmission in the Middle Bronze Age* (SIMA 88). Partille: Paul Åströms Förlag.

Weingarten, J. 1992. 'The multiple sealing system of Minoan Crete and its possible antecedents in Anatolia', *Oxford Journal of Archaeology* 11(1), 25–37.

Weingarten, J. 1994. 'Two sealing studies in the Middle Bronze Age. I: Karahoyuk, II: Phaistos', in Ferioli, P., Fiandra, E., Fissore, G. G. and Frangipane, M. (eds.), *Archives Before Writing. Proceedings of the*

International Colloquium Oriolo Romano, 23–25 October 1991. Rome: Ministero per i beni culturali e ambientali, 261–95.

Weingarten, J. 2010. 'Corridors of power: a social network analysis of the Minoan replica rings', in Müller, W. (ed.), *Die Bedeutung der minoischen und mykenischen Glyptik. VI. Internationales Siegel-Symposium, Marburg, 9–12 Oktober 2008* (CMS Beiheft 8). Mainz: von Zabern, 395–412.

Weingarten, J. 2015. 'The arrival of Bes[et] on Middle-Minoan Crete', in Mynářová, J., Onderka, P. and Pavúk, P. (eds.), *There and Back Again: The Crossroads II. Proceedings of an International Conference Held in Prague, 15–18 September 2014*. Prague: Charles University in Prague, Faculty of Arts.

Weingarten, J. and Sanavia, A. 2016. 'The transformation of tritons: some decorated Middle Minoan triton shells and an Anatolian counterpart', in Alram-Stern, E., Blakolmer, F., Deger-Jalkotzy, S., Laffineur, R. and Weilhartner, J. (eds.), *METAPHYSIS. Ritual, Myth and Symbolism in the Aegean Bronze Age. Proceedings of the 15th International Aegean Conference, Vienna, Institute for Oriental and European Archaeology, Aegean and Anatolia Department, Austrian Academy of Sciences and Institute of Classical Archaeology, University of Vienna, 22–25 April 2014* (Aegaeum 39). Leuven and Liège: Peeters, 335–44.

Wengrow, D. 2010. 'The voyages of Europa: ritual and trade in the Eastern Mediterranean circa 2300–1850 BC', in Parkinson, W. and Galaty, M. (eds.), *Archaic State Interaction: The Eastern Mediterranean in the Bronze Age*. Santa Fe, CA: School for Advanced Research Press.

Wengrow, D. 2011. 'Cognition, materiality and monsters: the cultural transmission of counter-intuitive forms in Bronze Age societies', *Journal of Material Culture* 16(2), 131–49.

Wengrow, D. 2013. *The Origins of Monsters: Image and Cognition in the First Age of Mechanical Reproduction*. Princeton, NJ: Princeton University Press.

Whitelaw, T. 1983. 'The settlement at Fournou Korifi Myrtos and aspects of Early Minoan social organization', in Krzyszkowska, O. and Nixon, L. (eds.), *Minoan Society: Proceedings of the Cambridge Colloquium 1981*. Bristol: Bristol Classical Press, 323–45.

Whitelaw, T. 1992. 'Lost in the Labyrinth? Comments on Broodbank's "Social change at Knossos before the Bronze Age"', *Journal of Mediterranean Archaeology* 5(2), 225–38.

Whitelaw, T. 2012. 'The urbanisation of Prehistoric Crete: settlement perspectives on Minoan state formation', in Schoep, I., Tomkins, P. and Driessen, J. (eds.), *Back to the Beginning: Reassessing Social and Political Complexity on Crete during the Early and Middle Bronze Age*. Oxford: Oxbow, 114–76.

Whitelaw, T. 2017. 'The development and character of urban communities in Prehistoric Crete in their regional context: a preliminary study', in Letesson, Q. and Knappett, C. (eds.), *Minoan Architecture and Urbanism: New Perspectives on an Ancient Built Environment*. Oxford: Oxford University Press, 114–80.

Whitelaw, T. 2018. 'Recognising polities in prehistoric Crete', in Relaki, M. and Papadatos, Y. (eds.), *From the Foundations to the Legacy of Minoan Archaeology: Studies in Honour of Professor Keith Branigan* (Sheffield Studies in Aegean Archaeology 12). Oxford: Oxbow, 210–55.

Whitelaw, T. 2019. 'Feeding Knossos: exploring economic and logistical implications of urbanism on Prehistoric Crete', in Garcia, D., Orgeolet, R., Pomadère, M. and Zurbach, J. (eds.), *Country in the City: Agricultural Functions in Protohistoric Urban Settlements (Aegean and Western Mediterranean)*. Oxford: Archaeopress, 88–121.

Whitelaw, T., Bredaki, M. and Vasilakis, A. 2018. 'The long-term dynamics of Knossos in context', in *Proceedings of the 12th International Congress of Cretan Studies*. Heraklion: Society of Cretan Historical Studies (https://12iccs.proceedings.gr/en/proceedings/category/39/35/816) [Accessed 27 December 2021].

Whitelaw, T., Day, P., Kiriatzi, E., Kilikoglou, V. and Wilson, D. 1997. 'Ceramic traditions at EM IIB Myrtos,

Fournou Korifi', in Laffineur, R. and Betancourt, P. (eds.), *TEXNH. Craftsmen, Craftswomen and Craftsmanship in the Aegean Bronze Age. Proceedings of the 6th International Aegean Conference, Philadelphia, Temple University, 18–21 April 1996* (Aegaeum 16). Liège: Université de Liège, 265–74.

Whittaker, H. 2016. 'Horns and axes', in Alram-Stern, E., Blakolmer, F., Deger-Jalkotzy, S., Laffineur, R. and Weilhartner, J. (eds.), *METAPHYSIS. Ritual, Myth and Symbolism in the Aegean Bronze Age. Proceedings of the 15th International Aegean Conference, Vienna, Institute for Oriental and European Archaeology, Aegean and Anatolia Department, Austrian Academy of Sciences and Institute of Classical Archaeology, University of Vienna, 22–25 April 2014* (Aegaeum 39). Leuven and Liège: Peeters, 109–14.

Wiencke, M. 1981. 'Typology and style of Prepalatial seals', in Niemeier, W.-D. (ed.), *Studien zur minoischen und helladischen Glyptik* (CMS Beiheft 1). Berlin: Gebr. Mann Verlag, 251–61.

Wilkens, B. 1996a. 'Faunal remains from Italian excavations on Crete', in Reese, D. S. (ed.), *Pleistocene and Holocene Fauna of Crete and Its First Settlers* (Monographs in World Archaeology 28). Madison, WI: Prehistory Press, 241–61.

Wilkens, B. 1996b. 'I resti faunistici di Haghia Triada (Creta) in età neo e postpalaziale. Nota preliminare', in De Miro, E., Godart, L. and Sacconi, A. (eds.), *Atti e memorie del secondo Congresso internazionale di micenologia, Roma-Napoli, 14–20 ottobre 1991*. Rome: Gruppo Editoriale Internazionale, 1511–20.

Wilkens, B. 2003. 'Hunting and breeding in ancient Crete', in Kotjabopoulou, E., Hamilakis, Y., Halstead, P., Gamble, C. and Elefanti, P. (eds.), *Zooarchaeology In Greece: Recent Advances* (BSA Studies 9). London: British School at Athens, 85–90.

Williams, R. 1983. *Keywords: A Vocabulary of Culture and Society, rev. edn*. London: Fontana.

Willis, R. 1975. *Man and Beast*. St Albans: Paladin.

Willis, R. 1990. 'Introduction', in Willis, R. (ed.), *Signifying Animals: Human Meanings in the Natural World*. London: Unwin Hyman, 1–24.

Winder, N. 1991. 'Interpreting a site: the case for a reassessment of the Knossos Neolithic', *Archaeological Review from Cambridge* 10(1), 37–52.

Winkler-Horaček, L. 2011. 'Wege der Sphinx: Von Ägypten und Vorderasien nach Griechenland', in Winkler-Horaček, L. (ed.), *Wege der Sphinx. Monster zwischen Orient und Okzident*. Rahden: Marie Leidorf, 99–115.

Wolch, J. and Emel, J. 1995. 'Bringing the animals back in', *Environment and Planning D–Society & Space* 13(6), 632–6.

Wolf, D. 2019. *Monsters and the Mind: Composite Creatures and Social Cognition in Aegean Bronze Age Glyptic* (Daidalos 9). Heidelberg: Propylaeum.

Wolfe, C. 2003. 'In the shadow of Wittgenstein's lion', in Wolfe, C. (ed.), *Zoontologies: The Question of the Animal*. Minneapolis: University of Minnesota Press, 1–57.

Xanthoudides, S. A. 1924. *The Vaulted Tombs of Mesara: An Account of Some Early Cemeteries of Southern Crete*. London: Hodder and Stoughton.

Xirouchakis, S. 2015. *Birdwatching in NATURA 2000 Network Areas on Crete: A Birdwatching Guide*. Heraklion: Natural History Museum of Crete.

Yannouli, E. 2003. 'Non-domestic carnivores in Greek Prehistory: a review', in Kotjabopoulou, E., Hamilakis, Y., Halstead, P., Gamble, C. and Elefanti, P. (eds.), *Zooarchaeology in Greece: Recent Advances* (BSA Studies 9). London: British School at Athens, 175–92.

Younger, J. G. 1976. 'Bronze Age representations of bull-leaping', *American Journal of Archaeology* 80(2), 125–37.

Younger, J. G. 1984. 'Aegean seals of the Late Bronze Age: masters and workshops III', *Kadmos* 23(1), 38–64.

Younger, J. G. 1988. *The Iconography of Late Minoan and Mycenaean Sealstones and Finger Rings*. Bristol: Bristol Classical Press.

Younger, J. G. 1989. 'A large stylistic group of the late XVth century', in Pini, I. (ed.), *Fragen und Probleme der bronzezeitlichen ägäischen Glyptik. Beiträge zum 3. Internationalen Marburger Siegel Symposium, 5–7 September 1985* (CMS Beiheft 3). Berlin: Gebr. Mann Verlag, 339–53.

Younger, J. G. 1993. *Bronze Age Aegean Seals in their Middle Phase (ca. 1700–1550 B.C.)* (SIMA 102). Jonsered: Paul Åströms Förlag.

Younger, J. G. 1995. 'Bronze Age representations of Aegean bull-games, III', in Laffineur, R. and Niemeier, W.-D. (eds.), *POLITEIA. Society and State in the Aegean Bronze Age. Proceedings of the 5th International Aegean Conference, University of Heidelberg, Archäologisches Institut, 10–13 April 1994* (Aegaeum 12). Liège and Austin: Université de Liège and University of Texas at Austin, 507–45.

Younger, J. G. 1999. 'The Cretan Hieroglyphic script: a review article', *Minos* 31–32, 379–400.

Younger, J. G. 2000. 'The Spectacle Eyes Group: continuity and innovation for the first Mycenaean administration at Knossos', in Müller, W. (ed.), *Minoisch-mykenische Glyptik: Stil, Ikonographie, Funktion. V. Internationales Siegel-Symposium, Marburg, 23–25 September 1999* (CMS Beiheft 6). Berlin: Gebr. Mann Verlag, 347–60.

Younger, J. G. 2005. 'Cretan Hieroglyphic wool units (LANA, double mina)', in Perna, M. (ed.), *Studi in onore di Enrica Fiandra. Contributi di archeologia egea e vicinorientale* (Studi egei e vicinorientali 1). Paris: De Boccard, 405–9.

Younger, J. G. 2013. 'Notes to the Hieroglyphic Signs' (http://people.ku.edu/~jyounger/Hiero/SignNotes.html) [Accessed on 2 June 2021].

Younger, J. G. and Rehak, P. 2010a. 'Minoan culture: religion, burial customs, and administration', in Shelmerdine, C. W. (ed.), *The Cambridge Companion to the Aegean Bronze Age*. Cambridge: Cambridge University Press, 165–85.

Younger, J. G. and Rehak, P. 2010b. 'The material culture of Neopalatial Crete', in Shelmerdine, C. W. (ed.), *The Cambridge Companion to the Aegean Bronze Age*. Cambridge: Cambridge University Press, 140–64.

Yule, P. 1980. *Early Cretan Seals: A Study of Chronology (Marburger Studien zur Vor- und Frühgeschichte 4)*. Mainz: von Zabern.

Zeimbekis, M. 1998. *The Typology, Forms and Functions of Animal Figures from Minoan Peak Sanctuaries with Special Reference to Juktas and Kophinas*. Unpublished PhD thesis, University of Bristol.

Zeimbekis, M. 2004. 'The organisation of votive production and distribution in the peak sanctuaries of state society Crete: a perspective offered by the Juktas clay animal figurines', in Cadogan, G., Hatzaki, E. and Vasilakis, A. (eds.), *Knossos: Palace, City, State* (BSA Studies 12). London: British School at Athens, 351–61.

Zeimbekis, M. 2006. 'Grappling with the bull: a reappraisal of bull and cattle-related ritual in Minoan Crete', in *Proceedings of the 9th International Congress of Cretan Studies, A2*. Heraklion: Society of Cretan Historical Studies, 27–44.

Zervos, C. 1956. *L'art de la Crète néolithique et minoenne*. Paris: Éditions 'Cahiers d'art'.

Zeuner, F. 1963. *A History of Domesticated Animals*. London: Hutchinson.

Zouzoula, E. 2006. *The Fantastic Creatures of Bronze Age Crete*. Unpublished PhD thesis, University of Nottingham.

INDEX